Resources
and Energy

Resources and Energy

An Economic Analysis

Ferdinand E. Banks
University of Uppsala

LexingtonBooks
D.C. Heath and Company
Lexington, Massachusetts
Toronto

Library of Congress Cataloging in Publication Data

Banks, Ferdinand E.
  Resources and energy.

  Includes index.
  1. Energy industries. 2. Petroleum industry and trade. 3. Mineral
industries. 4. Power resources—Economic aspects. 5. Raw materials—
Economic aspects. I. Title.
HD9502.A2B35   1982   333.79   81–47967
ISBN 0–669–05203–5   AACR2

Copyright © 1983 by D.C. Heath and Company

Published simultaneously in Canada

Printed in the United States of America

International Standard Book Number: 0–669–05203–5

Library of Congress Catalog Card Number: 81–47967

For my teachers at Illinois Tech, Roosevelt University, and The University of Stockholm; and also at Jackson and Orr.

For my students at Stockholm, Uppsala, Dakar, and Sydney; and also at Lewis, Gifu, and S. Gmuend

Contents

Figures

Tables

Preface and
Acknowledgments

I began thinking about writing this book when I was a visiting professor at the Centre for Policy Studies, Monash University (Melbourne) and I want to thank Richard Clarke, Doug Hocking, and Mike Porter for the many enjoyable conversations I had with them on some of the subjects taken up in this book.

I am especially indebted, however, to Upplands Bank of Uppsala, Sweden, and the Tore Browaldh Foundation for the grants that made possible my travel and research in 1981, during which time I was able to sort out my priorities and determine just what kind of book needed to be written. In addition to basic material on oil, coal, gas, and uranium, I try to provide a concise introduction to such things as the refining of oil, petrochemicals, futures markets, inventories, capital costs, tin, stock-flow models, and certain other topics that are not usually handled in most economics text and reference books. There is also a short but comprehensive survey of iron and steel. Chapter 8 provides a brief summary of the book.

In the course of my research in Germany I received especially valuable assistance at Kiel and Berlin. In Australia I would like to thank Mike Folie of Shell, Australia; Professor Greg McColl and Tom Mozina of the University of New South Wales; Dr. Duncan Ironmonger and Dr. Phyllis Rosendale of the University of Melbourne; Professor Geoff Taylor of the Australian National University; and Professor Ken Blakey of Wollongong University for permitting me to lecture and lead seminars in that country. I also owe a debt of gratitude to my audiences at Aix-en-Provence, Zurich, Lund, Oslo, Stockholm, and Cambridge University; and an even greater debt to Franz Ettlin of the Universities of Stockholm and California (Berkeley), for his assistance with an econometric problem.

Finally, I would like to express my appreciation to the editors and publishers of the *OPEC Bulletin* and the *Chemical Economy and Engineering Review* (Tokyo) for permission to use material I published in those journals; and also *Resources Policy* for permission to reproduce large parts of my articles titled "Australian Energy Resources" (June 1981) and "Fixed Capital and the Optimal Rate of Mineral Extraction" (December 1981). Furthermore, I would like to thank Mr. Åke Wennmann of Svenska Handelsbanken for his help with both this book and my current work on the world coal market. I am also extremely grateful to the many readers (and also reviewers) who have written to me about my book *The Political Economy of Oil,* and to the people at Lexington Books for their assistance with the changes and corrections in the second printing of that volume.

And, although they may be surprised to hear it, I owe a major debt of gratitude to the students in my courses in international economics who have seen to it that I received the mental gymnastics so important for this kind of activity. A similar service was rendered by the chief prosecutor of Uppland, Sweden, and people in Gothenberg who almost, but not quite, got their work into circulation.

Resources
and Energy

1 Introduction

This book features two long core chapters, on oil and on nonfuel minerals; and a reasonably complete exposition of the econometrics of primary commodities that, I think, contains enough materials to give most interested people a satisfactory insight into the basic techniques and their use.

There are also chapters on coal, gas, and uranium, which include a compact overview of the Soviet energy sector, and the Australian coal industry. Given the purpose of this book, these materials are kept at a very rudimentary level. Chapter 8 provides a summary of the book that contains a brief review of the latest developments on the energy and nonfuel minerals front.

The chapter on oil extends some of the most important topics examined in my book, *The Political Economy of Oil,* and I want to point out that the reader will find here more detailed descriptions of such things as refining, petrochemicals, sea transportation, and stocks. Moreover, the appendix to that chapter presents a thorough introduction to oil and the futures markets. I have also been very careful to emphasize the relation between oil (and energy) consumption and certain macroeconomic phenomena such as unemployment and inflation. The spectacle of economists congratulating industry, governments, and themselves for saving energy while ignoring the appalling fact that most miracles of conservation are taking place against a background of growing social and economic insecurity is one of the most distressing shortcomings of my profession—which happens to be a profession where certain types of shortcomings threaten to become the rule, rather than the exception.

Considerable effort has been made in this book to lay the groundwork for more extensive research in the economics of coal and gas, and to provide the reader with a brief but useful analysis of the very important iron and steel industry. Regrettably I had the desire, but not the ability, to provide an elementary introduction to the economics of population; because there is no doubt in my mind that the population explosion is the greatest single threat to our civilization. But I *can* hope that the most dedicated and energetic individuals in the profession will attempt to find some answers to this deadly impasse, or at the very least bring the matter to the attention of their students.

In all except the chapter on econometrics, highly technical exposition is mostly kept separate from the main arguments, and I point out those sections that I consider a bit abtruse for elementary readers. Let me also note that a part of this book can be understood by readers without any formal training in economics, while a cover-to-cover tour requires only an elementary knowledge of microeconomics.

Appendix 1A:
Units and Equivalencies

Prefix	Symbol	Power	Meaning	Example
kilo	K	10^3	thousand	KW (kilowatt)
mega	M	10^6	million	MW (megawatt)
giga	G	10^9	billion	GJ (gigajoule)
tera	T	10^{12}	trillion	TJ (terajoule)
peta	P	10^{15}	thousand-trillion	PJ (petajoule)
exa	E	10^{18}	million-trillion	EJ (exajoule)

Next it should be noted that 1 t is the designation on one metric ton, or 1 *tonne,* which equals 2,204 pounds (lbs). We also have a short ton, which in some countries is simply called a ton. One short ton (ton) = 2,000 lbs; thus 1 t = 1.1023 tons. Finally, there is a long ton, which is 2,240 lbs.

When working with energy, we are often interested in heat, which in textbooks and the press tends to be measured in British thermal units (Btu), where 1 Btu is the amount of heat needed to increase the temperature of 1 lb of water by 1° Fahrenheit. One metric ton (1 t) of anthracite coal has an energy content of 28,000,000 Btu, whereas 1 t of crude oil has an energy content of 42,514,000. One thousand cubic feet (1 Kcf) of natural gas has an energy content of about 1,000,000 Btu. These, of course, are average values.

It has been said that Btus will soon be replaced by joules in most of the literature. This is not certain, but in any event joules are a member of the international system of units (SI units), and their abbreviation is J. The kilowatt-hour is also a very popular and well-known unit. A less common unit is the therm (= 100,000 Btu = 1.055×10^8 J). A short table of equivalencies is thus:

	Joules (J)	Kilowatt-hours(kWh)	Btu
1 joule	1	0.278×10^{-6}	0.948×10^{-3}
1 kWh	3.6×10^6	1	3.412×10^3
1 Btu	1.055×10^3	0.293×10^{-3}	1

It is worth remembering that one million tonnes of oil equivalent (Mtoe) can be converted to British thermal units should this unit be relevant to the discussion. Handy transformations are 1 bbl oil = 5,800,000 Btu, where bbl signifies barrels. We also have 7.33 bbl = 1 t. The most popular unit for measuring both the consumption and production of oil is barrels per day (bbl/d). For example, 19 million barrels of oil per day (19 Mbbl/d)

3

is the present output of the Organization of Petroleum Exporting Countries (OPEC). This can be turned into another well-known unit, millions of tonnes per year (Mt/y) by multiplying by 50.

Power is defined as the rate of doing work. The best-known units for measuring power are the watt (W), which is equal to one joule per second, and the horsepower (hp). Once again we can present a table of equivalencies:

	Watts (W)	Horsepower (hp)	Btu/Hour
1 watt	1	1.341×10^{-3}	3.41
1 hp	0.746×10^3	1	2.54×10^3
1 Btu/hour	0.293	0.393×10^{-3}	1

Finally, in summary form, some useful equivalencies are

1. 1 barrel crude oil = 42 U.S. gallons and weighs 0.136 metric tons (tonnes)
2. 1000 cubic feet (1 Kcf) of natural gas = 28.3 cubic meters, where 1 cubic meter = 35.33 cubic feet = 35.33 ft^3
3. 1 kilowatt-hour (kWh) of electricity = 3,411 Btu = 860 kilocalories (kcals)
4. 1 metric ton bituminous coal = 27,700,000 Btu
5. 1 metric ton anthracite coal = 28,000,000 Btu = 7.06×19^9 calories
6. 1,000 cubic feet natural gas = 1.035×10^6 Btu = 2.61×10^8 calories
7. 1 metric ton of hard coal = 4.9 barrels of crude oil
8. 1,000 cubic feet of natural gas = 0.178 barrels of crude oil
9. 1,000 kilowatt-hour of electricity = 0.588 barrels of crude oil
10. 1 Quad of energy = 23.5 million tons of oil equivalent

2

The World Petroleum Market

This chapter consists of four parts. The first three form a nontechnical survey of the world petroleum (or, less formally, the world *oil*) economy. In the first two parts of this introductory effort I borrow liberally from my previous work, especially *The Political Economy of Oil* and *The International Economy: A Modern Approach,* since I am interested in providing the reader with a concise but up-to-date overview of one of the most important issues facing the international economy. In the second part I present an extensive overview of the supply of oil. The reader should understand that although there is still a great deal of petroleum in the world—as well as astronomical amounts of resources that can substitute for this commodity—we are now in the process of leaving (at least temporarily) the era of cheap energy based on perhaps the most flexible of all energy resources: crude oil. This is a transition that has already had profound consequences for the life-styles and incomes of tens of millions of people, although, oddly enough, most of them seem to be unaware of this fact. Many economists, for example, refer to the smooth adjustment to the higher price of oil and pontificate about switching to alternative energy sources, but do not seem to realize that these options mean that the share of energy in world output might conceivably double. This share is a rough measure of the amount of resources (such as capital and labor) required to obtain that energy, not of the quantity of energy that will become available. Needless to say, had energy remained as accessible as it was a few decades ago, these resources could have been used to produce investment or consumer goods.

Since the last portion of the chapter is more technical than the preceding materials, I use the third section to survey the effect of rising oil prices on such things as inflation and economic growth. The last part of the chapter, which consists of several subsections, discusses the probable evolution of the oil price in this decade and is designed to shed some light on the intentions of certain OPEC countries to modernize their economies at a maximum rate. This seems to be a controversial topic: I have encountered many individuals who believe or at least hope that the OPEC countries will be incapable of sustaining a high rate of economic progress over a long period of time. I also attempt to make clear in these final pages that it is still the OPEC countries that call the oil-price tune. This will be increasingly, not decreasingly, the case as these countries become fully aware of the relationship between their finite endowment of energy resources and their future economic development. Relations within OPEC are considered since by the end of the century the share of oil from Saudi Arabia, Kuwait, and the United Arab Emirates (UAE) could rise to about 90 percent, compared with 67 percent in 1980. Finally, there is a long appendix in which, among other things, the reader is introduced to futures markets and their growing importance for oil and oil products.

The Background

Petroleum, or crude oil, is a substance that has been known to humankind for thousands of years; but it was not until 1859, near Titusville, Pennsylvania, that Edwin L. Drake drilled the first modern well, although the first oil well in North America dates from 1829 (in Kentucky). Drake's well set off a wave of exploration in the vicinity, starting a boom reminiscent of the California gold rush. As for Drake, he assumed the title of colonel and, deciding to deal in oil stocks instead of the real thing, opened a brokerage office on Wall Street in New York City. When he died in 1880, he was practically a pauper.

Petroleum was first exported in 1861 (from Philadelphia to London). Standard Oil, the forerunner of the majors, or seven sisters, as the leading oil companies came to be called in some circles, was incorporated in Cleveland, Ohio, in 1870. The complete roster of the majors includes Exxon (Esso), Texaco, Gulf, Standard of California, Shell, Mobil, and British Petroleum. (The original Standard Oil was broken up into three companies—Exxon, Mobil, and Standard Oil of California—by a U.S. Supreme Court antitrust ruling.) The next two of the sisters to appear on the scene were Gulf and Texaco. Underlying the occasion was the discovery of oil at Spindletop, Texas, in 1901, which heralded the tremendous future finds in the U.S. Southwest, particularly in eastern Texas thirty years later. The first important foreign company to break into this select clique was Royal Dutch Shell, in 1907. This firm was formed from Shell, founded by Marcus Samuel of London, and Royal Dutch, whose assets were mostly in Indonesia (then known as the Dutch East Indies). Royal Dutch was the brainchild of Henri Deterding. In the original amalgamation 60 percent of the capital was Dutch, the rest English.

The first of the major oil discoveries outside the United States came at Mashid-i-Salaman in Iran (then Persia) in 1908. Some other important dates in the appearance of non–United States oil are 1922, when oil was found near the shores of Lake Maracaibo in Venezuela; 1927, the year of the Kirkuk strike in Iraq; and 1933, which saw the registering of the Kuwait Oil

Company. Saudi Arabian oil was also ushered onstage during this period, although not without some light comedy. Esso was offered a chance to take an exploratory and exploitation concession on a huge tract of that country for the grand sum of $50,000—but refused. This refusal went down in oil history as the billion-dollar blunder. It was also in the late 1920s that the majors formally agreed to regulate competition in the Middle East and, presumably, anywhere else that a conflict could arise between the majors themselves or between the majors and intruders—regardless of whether these interlopers were other oil companies or the governments that happened to own the land on which the oil was being pumped.

In my book on copper (1974) I describe the mutations that appeared on this market in the years just after World War II. Similar winds of change blew across the world petroleum economy about the same time. In 1948 Venezuela introduced a tax law that divided oil profits 50–50 between the operating companies and the state. Saudi Arabia soon adopted this practice; thus a principle that probably has a biological basis was formulated: natural resources located within the boundaries of a country belong to that country, irrespective of the political or historical circumstances that led foreigners to become involved with them in the first place. Although this proposition is definitely not universally accepted today, it has become uncontroversial enough to be discussed in public. Forty or fifty years ago, however, ideas of this sort were not bandied about so freely.

In 1951 Iran nationalized the Anglo-Iranian Oil Company. The Iranian government offered no compensation for the properties being taken over, reasoning that they had belonged to the Iranian people in the first place. Despite the inconvenience felt by the oil-producing fraternity, the companies did not make a spectacle of themselves by demanding restitution, nor was it necessary. In the two years that followed the nationalization, Iranian revenues were less than a single day's royalties under the old system. What happened was simply that the production of crude oil was increased in other oil-producing countries, particularly the United States. After a short period during which the output of crude was marginally under the desired level, enough supplies reached the market to satisfy virtually all categories of consumer. Under the circumstances, the operating companies did not expend a great deal of effort trying to negotiate a settlement, but simply sat back to await the inevitable.

The inevitable arrived in 1954, when the Iranian government was overturned in a coup organized by the CIA. A Western consortium then took over operation of Iran's oil-producing assets, with British Petroleum (BP) as the principal concessionaire, and continued to run it until the late 1970s, although beginning in the mid-1960s, its main function was to take and carry out orders from the Iranian authorities. One reason for this was that formally the nationalization of Iran's oil industry stayed on the books, and eventually both sides came to see that the master-servant relationship established in 1954 had been nullified. In 1954, however, other oil-producing countries, observing the details of the drama in Iran, called off whatever nationalization plans they might have been making and settled for the Venezuelan–Saudi Arabian formula of a 50–50 division of profits, with the profits themselves very strongly influenced by the so-called *posted price*. (This usage will be explained later.)

Before we scrutinize one of the more interesting oil-related problems of that period, it should be emphasized that the key element in upsetting the Iranians' nationalization attempt was the ability of the larger oil companies to bring about an increase in production elsewhere in the world, from either their own installations or those of independent producers, and not the interference of gunboats or marines (despite some talk of eventually resorting to these). Given an exponential increase in the world consumption of oil at that time of about 8 percent a year, with a large part of this increase scheduled to come from the huge reserves of the Middle East, anyone who gave the matter a few seconds' thought would have realized that in a decade or so world demand for oil would be so large that, in the event of a nationalization attempt by one or more of the leading Middle East producers, it would be quite futile to invoke the practice of increasing production elsewhere. This is so for two reasons: (1) for all practical purposes there was not enough oil elsewhere—except possibly in the United States—to compensate for a drop in Middle East production; (2) not only would U.S. production have to be raised, but also consumption in that country would have to be lowered in order to provide for consumers elsewhere. Given the growing cost of U.S. oil and the psychology of the U.S. automobile owner, this program was unfeasible in the short run and probably also in the long run. Thus, as early as 1954 the handwriting was on the wall; but, for reasons that are not yet entirely clear, the directors of the major oil companies chose to ignore it.

In March 1969 Iran demanded $1 billion from the operating companies on an "any way you can raise the money" basis, pointedly ignoring such things as the current flow of oil-company revenues, profitability, and so on. At the petroleum prices existing at that time, this injunction meant that the companies would have to lift and sell 15 percent more oil than they planned on handling. Needless to say, this was terrible news for these firms because oil production was already running slightly ahead of consumption. Also, giving in to the Iranians would tend to place the competence of the directors of the operating companies in a bad light. Even worse, if Iran succeeded with this challenge, the other producing countries would be tempted to try the same thing.

As we now know, on the latter point the operating companies were correct. Iran received its $1 billion, and soon afterwards the Libyan government announced a decrease in the production of oil, and Algeria increased the price of its crude. On top of this, Iran notified the management of the consortium that its royalties would have to come to $5.9 billion over the next five years. The oil companies' response to this shock was to notify anyone who would listen that this was more than the market could bear. Where economics was concerned, however, the directors of these companies and their economic advisors were badly mistaken. Almost ten years earlier the demand for oil had fallen considerably; in addition, there had been an increased supply of crude reaching world markets from Russia and from so-called independent oil companies operating in Algeria and Libya. As a result Exxon cut the posted price of its crude by 14 cents (without bothering to inform the Middle East oil producing countries of its decision); and eventually the majors fixed the official world oil price of $1.80 per barrel, where it languished for most of the decade. Despite the lamentations of the oil-company directors about the doldrums into which the world oil market had fallen, it should be remembered that in those days the supply of oil was considered virtually infinite; and it did not escape a few observers of the world energy scene that this low price helped ensure that the energy base for the equipment being designed and installed in the major industrial countries (especially in rapidly growing Japan and Germany) would be oil—not coal, gas, nuclear energy, or even conventional oil from new high-cost deposits. Thus, when the latest demands came along, there was nothing to do except let the price of oil increase. Since demand had almost caught up with supply, there were only marginal decreases in the overall demand for oil; many of the companies producing oil ended up with more revenue and profit rather than less. It is true, however, that these higher prices caused petroleum users and producers to begin thinking about such things as uranium, sizable increases in offshore exploration, and so on. Otherwise, for certain countries, the aftermath of the 1973–1974 oil-price increases might have assumed the proportion of a catastrophe.

As of 1970, the oil-producing countries were definitely on the offensive; and in truth, the oil-price increases they imposed a few years later probably could have been introduced earlier. There is no point in conjecturing why this was not done however, because as became clear later there were plenty of people around the Gulf who knew at least as much about energy economics as the executives of many of the big oil firms, their economists, and the experts they occasionally called in from the academic community. What must be made clear here is that the problem with the management of some (but not all) of the oil companies was not that they underestimated the strength of the producing countries or their ability to administer their own resources, but that they occasionally failed to comprehend that public profit does not always follow from private profit, as exemplified by the belief that the oil companies always knew best where oil was concerned—which was mostly true, but occasionally false.

OPEC: The Organization of Petroleum Exporting Countries

OPEC (the Organization of Petroleum Exporting Countries) was formed in 1960, and at first it was overlooked by the world press or regarded as just another international talk shop. At its sixteenth meeting, in 1968, however, the OPEC assembly adopted a Declaratory Statement of Petroleum Policy in Member Countries, which in broad outline stated that member countries would be assuming responsibility for their own resources as soon as possible. (Member countries and their estimated revenues in billions of U.S. dollars in 1980 are Saudi Arabia ($104.2), Iraq ($26.5), Libya ($23.2), Nigeria ($20), United Arab Emirates ($19.2), Venezuela ($18.9), Kuwait ($18.3), Algeria ($11.7), Iran ($11.6), Indonesia ($10.5), Qatar ($5.2), Gabon ($1.6), and Ecuador ($1.2). Some people took this declaration seriously, but many recognized experts on the oil industry preferred to believe that the producing countries did not possess the technical and managerial expertise to realize this ambition in the foreseeable future.

The reason for this latter attitude is simple: for anyone close to the oil industry in industrial countries, an OPEC takeover of the oil-producing assets on its members' territories was too horrible to contemplate; in fact the OPEC price rises of 1973–1974, which are what the Declaratory Statement was all about, represented a major defeat for the industrial world—at least in the way they came about. On the other hand, considering how the demand for oil has been growing since World War II, this defeat may have been inevitable. As table 2-1 indicates, the amount of oil—the lifeblood of world industry—under the control of OPEC and other Third World countries is large enough to put them in a very strong position for the indefinite future. By way of contrast, the energy picture can be whatever the leadership of the industrial countries wants it to be (as the OPEC directorate is constantly reminding its customers), since for energy the most important ingredient is scientific knowledge.

OPEC first became aware of the strength of its position in December 1970, at an OPEC meeting in Venezuela. There the members collectively came to understand that the competitive position of oil was secure and that there was no longer any reason to hold back in applying pressure on the oil companies. Among the surprises concocted for the companies at that meeting and at a later meeting in Teheran was an upward adjustment in the profit share going to the producing countries, which was increased to 55 or 60 percent. In addition, posted prices were increased by several percent to

Table 2-1
OPEC Countries' Oil Exports and Export Incomes, Oil Demand and Supply for Noncommunist Countries, and 1978 Reserves for Sixteen Countries

	OPEC Countries' Exports[a] and Export Revenues[b]					
	1978		1979		1980	
Saudi Arabia[c]	8.1	(34.6)	9.2	(55.5)	9.6	(104.2)
Iraq	2.4	(9.6)	3.3	(20.3)	2.4	(26.5)
Iran	4.5	(20.9)	2.6	(20.8)	1.1	(11.6)
Libya	2.0	(8.6)	2.0	(14.8)	1.7	(23.2)
Nigeria	1.8	(8.2)	2.2	(16.1)	1.9	(20.0)
Kuwait[c]	2.1	(8.0)	2.5	(16.0)	1.6	(18.3)
UAE	1.8	(8.0)	1.8	(12.4)	1.7	(19.2)
Venezuela	1.9	(5.6)	2.1	(12.0)	1.8	(18.9)
Algeria	1.1	(4.6)	1.0	(7.2)	0.9	(11.7)
Indonesia	1.4	(4.8)	1.3	(8.1)	1.2	(10.5)
Qatar	0.5	(2.0)	0.5	(3.1)	0.5	(5.2)
Ecuador	0.1	(0.4)	0.1	(0.8)	0.1	(1.2)
Gabon	0.2	(0.5)	0.2	(0.9)	0.2	(1.6)

Oil Demand and Supply[d]		
	1980	1981
Demand, excluding OPEC	47.6	44.1
U.S.	17.0	16.2
Japan	5.0	4.7
Europe	13.7	12.7
Other industrial	2.9	2.8
Non-OPEC LDCs	8.7	8.5
Consumption	47.3	44.9
Stock changes	0.3	−0.8
Non-OPEC supplies		
Industrial countries[e]	16.5	16.4
Mexico	2.1	2.5
Other LDCs	3.6	3.7
OPEC exports	24.5	21.5
OPEC domestic consumption	2.3	2.6
OPEC production	27.7	24.1

Reserves[f]			
USSR	58,438	Nigeria	12,273
U.S.	27,804	Kuwait	71,400
Saudi Arabia	113,284	UAE	31,904
Iran	44,966	Indonesia	7,824
Iraq	34,392	Canada	5,784
Venezuela	18,288	Mexico	28,407
China	20,025	Algeria	9,575
Libya	27,204	U.K.	10,191

Source: Shell Briefing Service; International Energy Agency.
[a]In millions of barrels per day (Mbbl/d).
[b]In billions of dollars, shown in parentheses.
[c]Including 50 percent of the neutral zone.
[d]In Mbbl/d.
[e]Including net exports to the West by centrally planned economies.
[f]In millions of barrels. Figures are for mid-1980.

compensate for higher prices paid by OPEC for the items they were import-ing from the industrial world. (Posted prices might also be called *reference prices,* and they were used to determine the taxes and royalties accruing to the producing countries. There was also a free market for oil that was patronized by most of the oil companies, and OPEC economists maintained that between February 1971 and June 1973 the price on the free market increased almost twice as fast as the posted price.) Considerable interest was also shown in a proposal that the petroleum authorities in Libya had presented an oil company in that country—namely, that the company should invest 25 cents in the country for every barrel of crude oil exported. Perhaps for the first time the OPEC directorate indicated openly that it was aware of the considerable fall in the real price of petroleum during the 1960s, with *real price* defined roughly as the money or nominal price of a good divided by the average price (in index form) of the products that good has to buy. In the case of oil, the real price can also be termed the *purchasing power* or the *terms of trade* of oil.

The role of Libya in the oil drama deserves further illumination. In 1970 Colonel Muammar al-Qaddafi sent for the chiefs of one of the operat-ing companies in Libya, Occidental Petroleum, and informed them that Occidental would be raising its posted price by 30 percent and that the tradi-tional 50–50 profit split between producing countries and oil companies would be altered to 58–42, though not in Occidental's favor. At first the oil men did not get the message, but when they learned that Qaddafi had dispatched soldiers to occupy their properties, they decided to go along with the new arrangement, at least temporarily. Then a bulldozer, accidentally or otherwise, cut the trans-Syrian tapline, interrupting a daily flow of 500,000 barrels of Saudi Arabian oil to Mediterranean tanker terminals and, from there, to world markets. In addition, for some inexplicable reason, the Syrian government decided that the line did not need repairing at that time. Whatever plans the oil companies in Libya, including the majors, may have had to trifle with Colonel Qaddafi were quickly canceled, and they buckled down to make the best of a bad situation.

The history of oil prices since 1970 is shown in figure 2–1. What this sketch does not show is that the nominal price of petroleum had not been behaving particularly well in the postwar period from the point of view of the producing countries. It may be that over the twenty years from 1950 to 1969, the real price of oil decreased by approximately one-half. As far as I can tell, however, it was Exxon's reduction in the posted price of crude that provided the principal impulse for the formation of OPEC, since the in-comes of the producing countries were determined by the posted price, and in general these countries adjusted their expenditure to this income. Saudi Arabia's income fell by $30 million for the year 1960–1961, causing Sheikh Abdullah Tariki to rush to Baghdad to express his solidarity with the gov-

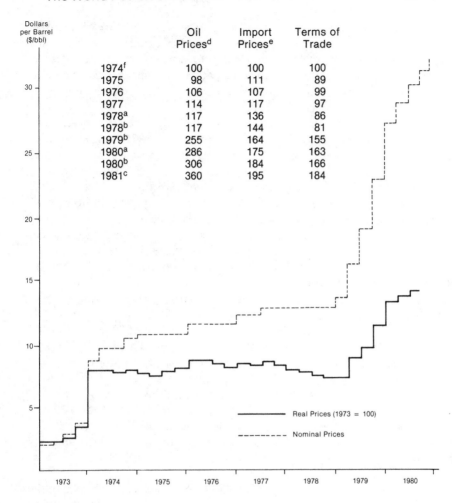

	Oil Prices[d]	Import Prices[e]	Terms of Trade
1974[f]	100	100	100
1975	98	111	89
1976	106	107	99
1977	114	117	97
1978[a]	117	136	86
1978[b]	117	144	81
1979[b]	255	164	155
1980[a]	286	175	163
1980[b]	306	184	166
1981[c]	360	195	184

——— Real Prices (1973 = 100)

- - - - - - - Nominal Prices

[a]Yearly average.

[b]December.

[c]Estimated.

[d]OPEC marker crude through December 1978; Effective OPEC price afterward.

[e]Wholesale prices of nonfood manufacturers in twelve industrial countries in dollar terms, weighted by OPEC imports.

[f]Index: 1974 = 100.

Figure 2–1. The Real and Money Price of Saudi Arabian (Marker) Crude and the Real Price (Purchasing Power) of Oil to OPEC between 1974 and 1981

ernment of Iraq (which felt itself to be under pressure because of its threat to revise producing-company concessions in that country). A short time later Juan Perez Alfonso of Venezuela flew to Baghdad, where he urged the Arabs and the Iranians to band together with each other and, for that matter, with any oil-producing countries desiring to construct a common front against the oil companies and their protectors.

Another thing figure 2–1 does not indicate is that the price increases of 1973–1974 were not entirely arbitrary. From early 1971 the intention of OPEC directorate was to get the price of Persian Gulf oil up to $5 or $6 a barrel as soon as possible. In line with this goal, OPEC began imposing various restrictions and inconveniences on the oil companies, hastening the day when the companies would find themselves completely without privileges, and perhaps without properties, in the producing countries. Among the first suggestions forwarded by OPEC was that the operating companies should begin considering a transfer of 20–25 percent of their assets to all producing countries—except, of course, those countries that were already asking for or taking more.

The oil companies naturally were not overjoyed at having to contemplate arrangements of this nature. The U.S. Department of Trade valued oil-company assets overseas at $1.5 billion in 1970 and estimated these as having a yield (or quasi-rent) of $1.16 billion. These figures imply a percentage yield on capital of about 77 percent, considerably above the 15–20 percent average that most foreign investments by U.S. firms were bringing in. (Even if this reported quasi-rent figure were in fact revenue, however, I still calculate a before-tax return on capital very near to 50 percent for the period taken up by the Department of Trade report.) The operating countries were prepared to surrender any part of these uniquely valuable assets only if they had no alternative, which in fact was the situation.

The present discussion will be concluded by saying something about the arrangement eventually worked out by the producer countries for the distribution of oil between the producer companies and their hosts. First, some of the oil produced in a given country falls in the category of *equity oil* or *equity crude*. This is the portion of the output produced that belongs to the oil companies in the sense that they can be reasonably sure of getting this amount at a reasonably stable price. The exact pricing details are explained at some length in *The Political Economy of Oil,* but in essence the price an operating company pays for equity oil is effectively equal to the cost of extraction *plus* the royalty per barrel *plus* the income tax per barrel, where this latter entry is calculated on the basis of the posted price. It is the posted price that largely determines the cost of oil to the firms, although any cost can be adjusted by attaching various fees in the form of surcharges, levies, or even exploration fees if a company does not do enough exploration. Generally, equity oil comes to between 30 and 50 percent of output. The rest of the output (total output minus equity oil) is called *participation oil* and belongs to the producing country. A portion of this participation oil, called *buy-back oil,* is sold by the countries to the oil firms at a price called the *buy-back price* or, sometimes, the *participation price.* This price is supposed to be determined on the basis of negotiations between seller and buyer. The usual arrangement is for the oil firms to commit themselves to buying a certain amount of oil from the producing countries, thus assuring OPEC that it will have a market for a large part of its oil even during periods of oversupply. Although the oil companies have tried, on at least one occasion, to avoid buying the amount of buy-back oil that the countries insisted they buy (and that, legally, they had contracted to buy), the general practice is to honor commitments. This is a question not merely of business ethics, but also of being assured of deliveries of crude when oil is in short supply. Without these deliveries the oil firms jeopardize the profitability of their refineries. (At present the companies are buying more oil than they want to buy. Since the demand for refined products is sagging, they are suffering severe losses because they cannot resell this oil but, instead, must add it to inventories that are already excessive.)

An interesting trend that seems to be developing is the tendency of the producing countries to sell more oil themselves: both on the spot market, and directly to countries—from one state enterprise to another. In 1973 the private oil companies gained access to 92 percent of the total supply of world oil, but in 1979 this figure was only 58 percent. The exact reason for this change in marketing strategy is unclear, although some guesses seem in order. In 1979 Shell and Exxon alone had combined profits on the order of $10 billion, with a sizable part of this coming from their distribution and refining operations. The OPEC countries may be able to cut themselves in on a small part by the aforementioned behavior. It is also possible that there are many people and organizations in the oil-importing countries who would be happy to see OPEC countries with some of these profits, since it might make them less aggressive in raising oil prices. Furthermore, in line with an argument presented later in this chapter, this might be the start of a campaign to wean the oil companies away from a certain intake of oil for their refineries in order to persuade these companies either to close down some capacity or at least not to construct any more. Thus later on there would be more scope for OPEC refining installations.

Post-1973 Pricing

In the spring and summer of 1973, King Faisal of Saudi Arabia was hard at work trying to convince the United States to decrease its support for Israel. Since contact through conventional diplomatic channels produced no results, he turned to the oil companies, informing their directors that the relations of these companies with their host governments could hardly improve if they did not use their influence in Washington. The oil men duly interceded, but their messages fell on deaf ears. The agonies and ecstasies being experienced by the oil companies' leadership had begun to bore the decision makers in Washington, who for the most part were preoccupied with the decline of the dollar and the recovery from the Vietnam fiasco.

In August 1973 Sheikh Yamani announced that Saudi Arabia would under no circumstances increase production by more than 10 percent a year. Given the projections for desired consumption of petroleum over the coming years, this meant that the industrial countries might soon face an undersupply of his vital input. At almost the same time, Libya announced that it was unilaterally increasing the price of its crude to $6/barrel. This move caused President Nixon to initiate a futile attempt to persuade the government of Libya that it was making a mistake, reminding it of such things as the Iranian boycott of 1951–1954 and, at the same time, trying to convince his allies and well-wishers throughout the world that a little well-chosen rhetoric from the Oval Room of the White House was still a potent factor in international politics.

On 6 October, war broke out among Egypt, Syria, and Israel. On 8 October, at OPEC's regular meeting in Vienna, Sheikh Yamani demanded $6 for each barrel of oil that OPEC members turned over to the oil companies. The companies in turn bid $3.50/barrel. After a while, with OPEC's offer at $5 and the oil companies' at $4, the meeting was adjourned.

The directors of the oil companies, still imagining that they counted for something in setting oil prices, claimed that they had to consult their governments before rendering a final decision. The OPEC directors simply shuttled over to Kuwait where, without further ado, they fixed the oil price at $5. Now, however, with U.S. weapons pouring into Israel, the Israeli army across the Suez Canal, and the road to Cairo open, something more drastic seemed in order to the Arab members of OPEC. It came in the form of a boycott of those countries that were most active in their support of Israel.

The United States and Holland would get no oil, whereas other countries on the pro-Israeli list would have their deliveries cut by 10 percent at first and later by 5 percent a month. (Those countries not on the list could still buy oil, but only at the new high price, which was not much help.) This tightening of the oil screw, rather than the possibility of a major confrontation between the United States and Russia in the Middle East, showed the consumers of the major industrial countries just what kind of world they were living in. They reacted by making it clear that they would accept any settlement that would keep their automobiles on the road. In November

OPEC met again in Vienna and informed the oil companies that, since they were apparently short of constructive solutions to various problems of mutual concern, OPEC would take the matter of oil pricing into its own hands. Price formation via negotiation between producing countries and outsiders now belonged to history.

Shortly before the next OPEC meeting, OPEC economists advanced the opinion that an oil price of $17/barrel was called for by the existing state of supply and demand. This price may not seem excessive on the basis of today's oil price; but had it been imposed and maintained at that time, it would have undermined the noncommunist industrial world and probably brought on some kind of military intervention against OPEC. In the end Saudi Arabia refused to go along with this price, and OPEC settled for a tariff of $11.65/barrel. (According to J.J. Servan-Schreiber, the celebrated presidential adviser, Dr. Henry Kissinger, was indifferent to the level of oil prices, seeing them mainly as a tool for financing a counter-Soviet military buildup in the Middle East in general, and Iran in particular.) As for the oil companies, they confirmed their new status by maintaining a deep silence, although by way of contrast a few prominent economists began concocting fairy-tales about how easily the new price could be accommodated as long as a part of OPEC's revenues could be recycled to Wall Street and London.

The boycott ended in March 1974. Since that time there have been a number of minor upward adjustments in the oil price, as well as the second oil-price shock that followed the revolution in Iran and the falloff in supplies from that country. All in all the OPEC countries have shown a great deal of restraint; and I think that had the governments of the industrial countries been a bit more imaginative, the world energy picture could have been a great deal different from what it is today. One thing is crystal clear, however. Some of the members of the OPEC directorate have shown a profound understanding of the economics of the industrial world that has not been matched by the officers of many financial institutions; or by the indolent, incompetent, and perplexed functionaries of many international organizations; or by a surprising number of academics. Certainly, there is a general unawareness among these people and their political masters that the social and economic progress of the industrial world over the last few decades has been based on cheap energy. With energy rapidly growing more expensive, a number of important social and economic changes will have to take place in the near future unless the citizens of the United States and western Europe have developed a secret passion for higher rates of unemployment and inflation, and a great deal less social security. Moreover, they completely misunderstand or underestimate the intention of certain key OPEC states to conserve their priceless hydrocarbon resources and use them as inputs in domestic processing operations.

The Supply of Oil and Oil Products

With the preceding section as a background, we are now in a position to deepen and widen our discussion and, above all, to commence our examination of the very important topic of petroleum-supply economics. First, however, some general observations are in order. Just a year ago, prior to the present glut of oil, the industrial world was in a state of extreme agitation as the price of oil was adjusted up almost monthly by direct price increases as well as by levies and surcharges of various kinds. This disquiet is understandable since, for the world as a whole, crude oil accounted for more than 40 percent of the primary energy consumption; for the industrial countries this figure tends to be much higher.

It is also of some concern for the Western industrial countries that the majority of identified resources of petroleum that can be recovered at acceptable costs (that is, *reserves*) lie outside their immediate political orbit. Of approximately 650 billion barrels (650 Gbbl) of oil reserves worldwide, about 58 percent can be attributed to the Middle East. The USSR is also a major possessor of reserves, as well as being the world's largest oil producer. Considerable anxiety has resulted from the recent events in Iran and speculation as to whether the Iranian experience could be duplicated in other Middle Eastern settings; certainly, no one knows exactly what the future role of Saudi Arabia will be on the world oil scene. Several years ago is was widely held that the kingdom was on its way to a daily production of 18–20 million barrels of oil per day (18–20 Mbbl/d), but now the governments of the oil-importing countries consider themselves fortunate if the Saudis agree to keep production from dropping below the 8–8.5 Mbbl/d range.

There are of course huge amounts of petroleum whose location is unknown today, but which will at some point in the future be placed in the reserves category. Exploration maps make it clear that only the United States has really been given a complete examination; as will be made clear later, however, geologists feel that most of the big strikes have taken place and that those taking place later will involve high-cost oil. In fact, although it is not fashionable to discuss the subject, the cost of reduced dependence on OPEC oil (which for the most part can be lifted for an outlay of well under $1/barrel) will be a number of macroeconomic discomforts, of which the most prominent are lower rates of aggregate economic growth and levels of unemployment that may eventually compare with those of the 1930s.

The emphasis in this chapter, as well as in *The Political Economy of Oil,* is on the reduced availability of petroleum in relation to the continuing need for this resource. It is only fair, however, to point out that forecasts of drastically reduced supplies of oil and other natural resources are an old story on this planet. It also seems that no sooner are these jeremiads circu-

lated than announcements ring out about giant new discoveries or technological prodigies that accelerate the substitution of one resource for
another. Taking just oil, it is interesting to regard a few opinions rendered
for the United States by people who should have known better:

Year	Prophecy	Source	U.S. Output
1866	Synthetics available if oil production should end	U.S. Revenue Commission	3.5 Mbbl/y
1885	Little or no chance for oil in California	U.S. Geological Survey	30.0 Mbbl/y
1891	Little or no chance for oil in Kansas or Texas	U.S. Geological Survey	53.0 Mbbl/y
1914	Total future production of no more than 5.7 billion barrels	U.S. Bureau of Mines	240.0 Mbbl/y
1920	Domestic production almost at a peak	Director U.S. Geological Survey	400.0 Mbbl/y
1939	U.S. oil supplies will last only thirteen years	Department of the Interior	1,300 Mbbl/y
1949	End of U.S. oil supply almost in sight	U.S. Secretary of the Interior	1,900 Mbbl/y

On the other side of the ledger, it is easy to cite crank accounts of a
world in which 15 or 20 billion people toil and frolic in solid bourgeois comfort without having to worry about where they will get their energy. Her-
man Kahn paints this kind of picture in both his written work and his lec-
tures, and there seems to be a large audience for these notions.

Regardless of the fantasies of pseudoprophets, the present outlook is
for a world in which the demand for energy will almost certainly continue to
rise, though considerably under the pre-1973-1974 trend rate. In the short
run, life-styles cannot change; even if they could, there are always the twin
driving forces of population growth in the Third World and the need for at
least a small amount of economic growth in the industrial world. There is
no point in being concerned with population growth in less-developed countries (LDCs) in this book, because for all practical purposes it is already out
of control. This is going to cause future generations more agony than any
annoyance associated with petroleum. Similarly, though less dangerous,
should the tendency persist for economic growth in the industrial world to
fall well short of 3 percent, then we can expect an extension of such
unneighborly practices as protectionism in trade and a more widespread sec-
tarianism on the domestic front—to begin with. In addition, as has been
made clear in the United States recently, investments in such things as pollution control are being canceled or postponed. In almost every country out-

side the United States, the amount of funds available for investment in alternative energy technologies exhibits a yearly decrease in real terms.

Generally it has been the case that the growth in energy use has outstripped by a large margin the aggregate growth in economic activity, but now this pattern is perhaps being reversed. In some ways this is a good thing, but it should never be forgotten that the productivity of labor is a function of the physical capital with which it collaborates. In the short run, at least, since energy and capital are complementary, this will mean that on the average there will be a smaller amount of capital per employee. Although it might appear that this arrangement will favor employment, the fact is that it will have drastic repercussions for productivity, investment, and—unless real wages and salaries can move rapidly downward—the growth of employment opportunities.

The last item that will be taken up here has to do with the possibility of accidental interruptions of oil supplies from politically sensitive regions. These accidents involve such things as the blowing up or bombing of pipelines; the closing of the Strait of Hormuz or a waterway of similar importance as a result of a misunderstanding; the change in attitude of one or more governments in the Middle East toward one of the great powers, or even one of the not so great; the bombing of Arabian oil by a non-Arab country and, in particular, by Israel. This last nightmare, incidentally, is never even mentioned in diplomatic circles, although it is certainly high on the list of Israel's priorities in the event of a future war. We can also take note of such things as the 1977 fires in the Abqaiq oil-producing center of Saudi Arabia, the Ekofisk blowout, and the extensive damage suffered by the trans-Alaska pipeline in 1977.

Elementary Supply Economics

Contrary to general belief, petroleum is found in small open spaces (pore spaces) in permeable reservoir rock, and not in large open caverns. An oil field consists of one or more distinct accumulations of this nature, which, inappropriately enough, are called pools. If sufficient gas is dissolved in the oil and the confining pressure is high enough, a well driven into oil will release that pressure, and the expanding gas will drive the oil to the surface, sometimes in the form of a gusher. In this situation the oil does not have to be pumped, at least during the early life of the field; however, the rate of production that can be sustained from any particular reservoir tends to decline as the oil in it is depleted.

If the natural pressure of the field becomes insufficient, the oil is pumped out. The two arrangements mentioned here (natural release and simple pumping) are called *primary production*. In the United States in

1977–1978, the amount recovered by primary production came to 25 percent. In fields where primary production results in an inadequate yield, secondary methods are also employed. These involve either pumping water into the stratum below the pool, or pumping gas into the layer above, with the intention of flushing out the crude oil clinging to the reservoir rock. Between 10 and 60 percent of the original *oil in place* in various reservoirs in the United States is recovered by these two methods, with the cumulative recovery average being about 31 percent. For every barrel of crude oil produced, more than two must be left in the ground if recovery is dependent on primary and secondary recovery methods at their present level of efficiency.

Thus two-thirds remains either in the form of a film on the walls of the pores, or locked in the reservoir by droplets of water. The key to recovery at this (tertiary) stage is lowering the viscosity of the oil by such things as the injection of steam or the ignition of underground fires. Tertiary (or enhanced) recovery also includes injecting gases like carbon dioxide or nitrogen; using chemicals called surfactants to wash the droplets of oil from rock pores; and using polymers to thicken the water used in flushing. Everything considered, these methods are still relatively unimportant; and in the United States—where they are used more liberally than elsewhere—they have raised the recovery rate to 32 percent, compared with a worldwide average of 30 percent. There is, in fact, an important body of opinion that claims that these tertiary methods are doomed to remain uneconomical except in very special situations. A complete analysis of this topic can be found in Dafter (1981).

Although crude oil is a clearly defined commodity, it is still possible to distinguish a number of varieties. The most widely used criterion for distinguishing petroleum products and crude oils is the American Petroleum Institute (API) degrees gravity scale, which essentially converts the specific gravity of a liquid to an API number that declines as the specific gravity increases. Thus Venezuelan *conventional* crude has an API of 26, which means that it is very heavy, whereas Algerian crude has an API of 41, which means that it is very light. By way of contrast, asphalt, which has a specific gravity of 1.08, has an API number of 0, whereas gasoline—whose specific gravity is 0.74—has an API of 60. The principal fuels derived from crude oil are gasoline and fuel oil, and these are referred to as oil products. Oil products are normally obtained by refining crude, where the simplest operation in refining is the distillation that isolates the various components. Oil and oil products can be moved to markets in a number of ways, of which the two most prominent are large tankers (for transportation between seaports), and pipelines.

Next we can consider the important question of the oil resource base. Here *resources* are distinguished from *reserves* in that resources include oil in place that is undiscovered or, for that matter, is discovered but cannot be exploited for economic or technological reasons. Reserves, on the other hand, are resources that, given the prevailing state of science and technology, *can* be economically exploited. The consensus view of global reserves is that the crust of the earth originally contained about 2 trillion (2,000 billion) barrels of recoverable oil, of which about one-half has already been discovered and—according to so-called informed sources—the other half will eventually become available. Up to now somewhat more than 400 billion barrels have been extracted, and world proved reserves are almost 650 billion barrels. The geophysicist M. King Hubbert has combined estimates of petroleum resources with an analysis of historical data to plot the way world oil production should rise and decline. On the basis of oil-consumption rates in the mid-1970s, he concluded that output from the world's oil fields will reach a peak in the 1990s and will fall rapidly thereafter. The kind of logic used in this forecast is discussed in some detail later in this chapter, but an abbreviated exposition might be of some use at this point.

The basic building block for this analysis is the concept of the optimal rate of resource exploitation, which means that for economic reasons no more than a fraction of the reserves of a given oil field should be removed in any given year. This fraction is generally put at about one-tenth but is dependent on the characteristics of the field and, at least in theory, can be much smaller. If, for instance, a field contains 150 units of oil, and we desire to lift 10 units a year, we can do so for six years without violating the criterion expressed here—assuming that the *critical* reserve-production ratio is 10. This is so because in each of the first six years we are lifting less than one-tenth of the total extractable reserves of the field. Over this period the reserve-production (R/P) ratio falls from 15 to 10 (that is, 150/10, 140/10, 130/10, . . . , 100/10) as reserves fall by the production rate of 10 units a year. After the sixth year, however, if we continue to produce 10 units per year, the R/P ratio would become less than 10: that is, we would be lifting more than one-tenth of the field's remaining reserves in a year. Thus after the sixth year the R/P ratio takes over and determines production.

In the sixth year reserves were 100, and production was 10. In the seventh year reserves are 90 (100 − 10); with a desired or critical R/P ratio of 10, production is limited to 90/10 = 9 units. The following year, with reserves equal to 81 (90 − 9), production would be 81/10 = 8.1, and so on. This analysis was carried out for a given field, but in a typical oil-producing country there could be fields under development and not producing oil, even though they contribute to the total of proved reserves; the same is true for the world as a whole. Thus an appropriate global R/P ratio should be higher than 10, perhaps as much as 15. In the United States, including Alaska, the R/P ratio is now under 10, and in the lower forty-eight may be under 9 and close to what geologists consider the absolute minimum figure.

It should be emphasized, however, that the critical R/P ratio is basically an economic rather than a geological or technical concept. It is technical in that a too rapid depletion of an oil field physically damages the field, reducing the amount of oil that ultimately can be taken from a given field. The most important aspect of this situation, however, is that the reduction in reserves reduces the (expected) discounted cash flow from a field. By extension, in an emergency such as a war the R/P ratio can be pressed down to well under the critical value.

In the foregoing example, when production turned down, there were still 90 units of reserves in the ground out of an original 150—well over half the original reserves. This indicates that there is a sound theoretical basis for King's theory that a decline in oil output in this century is inevitable, even though known reserves are huge. Note also that in contrast to the foregoing example, the world demand for oil has not been static but, until a few years ago, was increasing at rates of up to 5 percent a year, or even more. Had these growth rates not decreased considerably because of the slowdown in world economic growth, world oil production might have peaked in the late 1980s or early 1990s, with as much as one trillion tons of recoverable oil in the ground at the time of peaking. Regardless of *when* output peaks, however, it is impossible to deny the fundamental logic of Hubbert's analysis, which ends up with a construction similar to the one in figure 2-2.

Now we can discuss the possibilities for moving the peak of the foregoing diagram to the right—that is, into the more distant future. Principally, there are two ways to go. The first is to find more oil, either in conjunction with existing fields, or in new fields. In the United States, for the first time in the history of that country, the drilling boom that began after the 1973-1974 oil-price rises did not succeed in finding appreciable amounts of new oil, although the discovery rate may have been increased somewhat. The only thing this drilling boom made clear is that U.S. oil production has already peaked. The problem is that it is not sufficient to find *more* oil, but oil must be found in enormous quantities and not dribbles. Thus when oil geologists speculate on finding enough oil to reverse the present tendency to run down reserves, they are thinking in terms of giant or supergiant fields; and the general belief is that there are very few of these left. Richard Nehring of the Rand Corporation has argued that the problem of estimating the global availability of oil can be greatly simplified by concentrating only on giant and supergiant fields—those containing more than 500 million and 5 billion barrels of oil, respectively. In this century 33 fields in this category have been located worldwide, and Nehring estimates that outside the United States 75 percent of all ultimately recoverable oil lies in giant and supergiant fields. The corresponding figure is about 25 percent for the United States. Unfortunately, however, despite one of the most rapid oil-price increases in history, the 1970s was the first decade in sixty years when no giant or super-

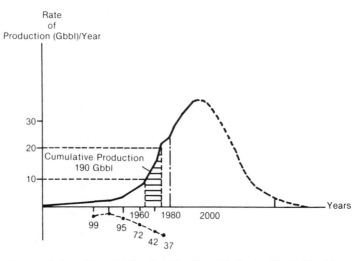

Reprinted by permission of the publisher from Ferdinand E. Banks, *The Political Economy of Oil* (Lexington, Mass.: Lexington Books, D.C. Heath and Company, 1980), p. 49.

Figure 2-2. Cumulative World Oil Production and the World R/P Ratio

giant field was discovered. One reason for this may be that there are few of these left to find.

The other possibility involves reducing the demand for conventional oil by utilizing the copious supplies of other oily substances in the earth's crust (shale oil, heavy oils, tar sands, and so on). This is possible, of course, but it is expensive. In a 1977 briefing paper British Petroleum estimated that there are at least 400 billion barrels of heavy oil and oil from tar sands, and 200 billion barrels of shale oil, that can eventually be recovered; many reputable estimates are even more generous. Moreover, a large part of these unconventional oils are located in the United States and Canada, with the USSR and Venezuela also major repositories of these materials. Table 2-2 shows some of the latest cost estimates for energy resources.

These figures, though among the best available, are approximate and are subject to rapid upward adjustment. It should be appreciated also that the cost of liquids from shale given previously does *not* cover rectifying the enormous environmental disturbances that could be incurred in exploiting this resource. The same thing is true of coal, though on a smaller scale. It may be for this reason that certain very large oil companies are closing down their shale operations.

Finally, some attention should be paid to conservation. Although I prefer to believe that the major force causing the present stagnation in oil

Table 2-2
The Cost of Some Energy Sources
(dollars per barrel of oil equivalent)

Energy Source	Production Cost[a,b]
Middle East oil from existing fields	1–3
North Sea oil from existing fields	5–20+
Liquids from tar sands and/or shale oil (North America)	15–35
Nuclear input (fossil-fuel equivalent)[c]	7–20
Coal in the United States	4–8
Imported coal, northwest Europe	10–14
Domestic coal, western Europe	9–20
Liquefied natural gas imports	25–35
Synthetic natural gas from coal (U.S.)	35–50
Liquids from imported coal (Europe)	45–65
Biomass	45–80+
Solar heat	120+
Electricity from solar, wind, tidal	120+

[a]All figures in 1980 dollars, and all energy forms taken as the thermal equivalent of a barrel of oil.

[b]Production costs only.

[c]To obtain the electrical-energy equivalent of a barrel of oil requires x amount of nuclear fuel, which in turn costs $7–$20 (as shown).

use is the drastic slowdown in economic growth in the major industrial countries, it must be admitted that improved energy efficiency has played a very important role in reducing energy consumption, and perhaps can do even more in the future. This is going to be a long-run prospect, however, since it involves the substitution of capital for energy, at least to a certain extent, as well as a great deal of innovation. Moreover, much of this activity is going to have to be carried out by individuals rather than corporations; and, as we know, the capacity of many individuals for evaluating energy-saving investments is not very great. Even so, some analysts claim that by the end of the century, energy-saving innovations alone may be capable of reducing total energy use to under the 1973 level. My opinion is that although this may be true for a few countries under certain circumstances, for the world as a whole energy use will continue to climb, though at a declining rate of increase. (This topic is treated in more detail in chapter 8.)

Petroleum Refining

Petroleum refining can be characterized as a variable-proportions, joint-product industry. The principal input is crude oil that is converted into dif-

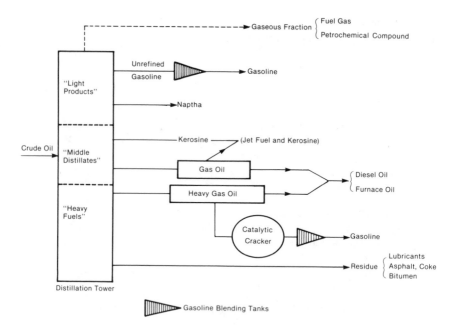

Figure 2-3. The Output Pattern of a Typical Refinery

ferent *slates* of refined products (or oil products). The exact composition of the slate depends on the demand for outputs as expressed through their price. At present the demand profile is heavily weighted in favor of the transport fuels and petrochemical raw materials, instead of the fuel oil and other heavy-oil products that were so desirable a few years ago; as a result, oil companies have had to invest heavily in the equipment needed to satisfy this type of demand. The point here is that the cost of each slate is different since they differ in capital intensity; for example, gasoline-heavy slates require more capital equipment per barrel of output than do others. Diagrammatically, the refinery input-output scheme looks as shown in figure 2-3.

This arrangement functions as follows. Crude oil is heated and pumped into a tall distillation tower that is pressurized and is hotter at the bottom than at the top. The various oil products have different boiling points, with those that are the lightest having the lowest. When the crude enters the tower, the heaviest part remains in liquid form and falls to the bottom. The rest is vaporized, but the various constituent products return to liquid form as they reach the lower temperatures higher up the column. As a result, they can be piped away.

The amount of each product that can be obtained from a refinery can be altered somewhat by changing the temperature and pressure in the tower, but the difficulty with the basic arrangements shown in figure 2-3 is that the refinery will always produce a certain quantity of each product, to include a fairly high percentage of fuel oil. (This is true even with North Sea crudes that are rich in lighter products.) Major changes that involve getting more light products require investment in upgrading equipment such as catalytic crackers. These crackers, together with a catalyst, crack apart the chains of molecules in one or more of the heavier products to form lighter products. Naturally, putting in these crackers is an expensive proposition, particularly in the face of a falling demand for all oil products.

To indicate the change in demand patterns, Esso (UK) has estimated that in 1973 about 46 percent of the total demand for oil products was for the heavier products, in particular fuel oils. Demand for middle distillate (such as industrial heating oil) was 28 percent of the total, and demand for the lighter products was 26 percent. By 1985 it is believed that these figures will be 29, 35, and 36 percent. With total demand down, refinery capacity is in oversupply in western Europe, and a number of refineries are scheduled to close. It is said, for example, that 30 percent of Europe's refining capacity will have to go if the present capacity-utilization figure of 58 percent is to be improved on. An interesting irony involving refineries has recently surfaced. In 1980, refineries were reducing their capacity because they were being squeezed between the rising price of crude and the falling demand for various oil products. Now it is the oversupply of crude that is causing some refiners problems, particularly in the United States. In that country the major refiners have access to so much inexpensive Saudi Arabian oil that they must either store it at very high charges or sell it at its original cost. Selling it, however, often means selling it to competitors who were in dire straits because they did not have access to cheap crude.

Later in this chapter the reader will be given some insights into OPEC intentions to play a major role in the great world of petrochemicals, but similar plans also exist for expanding OPEC's refining capacity. As has been made clear, the world refining industry is suffering from severe overcapacity; as a result, profit margins have fallen sharply in Europe and, for a number of refiners, in the United States. Even so, OPEC producers are determined to own a larger share of the world's processing facilities. In 1980 alone their refining operations expanded by 17.5 percent to about 6.3 Mbbl/d, and now make up 7.5 percent of the world total. Present intentions are for another jump in output of 50 percent in the coming four years, bringing total capacity up to 9.3 Mbbl/d by the beginning of 1985. A summary of projections for OPEC countries is given in table 2-3 on a country-by-country basis.

The country with the most ambitious plans is Saudi Arabia, where everything possible is being done to supplant Venezuela as OPEC's largest

Table 2-3
Existing and Projected Refinery Capacity of OPEC Countries
(thousands of barrels per day)

	1981	Under Construction	Additional Planned	Total Expected (1986)
Algeria	442	—	334	786
Equador	86	—	108	195
Gabon	20	—	—	20
Indonesia	486	186	265	937
Iran	1235	—	—	1235
Iraq	249	140	—	389
Kuwait	554	58	154	766
Libya	142	220	—	362
Nigeria	260	—	—	260
Qatar	14	47	—	62
Saudi Arabia	717	734	466	1917
U.A.E.	126	56	172	354
Venezuela	1349	—	150	1499

Source: OPEC Bulletin; OAPEC Bulletin, various issues.

refiner. Iraq was in second place where ambition is concerned, with plans to eventually expand refined output by a factor of three to approximately 1.0 Mbbl/d; but it must be accepted that Iraqi plans may be delayed by the war with Iran, just as the plans of OPEC as a whole might be held up by a faltering world market. The second-largest OPEC refiner at present is Iran. That country has given no indication of a desire to increase its capacity in the near future, however, although it can be expected that immediately after the conclusion of the war with Iraq, the refinery at Abadan will be repaired as rapidly as possible.

It is interesting that several countries that may be reducing their output of crude oil by a sizable amount in the coming ten or twelve years (Algeria and Indonesia) will have important additions to their existing refinery capacity coming on stream by the end of 1982 or the beginning of 1983. One conclusion might be that these countries will decide to reduce their production of crude in order to ensure that they have enough petroleum to supply their refineries over the lifetime of these installations. On the other hand, it could be that these refineries are intended to refine imported as well as domestic crude, since even countries such as Sweden—with no local oil production—maintain an important refining capacity.

At the time of this writing, refineries in the United States are closing at a record pace—faster than at any time in modern oil history, including the Great Depression. In 1980 fifty refineries, representing a daily processing capacity of one million barrels, terminated operations. About 25 percent of all independent refineries closed down, and the figure might be higher this year. As a result of the worst recession of the postwar period, Americans are buying less oil than ever, and refineries are functioning with a surplus capacity of about 6 million barrels per day. As indicated earlier, however, things are even worse in Europe, where surplus capacity may have reached 8 million barrels per day of a 20-million-barrel total capacity.

Some interesting economic questions have been raised by this alarming state of affairs. The most important of these probably has to do with the rate at which investment in new technology should take place. The Rock Island Refinery Company of Indianapolis, by investing heavily in technology that permitted obtaining more gasoline from a barrel of oil, was able to continue operating at about 80 percent of capacity, compared with a national average in the United States of 67 percent; their gain was somebody else's loss, however, since, given the overall development of the U.S. economy in 1981, there was no room for a sizable increase in gasoline consumption at that time. Thus the patent solution of mainstream economic theory—which is that everyone should have or should be investing in new technology—may not always be correct in these perilous times. In fact, the only thing an increase in new investment would mean today is a more rapid rate of refinery closings, which is precisely what is going to happen to European refineries when the new OPEC capacity starts coming on stream later in this decade.

Sea Transportation and Pipelines

Between the extraction and the refining of a large part of the world's supply of crude oil lies a very important economic activity: sea transportation. This activity also concerns refined products, which can be shipped in the same type of carrier as crude oil, though not always with the same efficiency.

Before 1973 sea trade in general grew at an average rate of 8 percent per year in volume and 11 percent in ton-miles. This discrepancy was largely the result of a faster-than-average growth in the shipping of crude oil. Oil has been the paramount sea cargo; until very recently oil tankers accounted for 50 percent of the world's shipping fleet, 40 percent of total tonnage, and over 50 percent of annual ton-miles. The amount of oil transported by sea increased from 250 million tons (Mtons) in 1954 to almost 2 billion tons (Gtons) in 1979. Over this period the world's tanker fleet increased from 3,500 ships to 7,000 ships and in total weight from 37 million deadweight tons (37 Mdwt) to 340 Mdwt. Despite an existing tanker surplus that came to 17 Mdwt at the end of 1979, tankers are still being built. For the most part these are smaller tankers (60,000–100,000 dwt); and one of the reasons for this is the deepening of the Suez Canal, which means that tankers can take the short route to European ports from the Gulf (which averages thirty-one days) rather than the long route around the Cape (which averages forty-one days). It was this long route that made large tankers (150,000 dwt and up) so economical when the Suez Canal was closed. Viewed over the last few years, the average cost of shipping a barrel of oil between the Middle East and Rotterdam has been about $1. These rates vary considerably, however. In the closing months of 1979, large tankers were getting freight rates of $1.22/barrel to $1.36/barrel, with their break-even cost being close to the top of this range. In early 1979 they could get only $0.137/barrel.

Generally freight rates are given in *worldscale flat rates,* which are established semiannually by the International Tanker Nominal Freight Scale Association and are the basic reference listing of tanker rates. The rate is given for each route and is the approximate cost of shipping a barrel of crude oil over the particular route on a tanker of standard size, which happens to be 19,500 long tons. Actual rates are then expressed in terms of the flat rate. For example, in February 1978 a 50,000-deadweight-ton tanker carrying crude from the Gulf to Japan quoted a spot (single-trip charter) rate of worldscale (WS) 80. Thus the actual rate was $0.80 \times \$1.26 = \$1.008$/barrel, where $1.26 is the listed or flat rate.

At the present time, a high proportion of the world's tanker fleet is lying idle. Here we find both smaller ships and the very large crude carriers (VLCCs) that were supposed to be the wave of the future in sea transport. There are now about 700 VLCCs in existence, but about 250 of these fall in the category of excess capacity and are to be found in uneasy retirement off Brunai or in the fjords of Norway, or are due to pay a visit to the scrap yards of Taiwan. Taiwanese buyers are having no problem at all buying ships that once cost up to $100 million for $4 million. Table 2–4 gives some idea of the size and ownership of the world's tanker fleet.

Today no one knows how the world tanker surplus is going to be alleviated. In the long run OPEC countries might be willing to buy some of the existing fleet; but by the same token they could decide that if they are going into the shipping business on a large scale, it might be better to invest in state-of-the-art equipment. This latter option seems quite possible just now, which is hardly good news for existing ship owners.

Before concluding this discussion it should be pointed out that although tankers are generally the cheapest form of transportation between two seaports, sometimes the tanker route is so long that pipelines are economical. Pipelines from the Gulf to the Mediterranean have been shown to be economical in the sense that they help reduce the cost of getting oil to Europe; and in the case of natural gas a seabed pipeline from North Africa

Table 2–4
Tonnage Structure of the World Tanker Fleet
(millions of deadweight tons)

	Total Tonnage	Tonnage over 175,000 Dwt	Tonnage on Order
Seven major oil groups	62.7	47.9	1.7
BP	7.0		
Exxon	16.9		
Gulf	4.1		
Mobil	5.7		
Shell	12.9		
Socal	9.4		
Texaco	6.7		
Other oil companies	68.2	30.8	4.1
Independents	191.6	110.6	13.2
Total	322.4	189.3	19.0

to Europe (Algeria to Italy) now exists. In the United States the high cost of operating ships between the Gulf of Mexico and east-coast ports has led to a heavy dependence on pipelines, particularly for carrying products that are refined near the major oil-producing centers of the Southwest.

Growing political collusion between Iraq and Saudi Arabia has resulted in an agreement that will allow Iraq to build a new pipeline across Saudi Arabia, probably terminating at or in the vicinity of the Red Sea coast seaport of Yanbu. The initial capacity of this outlet has been said to be 1 Mbbl/d, and it would be fed by oil from the Basrah deposits. Some plans apparently exist for the capacity of this pipeline to be brought up to 3 Mbbl/d, which is the amount Iraq was exporting prior to the initiation of hostilities with Iran. Just what economic considerations lie behind an extension of this magnitude are unfortunately not clear. Iraqi pipelines already carry oil from the Kirkuk region of Iraq to ports in Syria and Lebanon, although enmity between Syria and Iraq has apparently resulted in a greatly reduced flow through this line. There is also a pipeline from Iraq through Turkey that reaches the Mediterranean coast. The capacity of this conduit is 700,000 Bbl/d, and it seems to have been functioning more or less without difficulty. If a pipeline to Yanbu were built that could handle 3 Mbbl/d, it would give Iraq a pipeline capacity that was far above any production of oil that the present government of Iraq regards as conceivable.

Since about one-half of OPEC's production—which is approximately one-fourth of the supplies of the noncommunist world—is shipped through

Gulf ports, and therefore must pass the Strait of Hormuz (which, at least in theory, is one of the most easily blocked waterways in the world), then the oil-importing countries of Europe and the Americas, and Japan, have a vested interest in the pipelines that are being constructed in the Middle East. The new Saudi Arabian pipeline—sometimes called the Petroline—was built by Mobil at a cost of $2 billion. It will have a capacity of 1.85 Mbbl/d to begin with but may be expanded to 4 Mbbl/d later. Saudi Arabia also has a 1 Mbbl/d pipeline to the Mediterranean through Jordan and Lebanon, but it is closed at present because of unrest in Lebanon. For easily understandable political reasons, Saudi Arabia will probably prefer, in the future, to dispense with pipelines that traverse foreign soil.

In contrast to pipelines, the transportation of oil by tankers is a major source of environmental destruction. As many residents of Scandinavia know, the cleaning of tanker tanks and an assortment of accidents, many of them minor, have at times caused serious damage to Swedish resort beaches and to the water and beaches of both the Stockholm and Gothenburg archipelagoes. It is not unthinkable, however, that these problems can eventually be brought under control, since it is not particularly easy for the offenders to leave the scene of the crime.

Offshore Oil

Just now the search for offshore oil is accelerating, particularly around energy-deficient Europe. There is also a great deal of activity in the vicinity of the United States, where onshore supplies are definitely waning.

Offshore techniques are being brought to their finest flower in the North Sea. Recently the world's largest oil-drilling platform, the Statfjord B, was towed out to the world's largest offshore oil field (the Statfjord field), where it is expected to help generate a very large revenue for the Norwegian government. The cost of this platform was 12 billion Norwegian crowns (about $2 billion U.S.), but since it is intended to assist in the exploitation of a field containing 470 million tons of oil and 70 billion cubic meters of gas, very few people have been heard to complain about its cost.

At present, 280,000 barrels of oil per day are being obtained from a field adjacent to the one where the Statfjord B will be operating, and by 1984 there will be another field in operation in the general vicinity. Altogether, 600,000 barrels of oil per day will be lifted. This level of production could be held for thirty years.

The height of the Statfjord B platform is 271 meters, and its weight is 816,000 tons. It rests on four legs, which can also serve as storage tanks for oil. Although the Norwegian government is as much in control of its oil riches as any Middle Eastern government—perhaps even more so—this particular North Sea enterprise is being handled by the well-known oil major Mobil. Eventually, however, it seems likely that the Norwegians will assume a tighter control over oil production and exploration, which means that there may not be a warm welcome in that part of Scandinavia for the large oil companies. In fact, over the last few years taxes have been adjusted in such a way as to make the Norwegian sector of the North Sea an uninteresting playground for any firm in search of very high profits.

About 1,300 people were required to construct the Statfjord B. Its crew consists of 200 men who work in shifts of ten to twelve hours a day, for about 40 days out of 100. There is no problem in obtaining personnel for these platforms since salaries and working conditions are more satisfactory than those aboard most ships.

As shown in table 2–2, the cost of oil from existing offshore fields can often exceed $20/barrel. This is an important observation because if the oil price were to descend at the rate predicted by several so-called oil experts, certain offshore fields would cross the threshold of unprofitability, and investment in new offshore fields would all but cease.

The Rotterdam Spot Market

Appendix 2A will give the reader a basic but nontechnical introduction to spot, forward, and future markets. Since the Rotterdam spot market has become a household word, however, it deserves a brief introduction at this point in our exposition.

Rotterdam, Holland, is a huge refining center and the largest petroleum-handling port in the world. Conceptually, however, this has nothing to do with the Rotterdam Spot Market, since a spot market is a *way* (rather than a *place*) of doing business. For the most part oil is sold in bilateral deals between buyer and seller for future (forward) delivery. The exact price and quantity can be fixed in the contract, but it could also be stipulated in the contract that the price paid by the buyer at the time of delivery be other than that prevailing at the time the contract was drawn up. For instance, if the posted price of oil increases from the time at which the contract is entered into until the date of delivery, then the buyer might have to pay an appreciably higher price to get the quantity of oil specified on the contract. It may happen, however, that the purchaser of oil underestimates or overestimates his requirements when placing his contracts. If he underestimates, then he purchases the shortfall on the *spot* market (also referred to as the *free* market, where the price is set by short-run demand and supply). Obviously, if the buyer is in Marseille, his spot oil need not come from Rotterdam. Instead, it might arrive via a tanker that was in the Mediterranean and was redirected toward France when he gave his order (although in such a case there is a strong possibility that his order was handled by a Rotterdam oil broker).

Similarly, when buyers have too much oil, this oil can be sold on the spot market. There are other possibilities here. In Sweden several years ago, when it appeared that a small oil glut was in the offing, buyers placed fewer orders than usual with their customary suppliers (the large oil companies and the state companies of the producing countries). The glut did develop; and, since a great deal of oil was sold on spot markets by various countries and organizations, the spot price fell rapidly, and many Swedish buyers made some lovely profits. On the other hand, in 1979, when the price of oil was rapidly escalating and buyers worldwide were filling their storage tanks to the brim against the possibility of future shortages, those Swedish buyers who had not covered their requirements through long-term contracts had to pay very high prices for spot oil. Note the dilemma faced by both buyers and sellers when a commodity can be bought and sold on a forward market—where prices and quantities are specified—and/or on a spot market—where price formation depends on short-run supply-and-demand conditions. Should a buyer buy forward, and thus be assured of his requirements? Or should he take a chance and wait to buy on the spot market, hoping that the price will be low? In my courses in mathematical economics, I have often treated this kind of problem. As far as I am concerned, academic economics offers little toward its solution.

Lately, not only are the oil-producing countries selling more oil directly to the oil-importing countries, but they also have raised their deliveries to the spot market. Present estimates are that these countries now sell about 15 percent of their output on the spot market, compared with 10 percent during 1978 and slightly less earlier. The increased deliveries began in the middle of 1979, when the spot price of oil increased very rapidly relative to the posted price. Under the circumstances it made economic sense for anyone in a position to do so to divert supplies into the spot market. At the same time, however, it should be apparent that sellers generally would be unwilling to raise the amount sold on the spot market indefinitely, because in doing so they would have to accept the greater amount of uncertainty associated with the spot market (where future spot prices are unknown) as opposed to the forward market. Unfortunately there is little room for textbook behavior in situations characterized by the kind of uncertainty existing on the world oil market. It could be argued, however, that the problem is with the textbooks not with the markets.

Oil Stocks

In my books on copper, bauxite, and aluminum, I have emphasized the importance of inventories (stocks) and their role in real commodity markets, as opposed to textbook markets. What is true for copper and aluminum is doubly true for oil, although the problem is much more complex for oil than for these other commodities.

The first thing to distinguish here is a *supply chain* between the oil-producing facility and the final consumer. Each link, at any given time, is composed of a certain amount of more or less inert oil or oil products; even though these commodities are inert, however, a reduction in their level would eventually be reflected in a decreased amount reaching consumers—all other things being equal.

The chain is composed of stocks of crude oil in the oil-exporting countries, stocks at sea and in pipelines, stocks of oil products in refineries and depots, and so on. These supplies correspond to what in some economics texts is called working capital. On a world average the supply chain contains some sixty or seventy days of average consumption, or about 3 billion Bbl. A number of things have contributed to the gradual increase in the amount of stocks found in the supply chain, with one of the most noticeable being the increase in the size of oil tankers and the concomitant increase in storage capacity required in the vicinity of loading and discharging terminals.

Stocks of oil must also be held as a cover for seasonal variations. An alternative to building up stocks of, for example, heating oil in summer to provide for a higher demand in winter is to increase the flow of oil in winter (by utilizing more tanker and refining capacity). The economics of these options varies from location to location, however, and on balance it appears that an even use of tanker and refining capacity, with fluctuations in the level of inventories, is preferable to the opposite arrangement. It is also the case that stocks intended to accommodate seasonal variations in demand are generally held as close to the consumer as possible. For the reader intending to examine appendix 2A, it is useful to remember that with this category of inventory, considerable financial rewards can be gained by the stockholder who is a skilled weather forecaster or, by the same token, makes the same guess as his colleagues about the weather but is more competent in discerning and analyzing the other elements that influence future demand.

For the past few years the governments of the major oil-importing countries have also taken an interest in the level of stocks. In particular, they have been trying to ensure that, in the event of supply disruptions, vital economic activities in their countries can continue. Supranational organizations are also involved in this matter. European Economic Community (EEC) regulations require member countries to hold sufficient inventories to cover ninety days' consumption, based on the average demand for the previous year; the International Energy Agency (IEA) recommendation is for eighty days based on the previous year's imports. Although these stock levels may not seem excessive considering the types of interruption that could result from a major difference of opinions in the Middle East, it should be remembered that a world stock level of 3,000 million barrels is equivalent to an investment of more than $90 billion at present prices. Increasing stocks by an amount sufficient to get one extra day's coverage involves tying up at least another $1 billion in working capital.

Figure 2–4 shows world stocks (in all three of the foregoing categories) sufficient to provide about seventy days' cover for normal world consumption—assuming, of course, that there are no strange accidents in the storage areas of oil-exporting countries, no unforeseen barriers appearing across the straits of Hormuz, and so on. The figures associated with this diagram, however, do *not* include gasoline in the tanks of automobile owners or the storage facilities of gasoline stations, heating oil in the storage tanks of private homes, the often very large quantities held by private concerns and utilities, and the inventories in warehouses that are the property of speculators.

Lately a great deal of information is being made available about exotic plans to store crude oil in so-called strategic stockpiles. The United States has a long history of maintaining this kind of inventory and is now in the process of stockpiling oil in salt domes in Texas and Louisiana. By the end of 1981 more than 200 million barrels were in storage, and sometime before 1990 this inventory is scheduled to reach 750 million barrels. The purchase cost of this oil may reach $40 billion, and there will also be a kind of *imputed* interest cost associated with holding this oil rather than selling it at a price higher than its purchase cost and investing the revenue from those sales in such things as bonds or bank accounts. It is also possible that if things work out with the oil stockpile as they did with the tin stockpile, some of these strategic stocks may be made available to private consumers at a time that is most inauspicious for the commercial sellers of oil.

The Japanese government plans to construct eight large government tank farms in selected locations off the Japanese coast. About 50 million barrels, or thirteen days' consumption, are already available in oil tankers anchored off the Japanese coast; eventually Japan is to have a reserve of 190 million barrels. One-third of all Japanese tankers are now being used to store oil, and as yet costs are regarded as moderate—especially when the Japanese government contemplates the damage that could be caused the Japanese economy if imports of oil were cut off for an extended period.

Most European countries are holding stockpiles; one, Holland, has huge inventories of oil and oil products on its territory. In theory these stocks cannot be used in the event of an emergency since they are committed to international trade, but their presence is very reassuring to both Holland and its immediate neighbors. The IEA maintains that countries should run down stocks and relax stockpiling rules whenever there is even a slight disruption of supplies to the oil-importing countries; some question must be raised, however, as to whether this kind of strategy makes sense if there is even a moderate probability of a prolonged disruption. Certainly commer-

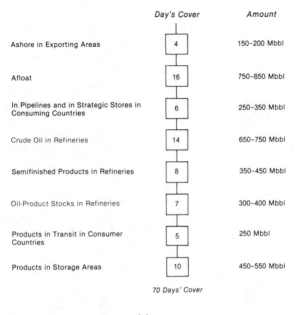

	Day's Cover	Amount
Ashore in Exporting Areas	4	150–200 Mbbl
Afloat	16	750–850 Mbbl
In Pipelines and in Strategic Stores in Consuming Countries	6	250–350 Mbbl
Crude Oil in Refineries	14	650–750 Mbbl
Semifinished Products in Refineries	8	350–450 Mbbl
Oil-Product Stocks in Refineries	7	300–400 Mbbl
Products in Transit in Consumer Countries	5	250 Mbbl
Products in Storage Areas	10	450–550 Mbbl

70 Days' Cover

(a)

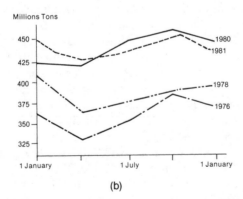

(b)

Source: IEA documents and Shell briefing reports.

Figure 2–4. (a) Stock Coverage Pattern, World Excluding USSR, Eastern Europe, and China; (b) Inventory Movements for Selected Years for IEA Countries

cial operators would find it distasteful to get rid of stocks knowing that, when restocking began, they might have to pay twice the price at which they sold their products. (This is a distinct possibility, as was made clear in the price run-up following the Iranian revolution). Somewhat more realistic-ally, the IEA has been active in promoting a scheme whereby oil-importing countries (the so-called IEA oil club) would share supplies in the event of an oil embargo or a precipitate falloff in the availability of oil due to a war or revolution in the Middle East.

At the moment total stock coverage may still be above the seventy days in the preceding example. Only a few months ago it was at least one-third more, although the unusually high level of interest rates common throughout the world today means that inventory carrying charges are probably at record heights. Still, many oil and oil-product consumers, remembering their difficulties during the 1979–1980 oil-price rises, are bear-ing this burden with a smile.

Oil and Macroeconomics

This section provides a transition between the preceding general discussion and the more technical materials to follow. First some attention will be devoted to various macroeconomic phenomena associated with the oil-price increases.

The first is inflation. Since oil is an input in most industrial processes, either directly or indirectly, and plays a prominent role in some of the most important consumer activities such as driving and the consumption of heating services, it would be difficult to argue that the 850-percent increase in the price of oil between October 1973 and January 1981 was not a major factor in raising the general price level. It was not only major, but crucial. Prior to the first oil-price shock, world prices were increasing at a rate of about 6 percent annually, and this was regarded as an aberration that would soon be brought under control. Today double-digit inflation is common in many industrial countries, and no knowledgeable person anywhere believes it possible to bring inflation rates lower than 1973 levels in the near future.

Even so, many economists contend that the oil-price increases are not responsible for the present world inflation, and that any normal economy is capable of adjusting to any type of external price shock. The adjustment they mean, of course, would take the form of a general decline in the stan-dard of living, as wages and salaries fall in phase with the increase in oil prices. Furthermore, these scholars insist, if employees are unwilling to go along with this arrangement, then it is the duty of the authorities to help them to conform by restraining the rate of growth of the money supply. An increase in oil prices will increase the costs of producers, who normally would raise product prices in order to obtain the revenues needed to cover these costs. If the money supply is not expanded, however, it will be impossible to buy, at higher prices, a large part of what has been produced. Thus employers will be forced to reduce costs by reducing wages and salaries. Should this be impossible for institutional reasons, they will be forced to close down some of their facilities and reduce their work forces. This is the theory. Although it sounds good, it has failed miserably on a number of occasions and it continues to fail. Since no one has bothered to give a rudimentary explanation of just why this theory does not work, however, I propose to do so here: it fails because the link between productivity and pay has been broken in many countries, so that inevitably many of the employees who are 'disciplined' by monetarism are in fact the most productive. For instance, in Sweden it is inevitably the least productive members of the work force who have the greatest job security. Textile and building workers periodically face long periods of involuntary unemployment, whereas pseudoscholars and aid agency employees, their clerical help and the people who arrange plane tickets and hotel reservations for them live off the fat of the land. Monetarism is likely to hurt aid agency drones far less than it does production workers.

Let us next consider what happens when authorities make an attempt to accommodate the impact of an oil-price rise. The analysis here employs a simple supply-demand scheme, as shown in figure 2–5a. Here we take the flow demand for output as a function of price, or $d = d(p)$. Aggregate output is an increasing function of price, or $y_s = y(p)$. It should be appreciated that we are dealing in real values. Now assume that the initial intersection of d and y_s takes place at full employment output, or y', which corresponds to an employment of N'. As noted earlier, an increase in the price of oil raises the cost of production, and shifts the aggregate supply curve to the left, so that output falls to y'', and there is an increase in the price level to p''. Note that y'' is below the full-employment level.

We can now examine what will happen if the authorities decide to counter the fall in output by monetary or fiscal expansion. In this case the demand curve is shifted to the right, causing a further rise in the price level and also a rise to a new full-employment output of y^*. Note that this comes about because, with money wages constant, real wages fall. Thus the full-employment labor force (N') still draws the same money wages, but they cannot buy as much; real gross demand may therefore be smaller. This is the kind of scenario Keynes would have used in his general theory had he been thinking about oil shocks. On the other hand, had money wages fallen (but not by less than the rise in oil prices), consumption would still have fallen (unless the people selling oil lent their revenues to domestic residents), but prices could have been kept from rising. This arrangement is shown in figure 2–5b. Note that the demand curve moves to the left under the

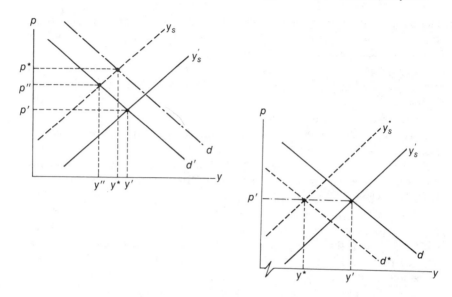

Figure 2-5. The Effect of an Oil-Price Rise on a Typical Economy

pressure of falling real incomes: money wages fall, but prices are constant. Note, however, that this is a very special case. It is, in fact, artificial.

We must now bring the oil-selling country into the analysis, but first the reader should be acquainted with the full significance of energy price increases on wages, rents, and the income of capital (or quasi-rents). In order to do this, let us consider a mythical land called Australia, where a great deal of tennis is played at night. An increase in the price of energy will mean that in some cases the total cost of operating a tennis club will exceed the cost of operating other types of facilities (such as parking lots) that make more efficient use of the production factors used to produce night-tennis services (land, electrical energy, and a small amount of labor). The proprietors of tennis clubs that operate only at night might, accordingly, experience a decline in the quasi-rent of their installation (where quasi-rent is defined as revenues minus operating expenses), and in some cases would conclude that, everything else remaining the same, they should trade in their nets for some parking meters. They will find it particularly difficult to stay in the tennis business when their courts deteriorate and they must consider the great expense involved in replacing them. Note, however, that should people be willing to pay more for their tennis, or should employees be willing to work for less, then it might still be profitable to continue operating. This example is similar to the preceding one in the sense that the supply curve for night-tennis services has shifted to the left.

Now let us consider day tennis and assume that in Australia tennis courts are used exclusively for the training of tennis players. The factors of production in this case are land, labor, capital in the form of ball machines, and fuel for the ball machines. The rise in the price of energy (fuel for the ball machines) does not mean that a club director will immediately dispense with ball machines. As in the previous example, however, he will have to contend with falling quasi-rents for this equipment. Thus, when the time comes to replace them or make expensive repairs, he may decide to sell the machines for their scrap value. In their place he hires more attendants, but since these attendants do not have the ball-projecting capacity of a machine (that is, their productivity without the machine is lower), their wages will also be lower. In this example energy and capital are clearly complements, whereas labor is a substitute for the energy-capital package. If the same amount of tennis-coaching services is to be provided at the same price as before the rise in fuel prices and the abandonment of ball machines, then the demand for labor will increase, but the remuneration of labor will decrease. If the wage of labor cannot fall, then the supply of tennis-coaching services will decrease in the sense that trainees will get three or four balls served up to them per minute instead of twenty or thirty. It should be clear, however, that if people are willing to pay more for their coaching, then they can continue to enjoy the presence of the ball machines. Specifically, they must be willing to pay most, or all, of the additional fuel cost. Similarly, the employees of the tennis club might be willing to take wage reductions large enough to make it attractive for the club director to keep the ball machines—particularly if they intend to continue working at the club, and if it turns out that without the machines their wages would be very low. (This last issue involves the productivity of the employees with and without the machines, as well as the exact amount of the increase in fuel costs.)

We can now consider the significance of this rise in fuel prices on real incomes in Australia, assuming that the citizens of that country use all their income for two things: consuming that ancient Australian-Swedish delicacy *sill* (which in Australia is sold in liquid form, mostly in cans), and/or consuming day tennis services. If the price of tennis services increases, many tennis players will devote more time to drinking sill. True, they may enjoy drinking sill; but for a large number this activity will be a poor second choice to tennis: as far as these individuals are concerned, their real income has unequivocally fallen.

Next let us take a slightly more complicated example. Suppose that the money value of Australian output is $1,000,000 per week, of which $600,000 is spent on 500,000 tennis lessons, and $400,000 on 400,000 cans of sill. Of the $600,000 spent on tennis, $100,000 is used to buy 100,000 units of oil from Manicura to use in the ball machines. In turn, the Manicurans spend this money for 100,000 cans of sill. I will also assume that Australia has 100,000 people in the work force, with 20,000 employed in the tennis sector and 80,000 at work producing sill. The money income of each employee is $10/week. I will also assume that they all have the same taste: each consumes 5 tennis lessons and 4 cans of sill per week. To begin, prices are $6/5 per tennis lesson and $1 per can of sill (= $500,000/500,000 cans).

Suppose now that one fine day the Manicurans decide that they want $200,000 for the 100,000 units of oil they have been exporting to Australia. The cost of a tennis lesson now rises to $7/5 since, on the basis of the figures given in the previous paragraph, each lesson consumes 1/5 of a unit of oil as an input to the ball machines. If to begin with we assume that each Australian wants to continue consuming 5 tennis lessons per week, then he or she can consume only 3 cans of sill as long as all consumption takes place out of income. Total Australian consumption is now 500,000 tennis lessons and 300,000 cans of sill. Obviously, the real income of all Australians has fallen. On the other hand, the purchasing power of a unit of oil has increased: one unit of oil will now buy two cans of sill. If the Manicurans desire, they can consume 200,000 cans of this delicious treat. What has happened here, quite simply, is that the real income of Australians has fallen because of the rise in the general price level with no change in money incomes. Even though the price of sill is unchanged while the price of tennis increases, however, the tastes of Australians are such that they want less sill and the same amount of tennis. Thus the reduction in their sill drinking makes possible an increased consumption of sill by Manicurans.

The foregoing should provide an adequate background for reviewing what happened to output, prices, and employment in the United States during the period 1973-1979. Following the oil-price shock of 1973-1974, prices in the United States increased by enough to restore the quasi-rents of capital goods to their 1972-1973 levels. With only marginal increases in money wages, real wages fell by enough to pay the increased oil bill (that is, we have essentially the same mechanism at work as in the fictitious example presented in the previous paragraph). Moreover, oil imports increased by a very large amount, which removed any barriers to the expansion of production that might have resulted from the physical unavailability of energy. It also happened that inflation and the depreciation of the dollar prevented the real price of oil from rising in the face of some minor upward adjustments of the money price of oil. By 1977 the United States had more or less recovered from the first oil-price shock. Both employment and output were increasing as a satisfactory rate, especially output; the principal difference between the economy in 1977-1978 and that in 1972-1973 was in the tendency for prices to move up faster after 1973.

Now let us return to the Australia of tennis lessons and canned sill, with a different scenario from the one just examined. This time let us assume that the rise in the price of oil, which causes a rise in the price of tennis lessons, results in a fall in the demand for these lessons. As a result, some of the employees of the tennis industry move into the sill industry, increasing the output of sill. Let us assume that demand falls by 250,000 tennis lessons, so that 10,000 employees leave the tennis industry and begin brewing sill. Less oil is now needed in the tennis sector, so imports of Manicuran oil decline. If by chance we have a linear economy, where production in all industries is proportional to the input of labor, then the fall in the demand for oil amounts to 50,000 units and, by the same token, an addition of 10,000 employees to the sill sector raises production in that industry by 62,500 cans of sill. (If 80,000 employees produced 500,000 cans of sill, then 10,000 will produce 62,500.) The total production of Australia is now 562,500 cans of sill and 250,000 tennis lessons, with Australians consuming 485,500 cans of sill and 250,000 lessons. This represents a clear fall in the real incomes of Australians because it is a second choice. (The price of sill is now 750,000/562,000 = $1.3333 per can.)

Another interesting scenario would be one almost identical to the last, except that the switch in demand from tennis to sill (which led to a total of 562,500 cans of sill being produced instead of 500,000) resulted in an increase in the price of sill (because, for instance, there were decreasing returns in the sill sector). Another plot might have tennis players and sill drinkers in Australia in possession of bank accounts as a result of some spell of frugality that they endured in the distant past. In this latter case they might decide that, despite the rise in the price of tennis lessons, they are going to continue consuming the same amount of both tennis lessons and sill—which they certainly can attempt to do simply by using their savings. This does not mean that they will be able to fulfill their intentions, however. Manicura still gets $200,000 for the oil they export to Australia. If they decide to spend all this money on Australian sill, while Australians also attempt to spend $400,000 on sill, then it is likely that the price of sill will increase because it is unlikely that sill manufacturers will be able to increase production in the short run. In these circumstances the Australian consumption of sill will still be less than 400,000 units. The interesting thing about this case, however, is that the rise in the price of sill has reduced the real price of Manicuran oil—in other words, its purchasing power: the $200,000 the Manicurans receive for their fuel will no longer purchase 200,000 cans of sill. Even so, the real consumption of Australians has decreased from what it was before the rise in oil prices; and the rise in the price of sill might cause the Manicurans to insist on even higher oil prices.

Two more examples follow, one simple and one complicated. In the simple case, the Manicurans increase the price of oil but do not spend their increase in oil revenue. Instead, they stash it in coconuts and bury the coconuts in the sand. Now, even though the price of tennis lessons increases from $6/5 to $7/5, it is still possible for Australians to take as many lessons as they did before the price rise, and to drink as much sill. All that is necessary is for the Australian government to print an extra $100,000 each week and give $1 to each Australian employee. As is easily verified, $11 will permit each of these employees to enjoy five tennis lessons (at a cost of $7/5 each) and four cans of sill. Observe also that this monetary expansion keeps the total demand high enough to maintain employment in Australia. A problem might arise later on, however, if, after many weeks of burying money in coconuts in the sand, a group of Manicurans with outsized thirsts dig them up and jet to Sydney in search of sill. Shortly after they arrive, many Australians will go to bed with parched throats.

At least they will still have their jobs, however. Things could be much more complicated if the Manicurans spent their oil revenues but, unlike in the previous examples, spent them in a fundamentally different way than Australians would. Here we could get both inflation and unemployment, depending on the ease or difficulty with which factors of production can be shifted from one industry to another. To generalize, let us talk in terms of oil-exporting countries (OEC) and oil-importing countries (OIC), and suppose that the residents of the OIC normally prefer the goods produced in industry A, but that the OEC prefers goods from industry B. Then an oil-price rise could lead to a lowered demand in industry A and an increased demand in B. Since prices in the real world go up more easily than they come down, however, we would expect price increases in industry B but hardly any movement in A. Accordingly, the net result would be a rise in the general price level. (This effect could be reinforced or weakened by a change in the spending pattern of residents of the OIC due to the price rise.) Moreover, there could be some problem in reallocating factors of production from one industry to the other. Some employees and fixed production factors located in industry A might no longer be needed in A or useful to B. Even if these employees were needed in B and theoretically could be transferred immediately, rigidities in the wage structure might not provide any incentive for them to move, particularly if the work in industry B were considered less pleasant than work in A, or involved changing localities. In these circumstances involuntary unemployment could easily result, particularly since most industrial countries have unemployment compensation schemes that foster wait-and-see attitudes.

Finally, it should be mentioned that the oil-price rises intitated by OPEC help some industrial countries—those that themselves possess energy resources. The sharp rise in oil prices has upgraded the value of Australian coal, gas, and uranium reserves. At least in theory, rising oil prices could have made England—with its temporary self-sufficiency in oil—the most important industrial country in Europe. Apparently, however, someone or something deemed otherwise. On the other hand, it could be argued that oil-price rises that theoretically favor countries like England and disadvantage highly productive economies like Japan are harmful to world economic welfare.

The Petrodollar-Recycling Problem

One of the most important aftermaths of the first oil price shock (1973–1974) was a rise in OPEC surpluses, which for our purposes are defined as OPEC export revenues minus the money value of OPEC imports. In the period named these came to about $60 billion, and there was considerable anxiety in various financial circles about the ability of the world banking system to recycle these petrodollars. As it happened, however, most of this anxiety was misplaced, since the financial institutions making up the Euromarket were well able to handle those funds that could not reach borrowers through more direct channels. By 1978 OPEC's surplus had fallen to $5 billion because of a decline in oil consumption in some of the major oil-importing countries and because the fall in the real price of oil palpably boosted the steadily rising export spending of the OPEC countries. (Much of this decline in purchasing power of oil could be traced to the depreciation of the dollar, which is the currency unit in which oil is purchased. See *The Political Economy of Oil.*) It was therefore presumed that the recycling problem, which had not really been a problem after all, could be solved within the framework of existing institutions.

Since the second oil-price shock, however, these surpluses have once again become huge. In 1980, for example, the total OPEC surplus came to more than $100 billion. In comparison with 1974–1975 the major industrial countries are interested in borrowing a large amount of this money. As in 1974–1975, however, the LDCs are anxious to claim their share.

According to some analysts, they have already claimed too much. The debt burden of these countries to private lenders is now well in excess of $300 billion, and their credit-worthiness is descending rapidly. The average debt ratio (foreign debt/gross national product) for LDCs averages out at close to 20 percent, which means that in some countries it is higher. It may be only a matter of time before private banks decide that they can no longer lend to countries with debt ratios of this magnitude (or, as bankers would say, refuse to reschedule their debts.) The exact significance of such an eventuality is impossible to predict, although it is clear that nonprivate institutions such as the International Monetary Fund (IMF) or a consortium of Western countries will in general be prepared to ensure that virtually no Third World country is put in the position of an individual who declares bankruptcy and then has his assets confiscated and sold in order to pay creditors. On the other hand, the credit institutions that carry the debt of semi-insolvent countries on their books are in a more delicate situation, since no one really knows whether their governments or central banks are as solicitous of *their* own welfare as of the peace of mind of politicians half a globe away. At one time there was a widespread belief that if a crisis of financing led to a bank failure or bank scandal, the entire international financial system would be placed at risk. In these circumstances, as shown in the affair of the Franklin National Bank, central banks are willing to go to a great deal of effort and expense to offset the result of a bank management's poor judgment; but in the future they might feel otherwise.

This brings us to the crux of the analysis. The size of the OPEC surpluses is a natural consequence of the level of OPEC's oil production and of the inability of the oil exporters to spend all their oil revenues on current goods and services. If the surpluses cause too much trouble because they cannot be lent, or can be lent only at a very low rate of return, or if a great deal of money must be advanced to countries whose borrowing jeopardizes the health of the world financial markets, then it is a simple matter for oil exporters to eliminate these surpluses by simply reducing their oil production to a level sufficient to pay for their imports and to make a few high-quality loans. Unfortunately for the oil-importing countries, however, this cure is worse than the disease. Thus the lending of billions of dollars to start or keep afloat projects that should never have been considered must continue. Only now—since the major lenders in OPEC have become much more insistent on having the real value of their revenues from these financial placements protected—borrowers are paying record rates of interest. There may be a number of ways to categorize the situation being treated here, but according to this economist it cannot be called a monetary problem. Instead, it turns on the irony that in a world of steadily rising unemployment—and in some countries deprivation—there is a shortage of viable investment projects capable of absorbing the OPEC surpluses. In industrial countries one of the reasons for this shortage is the existence of higher energy prices in concert with a general reluctance of employees to accept the lower real incomes implied by falling productivity and demand.

As figure 2–6 indicates, many OPEC financial placements are directed toward the industrial countries, as are almost all their nonportfolio investments. With respect to this latter category of investment, it should be emphasized that the governments of OPEC countries resent the coolness with which their direct (equity) investments are greeted in the industrial countries. Kuwait owns about 14 percent of Daimler-Benz and 10 percent of Krupp, but the Germans have made it clear that they would not be happy to see a major German corporation pass into the control of an OPEC country. In the United States OPEC investments have been questioned openly and aggressively in Congress; further, the significance of the blocking of Iranian

OPEC CURRENT ACCOUNT (1980)

Exports of Goods and Services	304
of which oil revenues	277
Imports of Goods and Services	− 202
Net Investment Income	9
CURRENT ACCOUNT	111
TRANSFERS	− 8[c]
CURRENT ACCOUNT AFTER	
TRANSFERS	103

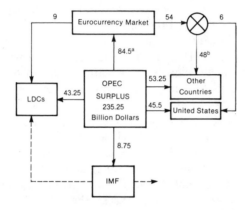

[a]All Eurocurrency *lending* not accounted for.
[b]Other countries: rest of world *minus* United States *minus* LDCs.
[c]Mostly aid.

Figure 2–6. Disposition of the Total OPEC Surplus for the Years
1974–1979, and the OPEC Current Account, 1980

assets at the time of the Teheran hostage incident has not been lost on
OPEC governments.

It should be emphasized that increasing portions of the OPEC surpluses
are being parceled out to the industrial countries. Thanks to its growing
deficit, a country like the United States is now borrowing tremendous
amounts, which is one of the reasons that world interest rates have reached
their present level. Also, Brazil's so-called spectacular borrowing is at present on a much lower level than the borrowing of Mexico and Venezuela; in
general, the oil-exporting LDCs borrow more than do the LDCs without
oil—although, obviously, they can afford to borrow more. International
institutions such as the IMF have also started borrowing from OPEC coun-

tries, and some interesting face-to-face deals have been arranged between OPEC countries such as Saudi Arabia and major industrial countries such as Germany. OPEC has also given a great deal of financial aid to nonoil LDCs.

Readers with a background in international trade probably comprehend that certain OPEC financial institutions may someday enjoy the same privileges held by their British counterparts in the nineteenth century and the first part of the twentieth, as moneylenders to the high and mighty of this world. Since the OPEC banks and lending agencies referred to here are obviously associated with the Gulf states, a great deal will depend on the political stability of that region—something no one can take for granted. At present there is a slow but steady movement of OPEC money and influence into the financial institutions of the non-OPEC world; this may represent a first step in altering the center of gravity of the world's capital markets. One thing is certain, however. Had it not been for the civil war in Lebanon, which resulted in a drastic change in the status of one of the world's most important financial pivots, Beirut, this process would be much farther along today.

At the time of writing it appears that in the next year or so OPEC surpluses may fall to a record low level, and some OPEC countries may enter the market as borrowers. This means that with the record budget deficits (excesses of spending over receipts) now predicted for the U.S. government, which are scheduled to be financed by borrowing instead of by printing money, interest rates could exceed the incredibly high levels of 1981. As a result, not only could the industrial countries be thrown deeper into recession, but a number of Third World countries (and perhaps even one or two in eastern Europe) that must borrow just to meet their interest payments would be forced finally to admit their complete lack of viability and, in one way or another, disown their debts. The effect of this type of crisis on the international financial system could be disastrous.

The Price of Oil in the Medium Term

[e]ven if they [OPEC] cut their output to zero, they could not for long keep the world price of crude above 10 dollars a barrel. Well before that point the cartel would collapse.—Milton Friedman (1974)

No one can predict the future, but with the help of economic theory it may be possible to speculate on what should happen in the presence of a given background or environment and to get a fairly good estimate of what could happen, particularly if we consider medium-term situations in which relatively rational people have time to recognize the options, discern their best interests, and if need be correct previous mistakes.

I cannot tell what will happen on the world petroleum scene tomorrow, although the popular media seem to think that a few men in the Arab world can. On the other hand, the long-run situation for petroleum (crude or conventional oil) is clear, despite the optimism of such writers as Professor Peter Odell. For better or worse, the exciting age of petroleum will start drawing to a close in the next century—perhaps early in that century. (Professor Odell continues to insist that oil reserves should be put at 5 trillion barrels, rather than the 2 trillion that is the consensus estimate of various government and oil-company scientists. In his *Oil and World Power* (1971) he concluded that the OPEC producers had "already secured the greater part of the total advantage they can expect to get from exporting oil.")

What about the medium term, up to the end of the twentieth century? In my book *The Political Economy of Oil* (1982) I present a case for a growing scarcity of petroleum, which will be reflected in a steadily appreciating real price for this commodity. I see no reason to change my opinion now, although I appreciate that the people who lift and sell OPEC oil may fail to recognize that they are faced with an essentially unique opportunity for systematic economic development that might be dissipated by a too rapid exhaustion of their irreplaceable hydrocarbon resources. Moreover, it should be made clear at this point that there are some serious misunderstandings about the nature of the present (February 1982) surplus of oil on world markets. This oversupply—commonly called a glut—is a flow and not a stock phenomenon. Reserves, or known supplies under the ground, of oil are definitely becoming scarcer in relation to the demands that are likely to be placed on these reserves over the next few decades if world economic growth is to reach and maintain what has previously been considered an acceptable level.

In the remainder of this chapter, I first comment on present developments in the world petroleum market, and make clear that although the money price of oil has been stagnant since early in 1981, the real price or purchasing power of oil has continued to increase. I do not claim that this particular phenomenon can continue indefinitely, however. Then I turn to the physical availability of oil and indicate that, in ten years or so, production could have turned down in some of the most important petroleum-exporting countries had they attempted to maintain the production rates that were reached at the beginning of 1981. A few events of this character should have an important psychological influence on other oil producers and should also alter the distribution of power within OPEC in the sense that Saudi Arabia would assume an even more important influence on the price of oil.

Next I make a brief inspection of some work of Marian Radetzki. Dr.

Radetzki has drawn a number of provocative conclusions about the future of energy usage in the industrial world with which I disagree; I also believe he has overemphasized the significance of the real price of oil for individual industrialized countries. As indicated later in this chapter with a simple example, a lower real price of oil is not inconsistent with a decreased availability of this commodity, in which case substantial economic losses might result for a particular country. It is because of Dr. Radetzki that I have chosen to deal with the medium run (1985–1990) instead of the mid-1990s, when, I think, both the developing petroleum shortage and the correct long-run strategy of the major oil-producing countries in OPEC will be clear even to energy experts in Gothenburg. It could be argued that the correct strategy is already obvious: by the late 1980s incontrovertible proof should exist that the more development-minded of the petroleum-exporting countries have a future as large-scale refiners of oil and producers of petrochemicals. When that situation occurs, and the governments and people of these countries can contemplate hundreds of years of development and prosperity—compared with the relatively few decades they will be likely to enjoy if they continue to pump oil at 1980 rates—they should feel an overwhelming urge to turn down the oil tap. Whether they will respond to this urge is quite another matter, however, and cannot be taken up here.

I continue by examining some macroeconomic aspects of the oil price, which, under certain conditions, could be of a great importance for the entire international financial system. I conclude the analysis by taking up some odds and ends associated with earlier topics. In order to emphasize certain aspects of the discussion, some elementary algebra has been introduced. Readers not interested in this kind of presentation can pass it by without suffering any appreciable loss.

The conclusion I will advance—that the underlying pressure on the oil price is upward—is not a prediction but a forecast, which means that it is contingent on the occurrence of a certain scenario: aggregate economic growth in the industrial world must reach, and maintain, the levels recorded in the late 1970s, or about 3 percent. This rate of growth, though well under the average trend rate of growth of 5.3 percent experienced by the seven largest countries in the Organization for Economic Cooperation and Development (OECD) during the period 1960–1973, should cause oil consumption eventually to close with oil production, since on the basis of existing evidence I interpret the latter as being on a downward course regardless of the state of the market. Although sufficient statistical evidence is not yet available, the opinion here is that the sharp decline in oil consumption over the past year or so is mainly the result of a deceleration in industrial output in the OECD countries that began in 1980 and cannot be reversed until 1983–1984 at the earliest. Some of this effect can be seen in table 2–5.

It can also be pointed out that in 1982 the economic situation for the

Table 2–5
The Relationship between the Growth of GNP and Oil Imports for the Seven Most Important OECD Countries

Year	Britain	Canada	France	Italy	Japan	United States	West Germany
Growth of GNP[a]							
1978	3.50	3.40	3.30	2.60	6.00	4.40	3.50
1979	1.70	2.90	3.20	5.00	5.90	2.30	4.40
1980	−2.25	0.25	2.00	3.50	5.50	−1.00	2.00
Oil Imports (Millions of Tons)							
1978	41.7	12.2	105.6	84.2	252.7	402.0	140.6
1979	19.5	82	126.0	99.3	286.0	411.6	145.6
1980	−2.3	9.0	109.0	98.1	243.9	321.2	127.0

Source: OECD, International Energy Agency.
[a]Percentage change from previous year.

countries mentioned in table 2–5 was less favorable than in 1980. If we focus on unemployment, approximately 30 million people are out of work in the twenty-four OECD countries. Predictions are that this figure will reach 32 million before 1984 if the governments of several of the major industrial countries do not make radical changes in their economic policies. As a result, it seems reasonable to expect that well before 1986 even the most incompetent politicians in these countries will be aware of the social and psychological dangers that could result from higher levels of unemployment, and that considerable efforts will be made to increase economic activity. In these circumstances oil consumption should increase considerably.

Short Comment on Real and Money Rates of Interest

In the presence of rationality and price increases, if the return on savings (the interest income) is not sufficient to protect against a fall in purchasing power, then it is likely that less will be saved. For instance, if $100 is saved, the rate of interest is 10 percent, the average price of consumption goods is unity (1), and there is no inflation, then the money yield of these savings is $10, whereas the real yield—the amount of consumption goods the money yield will buy—is 10 units of consumption goods (also 10 percent). If the rate of inflation is 5 percent, however, then the real yield falls. The money yield is still $10, since $100 put in a bank or used to purchase a bond will mean $110 at the end of a year; but this $110 will only buy $104.76 (≈ 105) units of consumption goods. The real interest rate is thus 5 percent, since 100 units of consumption goods sacrificed now yields 105 units in a year's time. It therefore seems plausible that the person willing to save $100 when the real rate of interest will decide to scale down his or her savings when the real rate of interest drops to 5 percent. (It should be recognized, however, that many people will save money even when the real rate of interest is negative).

    This discussion will now be put in algebraic form. Let us say that someone has M dollars that they can spend now, and as a result obtain M/p_0 units of real goods (where p_0 is the present price level). On the other hand, they can put this money in a bank and, after one year, obtain $M(1 + r_m)$ dollars. Then, if the price level after one period is p_1, the amount of goods that can be obtained at that time is $M(1 + r_m)/p_1$.

    The real rate of interest or real return—or what has been called on occasion the *commodity rate of return*—is the increased amount of real goods that can be obtained by waiting one year, measured as a percentage. If we continue with the present example, we get:

$$\frac{\dfrac{M(1 + r_m)}{p_1} - \dfrac{M}{p_0}}{\dfrac{M}{p_0}} = r. \tag{2.1}$$

This can be quickly simplified to

$$\frac{p_1}{p_0} = \frac{1 + r_m}{1 + r}.$$

In terms of p_0 and p_1, however, the rate of price increase can be defined as

$$\hat{p} = \frac{p_1 - p_0}{p_0} = \frac{p_1}{p_0} - 1 \quad \text{and} \quad \frac{p_1}{p_0} = 1 + \hat{p}. \tag{2.2}$$

From equations 2.1 and 2.2 we obtain

$$1 + r_m = (1 + r)(1 + \hat{p}). \tag{2.3}$$

If r and $\hat{p}$ are small, then $r\hat{p}$ is very small and the foregoing is approximately

$$r_m \approx r + \hat{p}.$$

Note from these expressions that as $\hat{p}$ increases, r falls; and when $\hat{p}$ exceeds r_m, r is negative.

The Oil-Price Yesterday and Today, and the Real Price of Oil

Before discussing the future price of crude oil, a clarification is required concerning this price in the recent past—especially the price during most of 1981 and 1982, when there was a visible surplus of oil on world markets. Although it may appear that the oil cartel has fallen on hard times, the simple fact is that the real price of oil (its purchasing power) has not been falling, even though the money price of oil (which can be taken as an average of the price on contracts signed by oil exporters and importers for the delivery of oil in the present and coming year) has dropped somewhat. The reason for this is that four-fifths of the imports of OPEC countries come from industrial countries other than the United States and countries whose currency moves in phase with the dollar; and the price of the dollar, which is the currency unit for which oil is sold, is appreciating relative to these other currencies. For example, in the first six months of 1981 the dollar rose by 18 percent against the German mark, 22 percent against the French franc, and 24 percent against the Italian lira, while the rate of inflation in the industrial countries ranged between 5 and 15 percent. Although OPEC imports from the United States cost more, the OPEC countries were still able to increase the purchasing power of their oil since their imports from other industrial countries cost less.

This being the case, the more enlightened of the oil-exporting countries can continue to sell a decreasing amount of petroleum for progressively higher real prices. Some countries naturally would like to sell more oil on the grounds that their development programs require a larger input of imports, but in my opinion most of these countries have more than enough revenue to support the actual (as opposed to the advertised) development they have been experiencing. Until recently the value of OPEC's equity and other investments abroad increased steadily. Table 2–6 gives an estimate of OPEC's holding of financial assets about the middle of 1981. To this can be added substantial holdings of real properties, particularly in the form of land, structures, and so on. Some of the better managed oil-producing countries, such as Kuwait, can probably finance a large part of their current investment from the yield on their foreign assets, since, according to some reports, the annual investment income of that country may be approaching $7 billion.

In light of the situation referred to at the beginning of this section, some have argued that OPEC should find it in its interest to help maintain the value of the U.S. dollar. Still, it cannot be denied that the financial and political muscle of OPEC was built up over a period when the dollar was depreciating on exchange markets, with increases in the real price of oil being brought about by occasional precipitate upward adjustments of its money price. There are undoubtedly several categories of exporting countries that prefer this arrangement. These include countries with relatively limited oil reserves that, for one reason or another, find it difficult to reduce their spending, and think that their total real receipts could be increased appreciably if the money price of oil were given a substantial boost. The same is true of countries with alternative energy resources (such as gas) that wish to see these resources appreciate in value so that it will be profitable to speed their exploitation. (There is also the interesting case of industrial countries that are energy independent and thus gain at least a theoretical advantage over their competitors every time the world price of energy rises.) Finally, as will be explained in detail later, there are countries with serious intentions to construct refineries and petrochemical installations, which may reason that increases in energy prices will slow down the establishment of these facili-

Table 2–6
OPEC Financial Assets (1981) and Estimated Oil Production Required by Budgetary Considerations

Country	Financial Assets (Billion Dollars)	Present and (Normal) Daily Production (Mbbl/day)	Oil Production Indicated by Budgetary Requirement[a]
Saudi Arabia	162	8.0 (9.90)	6.5
Libya	44	0.90 (1.83)	1.0
Kuwait	76	0.85 (1.76)	0.9
UAE	39	1.4 (1.70)	0.8
Qatar	16	0.4 (0.47)	0.6
Iran[b]	3	1.0 (1.47)	3.6
Iraq[b]	32	1.0 (2.64)	2.1
Nigeria	4.5	1.8 (2.05)	2.2
Algeria	3.8	0.7 (1.10)	1.2
Gabon	0.7	0.2 (0.17)	0.2
Venezuela	7.7	2.1 (2.16)	2.4
Ecuador	0.7	0.2 (0.20)	0.2
Indonesia	9.0	1.6 (1.57)	1.6[c]
Total	387.0		

Source: Morgan Guaranty; OPEC documents.
[a]Budget of the central government.
[b]Uncertain because of present conflict.
[c]Calculated as a residual.

ties in other parts of the world and perhaps eliminate some of the least profitable existing facilities.

One interesting point alluded to earlier is that since the beginning of 1981—for the first time in recent oil history—the real price of oil has been increasing even though the trade-weighted money price of oil has fallen slightly. Since this process turns on the appreciation of the U.S. dollar, the obvious question is how long this development can continue. I am among those expecting an eventual decline in the value of the dollar since, as things now stand, its value is out of proportion to the basic productivity of the U.S. economy in relation to the productivity of its major competitors. However, it should be understood that the United States is still extremely rich in energy and nonfuel resources in relation to these competitors. In the event of a political and sociological renaissance in the United States, including the installation of a government capable of organizing these resources, such things as oil-price shocks and other tumultuous departures from orthodoxy would favor the United States. For example, within a very short time after the first oil-price shock, the value of the dollar increased by 11 percent with respect to the German mark. In fact, everything else remaining the same, decreased supplies of OPEC oil work to appreciate the U.S. dollar relative to other currencies.

In conjunction with the theme of this section, several things should be made clear. The first is that the real prices we have considered are aggregates; and because the oil-exporting countries have considerable latitude to choose between different suppliers of imports as well as different import bundles, the purchasing power of oil may be higher for some oil-exporting countries than for others. (In an intertemporal sense it may be higher still for those countries that hold large amounts of financial assets denominated in an appreciating currency, or whose yield is guaranteed.) Put more simply, as the dollar appreciates relative to other currencies, they can carry on a smaller share of their trading with the United States, and thereby increase the purchasing power of oil. (Although the amount of oil being bought on spot or cash markets, as opposed to purchases made on a contractual basis, is increasing everywhere, I know of no country that buys more than 20 percent of its oil on spot markets, and this is probably a temporary phenomenon: historically, spot markets handle less than 10 percent of the trade in energy materials and metals. In addition, as the spot price of oil decreases, the oil-exporting countries will sell less to these markets.)

Next we can look at the situation from the point of view of individual purchasers of oil. A simple example might be illuminating here. Suppose that the price level in Sweden rises by 10 percent, while there are no changes in either the price of oil (in dollars) or exchange rates. If the oil-exporting countries are purchasing Swedish goods—and they must have these goods in the sense that no replacements are to be found elsewhere—then a unit of Swedish goods buys more oil: the real price of oil has decreased for Sweden, and Sweden has an unequivocal gain. Conversely, however, if the price of Swedish goods increases relative to the price of oil, then less oil may be made available to Sweden as the oil exporters go to other suppliers in order to obtain the products they were buying in Sweden.

Let us complicate this last situation somewhat and make it more realistic. Suppose Sweden produces export goods (such as machines) with a portion of its domestic supplies of capital and labor, and that these machines in turn are sold abroad to obtain dollars, which are then used to obtain oil. (In this example oil producers do not buy directly from Sweden.) Let us further assume that this oil is an intermediate product that is also used with the remaining supplies of Swedish capital and labor, and with rotten fish that wash ashore from the Baltic (and thus do not cost anything), to produce a single consumer good that we shall call *sill*. The oil is used to rinse the fish. A rise in the price of Swedish exports could cause a fall in the demand for these exports and, under certain conditions, could lead to a fall in the amount of revenue obtained for these exports (on condition that the demand for Swedish exports is elastic).

This would mean a fall in the quantity of oil that could be purchased. Next let us assume that the capital and labor freed by the fall in production in the export sector could move immediately into the sill sector, which means that production in this sector will be carried on with larger amounts of capital and labor, but smaller amounts of oil. If capital and labor are not very substitutable for oil, however, then production will fall in the sill sector. Assuming a (fixed) amount of capital and labor in the Swedish economy, a smaller production of the single consumption good means, other things being equal, that someone in Sweden loses—either labor, or the owners of capital, or both. Furthermore, if the factors freed in the export sector cannot move into the sill sector—which to some extent is usually the case in the real world—then this economy could suffer serious economic losses. Thus even though we assumed no change in the price of oil or in the exchange rate, the result of a decrease in the real price of oil for a single oil-importing country could mean a lowering of that country's standard of living due to a fall in the availability of oil to that country. For an individual country, movements in the real price of oil, considered in isolation from other economic variables, can have very little economic significance. (Germany, on the other hand, was able to benefit for many years from a fall in the real price of oil that resulted from the appreciation of the mark relative to the dollar. This appreciation, however, was initiated by steady increases in the demand for German exports).

An Algebraic Excursus on the Real Price of Oil

Two things are to be emphasized here: first, that the real price of oil is equivalent to the purchasing power of oil and therefore is expressed in physical units. If, for instance, a certain amount of oil q_o sells for a price p_o, the revenue from this sale is $p_o q_o = R$. If the price of, for example, machines to be purchased with this oil is p_m, then the amount of physical goods (machines) that can be obtained is

$$q_m = \frac{p_o q_o}{p_m}. \tag{2.4}$$

Note that the units of q_m are machines. We see immediately that with q_o given, if p_o increases or p_m decreases, more machines can be bought: the purchasing power of oil, or the real price, has increased. (Naturally, q_o can be set equal to 1 barrel).

Next we ask what happens when the oil exporters are purchasing more than one good, and several countries are involved. For example, one good is purchased for dollars (which is also the currency in which the price of oil is expressed), and another good is paid for in some other currenty—for example, kronor. In this case p_m becomes a price index that is weighted by the amounts of trade with the two countries; and, with $M_{1,2}$ signifying imports from the two countries, we get:

$$p_m = \frac{p_1 M_1 + ep_2 M_2}{M_1 + M_2} = \frac{M_1}{M_1 + M_2} p_1 + \frac{M_2}{M_1 + M_2} ep_2$$

$$= w_1 p_1 + w_2 ep_2. \tag{2.5}$$

In equation 2.5 e is the exchange rate between dollars and kronor, and has the dimension dollars/kronor. Thus we have, from 2.4 and 2.5:

$$q_m = \frac{p_o q_o}{w_1 p_1 + w_2 ep_2}. \tag{2.6}$$

From 2.6 we get immediately

$$\frac{\partial q_m}{\partial e} = - \frac{w_2 p_2 q_o p_o}{(w_1 p_1 + ew_2 p_2)^2}. \tag{2.7}$$

If e decreases—that is, the dollar appreciates relative to kronor—then q_m increases. In terms of the original definition, the purchasing power of oil has increased. It should also be clear in this example, that the effect of increases in e can be offset to some extent by importing more from country 1.

The Supply of Oil and the Optimal Depletion of an Oil Deposit

The next question concerns the physical availability of oil. Here, unfortunately, things are not as they appear. If we read, for example, that a country has reserves of 20 billion barrels of oil and is extracting one billion barrels per year (1 Gbbl/y), this does not imply that it will be able to continue extracting 1 Gbbl/y for twenty years, but rather that production should start declining after the end of the tenth year—at the latest.

To understand this issue it should first be appreciated that every oil field has a potential production rate that depends on the size of the field, its geological characteristics, and its facilities for lifting and transporting oil. In general it is uneconomical to produce more than 10–15 percent of the recoverable oil in a field in a single year, since producing more means that the amount of oil that can eventually be recovered is reduced. The analogy I use to explain this phenomenon is a machine, or a vehicle, that can be run at very high speeds if we desire, although to do so over extended periods means that we greatly reduce its usable life. Actually, a reserves-to-production (R/P) ratio of 10 is probably the absolute minimum that could be used for the world, since this would imply that all oil fields can simultaneously produce at the maximum rate, when in fact some fields will merely be under development, even though they contribute to the aggregate of proved reserves. This is the reason generally given for considering the critical R/P ratio to be greater than 10. I will discuss another reason later in this section.

A numerical example will now be constructed to illustrate how the R/P ratio influences production. Assuming initial reserves of 200 units, two cases will be examined. The first features a stationary production of 10 units/year; the second assumes that intended production grows at a rate of 10 percent a year (10, 11, 12.1, 13.3, and so on). To keep things simple the critical R/P ratio will be taken as 10. Figure 2–7 shows production over time for the two cases, with the relevant calculations just below the figures. If we observe the figure on the left, we note that production remains at the intended level of 10 until the R/P ratio reaches its critical value. Then this critical ratio (which is equal to 10) takes over and determines production. It should also be made clear that these calculations are somewhat more precise than those presented in *The Political Economy of Oil*. As will be explained, end-of-year reserves are used, and this somewhat complicates calculations when the critical R/P ratio is reached. As just mentioned, in the diagram on the left this takes place immediately after the tenth year is reached. From that point on production must be chosen so as to keep the R/P ratio at 10. This means that for each period t we must have

$$\frac{\text{End-of-period reserves } (t)}{\text{Production } (t)} = \frac{\text{Reserves } (t-1) - \text{Production } (t)}{\text{Production } (t)} = 10.$$

This can be simplified immediately to give

$$\text{Production } (t) = \frac{\text{Reserves } (t-1)}{11}.$$

For example, reserves at the end of period 11 are 91.9, so in order to maintain the R/P ratio at 10, production in period 12 must be

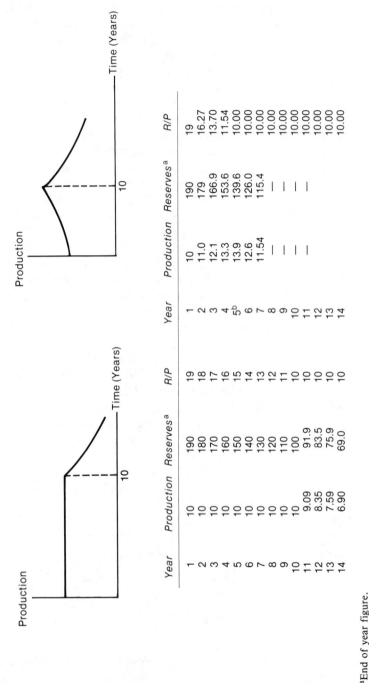

Year	Production	Reserves[a]	R/P
1	10	190	19
2	10	180	18
3	10	170	17
4	10	160	16
5	10	150	15
6	10	140	14
7	10	130	13
8	10	120	12
9	10	110	11
10	10	100	10
11	9.09	91.9	10
12	8.35	83.5	10
13	7.59	75.9	10
14	6.90	69.0	10

Year	Production	Reserves[a]	R/P
1	10	190	19
2	11.0	179	16.27
3	12.1	166.9	13.70
4	13.3	153.6	11.54
5[b]	13.9	139.6	10.00
6	12.6	126.0	10.00
7	11.54	115.4	10.00
8	—	—	10.00
9	—	—	10.00
10	—	—	10.00
11	—	—	10.00
12	—	—	10.00
13	—	—	10.00
14	—	—	10.00

[a]End of year figure.

[b]Note that in this period intended production was 14.6 units, but had production actually reached this level the critical R/P ratio of 10 would have been exceeded.

Figure 2-7. Production as a Function of Time, with a Critical R/P Ratio of 10

$$\text{Production (12)} = \frac{\text{Reserves (11)}}{11} = \frac{91.9}{11} = 8.35.$$

This results in end-of-period reserves for period 12 of $91.9 - 8.35 = 83.5$, and of course an end-of-period R/P ratio of $83.5/8.35 = 10$.

In the situation on the right in figure 2-8, production follows the intended path up to period 5, but from there on production must drop in order to keep the R/P ratio from falling below 10. Note also that in period 5 production has not increased by the intended 10 percent, because had it done so, production in that period have increased by 1.33 units to a value of 14.63 units. That would have resulted in end of period reserves of $153.6 - 14.63 = 138.97$, and thus an R/P ratio of 9.4989. This value is unacceptable since we have specified a critical R/P ratio of 10. (It is true, however, that there is no hard and fast rule for choosing the critical R/P ratio. Before Alaskan oil began to flow at its present rate, the aggregate R/P ratio for the United States fell under 10, although the general feeling in the U.S. oil industry seems to be that it should be kept above that figure.)

Next we shall consider the effect of differential field size on what might be called the aggregate critical R/P ratio. For instance, if we have a situation with two countries—or, for that matter, two different-sized fields in one country—where one country (or field) has reserves of 200 units, and the other reserves of 150 units; and if both are producing 10 units/year with the intention to maintain this output until the critical R/P ratio is reached, then we have an arrangement of the type shown in figure 2-8. (Once again, the relevant calculations are shown directly beneath the diagrams.)

As shown under the diagram on the right, production begins to fall in field 2 after the fifth year. At that time the aggregate R/P ratio is

$$\frac{R}{P} = \frac{R_1 + R_2}{P_1 + P_2} = \frac{150 + 100}{10 + 10} = \frac{250}{20} = 12.5.$$

We therefore see that although the critical R/P ratio for each field is 10, the aggregate R/P ratio when total production began to fall (from the value of 20 at which it had been maintained from the beginning) is appreciably higher.

Now let us apply some of the above observations to the real world, and in particular to those OPEC countries whose reserve position is not especially large relative to their production. The information in table 2-7 is for 1979.

Each of the countries mentioned had R/P ratios of less than 20 at the end of 1980. Together, in 1979, they produced a daily average of 8,341 million barrels of petroleum, or 27 percent of OPEC's output. Four of

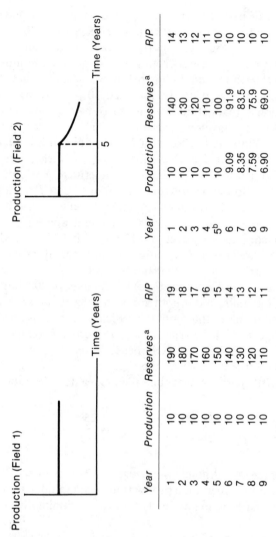

Year	Production	Reserves[a]	R/P
1	10	190	19
2	10	180	18
3	10	170	17
4	10	160	16
5	10	150	15
6	10	140	14
7	10	130	13
8	10	120	12
9	10	110	11

Year	Production	Reserves[a]	R/P
1	10	140	14
2	10	130	13
3	10	120	12
4	10	110	11
5[b]	10	100	10
6	9.09	91.9	10
7	8.35	83.5	10
8	7.59	75.9	10
9	6.90	69.0	10

[a]End of year figures.
[b]Year in which critical R/P ratio reached.

Figure 2–8. Production as a Function of Time in Two Fields with Different Initial Reserves

Table 2-7
Reserves, Annual Production, and *R/P* Ratio for Seven OPEC Countries

Country	Reserves[a]	Annual Production[a]	Average Daily Production[b]	R/P Ratio[c]
Algeria	8,440	421	1,154	20
Ecuador	1,100	78	214	14.10
Gabon	500	74	203	6.75
Indonesia	9,600	580	1,590	16.50
Nigeria	17,400	841	2,305	20.70
Qatar	3,670	185	508	20.30
Venezuela	17,870	860	2,367	20.80

Source: *OPEC Quarterly Review* (various issues).
[a]In Mbbl (for 1980).
[b]In thousands of barrels (for 1980).
[c]In years.

them—Indonesia, Algeria, Nigeria, and Venezuela—are major OPEC producers and very influential members of that organization; three—Indonesia, Algeria, and Nigeria—will have a desparate need for money many decades into the future if they are to fulfill even a fraction of their development goals. As these countries approach the date when, for the reasons given earlier, their production must begin to decline, and they become fully aware of the significance of this occurrence, they can be expected to do everything possible to raise the price of the oil they are selling. This means not only putting pressure on better endowed countries to decrease their output, but also finding it progressively easier to enter into production-cutting schemes themselves in order to support the price of oil.

At present, several of these countries are producing considerably less oil than is indicated in table 2-6. Algeria's production is less than 0.9 Mbbl/d and may have plunged as low as 0.7 Mbbl/d at one stage or another; Nigeria's output appears to have reached the range of 1.1–1.3 Mbbl/d or lower. Some say these countries are now spending more money than they are making; unless they can raise their production in the near future, may have to engage in considerable borrowing. In my opinion, given the extent of their oil reserves and the enormity of their requirements, it would be much more sensible to forget about the borrowing, leave their production at current levels, and concentrate on raising the efficiency of the economic activities taking place in their countries. In certain oil-producing countries, a little less corruption might also be helpful.

Moreover, another major OPEC producer, Iraq, had an *R/P* ratio of only 24 at the end of 1979; despite the contention by some people that exports of Iraqi oil have been reduced to a trickle because of the war between Iraq and Iran, the truth is that a sizable proportion of the prewar Iraqi exports of crude oil are still getting to market via Syrian, Lebanese, and Turkish pipelines. Eyewitness accounts also indicate that the pace of construction in Iraq has not slowed because of the war; and considering the seriousness of that country's development program, there is no reason to believe that the end of hostilities will result in the Iraqi government flooding the world market with oil in order to obtain the resources needed to rebuild its economy. In fact, helping to create a surplus of oil would be a way *not* to get these resources.

Strangely enough, the apostles of conservation in OPEC seem to be found in the ranks of the richest members. Libya has made a point of keeping its production at a minimal level while maintaining almost total independence from the industrial world; although Saudi Arabia engineered the glut that caused the downward pressure on the oil price in late 1981, that country has apparently ceased to invest in new facilities for crude-oil production; and there seem to be plenty of people in the kingdom who believe that Saudi social and economic ambitions would be better served if oil production were lowered to about 4.5–6.0 Mbbl/d. It also appears that the oil minister of Kuwait (Ali Kalifa al-Sabah) is the most articulate spokesman in OPEC for conservation and a production policy "based on a long-range vision for the economic and social transformation of OPEC societies." With the exception of a few states around the Gulf, I doubt whether this vision is widely shared; if it was, however, it would mean levels of oil production that are much lower than any being contemplated today.

Next I would like to examine a forecast by the U.S. Office of Technology Assessment of the supplies of oil that will be available to the noncentrally planned countries in 1985 (see table 2–8). From the middle column of this table we see at one extreme a situation in which the production of oil is at the very low level of 45 Mbbl/d. If things work out this way, there will definitely be a sharp increase in the price of oil around 1985, or even earlier. What about the opposite situation? I see no possibility for an output in the vicinity of 60 Mbbl/d unless there is a huge rise in the money price of oil, in which case countries like Iraq, Iran, Mexico, and the USSR might make extra efforts to increase their exports, and the production of relatively high-cost oil from small installations in the United States would be augmented considerably. My own estimates deal primarily with what I believe cannot happen. It is extremely unlikely, for example, that Saudi Arabia will raise its oil production to 11 Mbbl/d. I am also dubious about the output levels I show for Iran, Iraq, and Canada; for these countries, however, I prefer to overestimate rather than underestimate.

Thus the figure at the bottom of the last column should probably be less than 54 Mbbl/d. Assuming that 1979–1980 was the most recent period in which anything approaching normality existed in the world economy, the

Table 2-8
Supplies of Oil to the Noncentrally Planned Countries in 1979 and 1985 (Estimated)

	1979 (Actual)[a]	1985[a]	1985–1986 (My Estimate)[a]
United States	10.2	7.2–8.6	8.0
Canada	1.8	1.6–1.8	1.8
North Sea	2.1	2.8–4.0	4.0
Other developed countries	0.8	0.8	0.8
Total (developed countries)	14.9	12.4–15.2	14.6
Saudi Arabia	9.8	9.1–11.1	8.5
Iraq	3.4	2.7–4.5	4.0
Iran	3.0	2.0–4.0	4.0
United Arab Emirates	1.9	1.9–2.5	2.5
Kuwait	2.6	1.9–2.4	1.9
Other OPEC	10.7	9.5–10.5	10.0
Total (OPEC)	31.4	28.1–34.5	30.9
Mexico	1.6	3.0–4.0	4.0
Non OPEC LDCs	3.5	4.5–5.0	5.0
Total	5.1	7.5–9.0	9.0
Centrally planned countries	1.0[b]	1.9[c]–0.0	0.0
Total	52.4	45.1–58.7	54.5
Consumption, 1981 (estimated)		47.5–48.5	

Source: Report of the U.S. Office of Technology Assessment to the U.S. Congress.
[a]In Mbbl/d.
[b]This probably should be 1.5 Mbbl/d.
[c]Imports of petroleum.

downturn that arrived in the last part of 1980 was a replay of the downturn that followed the 1973–1974 oil-price increases; the upturn, when and if it appears—should also display characteristics similar to those of the 1975–1979 upturn. In these circumstances, if the upturn arrives before 1984, it should be possible for world petroleum demand to achieve a level of 54–55 Mbbl/d by the end of 1985 or soon after; in the subsequent adjustments on the world market (involving falling inventories and ad hoc production increases), the money price of oil will begin to climb once again.

A few more words about the structure of world petroleum demand are useful here. The consumption of oil has now stagnated and in some countries is falling. Despite the assurances of various academics and other personalities that this is a good thing, however, it has its negative aspects. In most respects oil is still the superstar of energy resources; although such things as natural gas are in many ways preferable, they cannot possibly be considered a short-run replacement for oil. Along with the declining use of oil—and possibly because of it—industrial production is falling to the extent that some people are speaking of the deindustrialization of large parts of the traditional industrial world. Economic growth is decelerating almost everywhere, the growth of productivity is decreasing, and unemployment is increasing. When I published my book *Scarcity, Energy, and Economic Progress,* the level of unemployment in the OECD countries was about 17 million. It has now passed 30 million and could reach 35 million by 1985 unless there is a substantial revival in investment and, concomitantly, economic growth. According to OECD estimates, the aggregate growth rate in the OECD countries must reach at least 3 percent to hold unemployment at its present level, and 5–6 percent if unemployment is to be decreased by a significant amount.

Of course, there is no guarantee that economic policies will change in the industrial countries or that the necessary steps will be taken to lift investment out of its present doldrums. The knowledge that an upsurge in economic growth could create the conditions for further major rises in the price of oil has now become part of the intellectual equipment of most—but not all—heads of state and their economic advisors; this is a major factor in the willingness of these individuals to tolerate economic conditions that could cause irreversible damage to the social fabric of their constituencies. On the other hand, it should be appreciated that there are a few countries that no longer possess any options in these matters; as time passes, there will be more. Denmark, with 20 percent of the people in its labor force between the ages of 18 and 30 unemployed, is one of these, and is rapidly approaching the point where production and employment cannot be permitted any further decrease. At this point the rate of growth of energy use will very likely increase.

It should also be remembered that energy requirements in the nonoil less developed countries (LDCs) can only rise, and that OPEC's own consumption is increasing. The World Bank has estimated that commercial energy consumption in nonoil LDCs will rise from 12.4 Mbbl/doe (oil equivalent) in 1980 to 22.8 Mbbd/doe in 1990, although indigenous producers in these countries will increase their output only from 7.8 Mbbl/doe to 15.2 Mbbl/doe. As things now stand, a significant portion of the incomes of LDCs is used to import energy; given their growing populations and lack of technological capabilities, there is literally no possibility for these countries to effect a rapid substitution out of oil and into some other relatively efficient energy medium. It is also interesting to note that a sizeable portion of the loans and aid received by these countries, including the aid they obtain from OPEC, is used to purchase energy materials; and many oil companies feel that, in the future, the most expansive market for petrochemicals will be in the Third World.

One topic that has not yet been touched on involves the various concepts of capacity. The first is called *facility capacity* and refers to the total installed capacity of gas-oil separating plants, main-trunk pipelines, and oil-loading terminals. In March 1979 the facility capacity of Saudi Arabia was reputedly 12.8 Mbbl/d. *Maximum sustainable capacity* is the maximum production rate physically sustainable for at least six months; this usually comes to 90–95 percent of facility capacity. For Saudi Arabia this came to 9.8 Mbbl/d at the date given. The last concept is *surge capacity,* the maximum output that can be produced for a short period such as a few weeks. As indicated by several production peaks, this level came to 10.5 Mbbl/d for Saudi Arabia in 1979.

Figure 2–9 provides a rough estimate of maximum sustainable capacity for the noncentrally planned world. Note that capacity is falling; thus, if there was a sudden surge in demand, maximum capacity might be encountered at a fairly early stage, bringing about a sharp increase in prices.

Ultimately, capacity seems to be tied to exploration and the augmentation of reserves. As noted earlier, more drilling has been taking place since 1973 than at any time previously; but up to now results have not been particularly impressive. In 1980 increases of more than 20 percent were recorded over 1979 in both number of wells drilled and in footage, and similar figures probably pertain to 1981. Still, there have been only trivial increases in the amount of oil that has been found, and certainly nothing approximating a giant or supergiant field. In fact, a number of statistical models of drilling activity indicate sharply decreasing returns in both oil and gas drilling. This trend could be turned around later for gas, when it becomes possible to drill much deeper and in different types of environments, but it is difficult to be optimistic about oil.

Calculating the Price Elasticity of Energy

For econometric reasons having to do with a general lack of price response prior to 1973 (as well as some multicollinearity problems), and a shortage of observations after 1973, formal econometric techniques do not seem to offer much scope for rigorous estimation of the price elasticity of demand for energy; but a simpler type of operation, using only the definition of elasticity, may be of some interest to the reader. First we define the demand for energy as a function of its price and some aggregate variable such as the GNP. Thus we have $q = q(p,y)$, and a total differentiation with respect to price yields

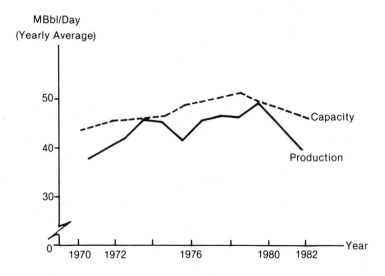

Figure 2-9. World Production Capacity (and Production), Excluding the Centrally Planned Countries

$$\frac{p}{q}\frac{dq}{dp} = \frac{\partial q}{\partial p}\frac{p}{q} + \frac{\partial q}{\partial y}\frac{dy}{dp}\frac{p}{q}\frac{y}{q}\frac{q}{y} = n_p + n_i\frac{p}{y}\frac{dy}{dp} \qquad (2.13)$$

Next, going over to finite movements in the variables, we can write 2.13 as

$$\left(\frac{\Delta q}{q}\right)_j = \left(\frac{\Delta p}{p}\right)_j n_p + n_i\left(\frac{\Delta y}{y}\right)_j$$

or
$$\hat{q}_j = n_p\,\hat{p}_j + n_i\hat{y}_j. \qquad (2.14)$$

In these expressions n_p and n_i signify the partial elasticities for demand with respect to price and the aggregate variable y. Now, for $J = 0, 1$, where 0 and 1 are averages covering several years, we get

$$n_p = \frac{(\hat{q}_0 - \hat{q}_1) - n_i(\hat{y}_0 - \hat{y}_1)}{(\hat{p}_0 - \hat{p}_1)}. \qquad (2.15)$$

Let us employ the following data: the percentage growth in energy prices in the United States in 1965–1973 was -0.1, and for 1973–1978 it was 6.8; the percentage growth in energy demand is 4.3 for 1965–1973 and 0.9 for 1973–1978. We also have the percentage GNP growth for 1965–1973, which was 3.7; for 1973–1978 it was 2.3. Insofar as n_i is concerned, this was taken as unity. Substituting these values in 2.15 gives

$$n_p = \frac{(4.3 - 0.9) - (3.7 - 2.3)}{(-0.1 - 6.8)} = \frac{2.0}{-6.9} \approx -0.3.$$

Normally it would be expected that the absolute value of the price elasticity of oil would be slightly larger than that for energy, since substituting for a particular energy type can take place much faster than, for example, substituting labor or capital for energy. Various estimates of the price elasticity of energy run from -0.2 to -0.8.

Oil Price and Energy Consumption after the First Oil-Price Shock

I would like to begin this section by discussing a contribution by Marian Radetzki (1981) on the demand for energy in 1985. According to Radetzki, energy consumption in the OECD countries, which he gives as 78.7 Mbbl/doe in 1979 will fall to 65.1 Mbbl/doe by 1985 if there is an aggregate increase in the gross national product (GNP) of 23 percent over the same period. Needless to say, GNP is not going to increase by 23 percent unless the world economic outlook brightens considerably; but it is still true that, on the basis of Radetzki's calculations, a 1-percent rise in GNP is accompanied by a 0.7-percent fall in the use of energy. At best this contention is unacceptable, but it does serve as a useful introduction to a review of energy consumption over the years following the first oil-price shock.

The technical difficulty here is that Radetzki has made an error in both his interpretation and his use of the various elasticities of energy demand (where an income elasticity is defined as the percentage change in the demand for energy resulting from 1-percent change in a variable such as GNP or domestic product and a price elasticity as the percentage change in energy demand if we have a 1-percent change in the price of energy). According to Radetzki, the relevant energy elasticity is unity, and the price elasticity is -0.5. Given these, he reasons as follows. A GNP increase of 23 percent between 1979 and 1985 would mean an increase in the consumption of energy from 78.7 Mbbl/doe to 96.8 Mbbl/doe. From this, however, Radetzki subtracts what he (incorrectly) estimates to be the residual effect of the 1973–1974 oil-price rise, which turns out to be 10 Mbbl/doe. This means that instead of a 1985 consumption of 96.8 Mbbl/doe, we now have 86.8 Mbbl/doe. Amazingly enough, it is to this 86.8 Mbbl/doe that Radetzki applies his price correction. Estimating that the 1979–1980 price rise to the final consumer was 50 percent, a price elasticity of -0.5 implies a 25-percent fall in consumption. Specifically, illegitimately multiplying the 1985 consumption of 86.8 Mbbl/doe by -0.25, he gets -21.7 Mbbl/doe as the consumption reduction due to price. His final figure for energy consumption in 1985 is thus $86.8 - 21.7 = 65.1$ Mbbl/doe.

No comment will be made on this particular result because I am certain that the reader will understand that both income and price elasticities should have been applied to the 1979 consumption figure. Even had this been done, Radetzki would still have obtained the wrong result because, as pointed out explicitly by Adelman (1980), it is a mistake to use conventional elasticities or incremental energy-income coefficients in these situations. Adelman promotes the use of average energy-income ratios, whereas many other investigators employ gross elasticities that take price effects into consideration. Since these gross elasticities are all positive, one might suspect that Radetzki stands almost alone in his vision of a world in which GNP growth is accompanied by a declining energy use.

Before leaving this topic let us note that had Radetzki, in his analysis of energy consumption in 1973–1979, not also made the basic mistake of subtracting his price correction from energy consumption after he used the income elasticity instead of before, he would have obtained the result that energy consumption in 1979 was lower than in 1973—a gratuitous absurdity in the light of the way things turned out over this period. Actual events indicate in fact, that the price effects of the 1973–1974 oil-price rise were extinct before 1978–1979. This will be immediately obvious to the reader who takes the time to examine some energy-consumption statistics presented later in this section; but it is also the interpretation of the flattening out of OECD oil and energy intensities (defined as oil or energy consumption divided by real GNP) during the years 1975–1977, and shown in the left-hand diagram in figure 2–10. I would also like to emphasize that it makes no sense at all from a scientific point of view to consider the long-run price effects of the 1973–1974 price rise while ignoring the long-run income effects, especially when the bulk of the GNP increase that Radetzki cites came later than the price rise.

As an afterthought, the reader should keep this in mind. Without being concerned about elasticities, long-run effects, and so on, it is possible to get an excellent picture of the place of energy in the scheme of things by simply considering the situation in England over the past seven or eight years. In that country, for some inexplicable but doubtless illogical reason, the price of energy has been permitted to move up in pace with the world price; at the same time industrial production (the most energy-intensive component of

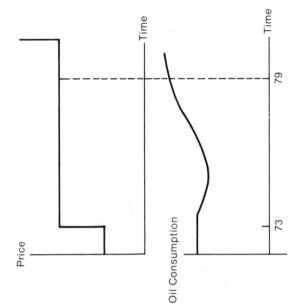

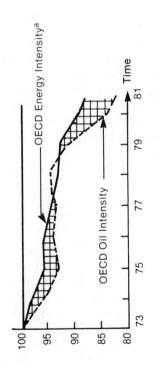

Note: The cross-hatched area indicates a net substitution of oil by other energy materials.

[a]Energy intensity equals total primary energy demand divided by real GNP.

Figure 2–10. OECD Oil and Energy Intensities, and an Idealized Portrayal of the Oil-Price Rise and Oil Consumption over the Period 1973–1979

national output) has been falling at a catastrophic rate. Even so energy consumption in Britain today differs only marginally from its 1973–1974 value; according to the British Institute of Energy it should rise by 0.5–1 percent a year in the coming year or two even if present economic conditions continue to prevail. All that is needed is for GNP to increase.

This portion of our discussion will be concluded by a brief examination of world oil and energy consumption over the period 1973–1978, with the idea of emphasizing the previous observation that the price effects of the 1973–1974 oil-price shock had essentially passed well before 1979, and to suggest that there might have been something in energy-consumption behavior during the 1970s that is relevant for the 1980s—or later. On the right-hand side of figure 2–10 I present an idealized portrayal of the 1973–1974 oil-price rise together with a similar representation of oil consumption between 1973 and 1979. This course of events reduced to almost a textbook situation in which the real price of oil increased by several hundred percent and then remained relatively constant for approximately five years. Over these five years the major industrial countries outside eastern Europe increased their consumption of energy (on the average) by about 0.5 percent per year, although world energy consumption climbed on the average by about 2.5 percent per year. The oil-price shock of 1979–1980 has of course reduced the demand for energy materials, mainly by helping to bring about a depression in the industrial world that was accentuated by the economic policies in certain industrial countries. Even so, in 1980–1981 there was definitely a growth in energy consumption in most countries registering real increases in GNP, although we are witnessing a shift between energy resources: less oil is being used, and more coal, nuclear, and so on. Just as there was an increase in optimism about oil supplies in 1975–1977 that caused a rise in oil intensities in most of the OECD countries, however, the same tendency has been noted in the United States recently as a result of the halt in the rise of the oil price.

The energy-consumption figures for 1973–1978 are given in table 2–9. Note in particular the consumption of liquid fuels, since these had almost recovered to their 1973 level by the beginning of 1976, even though the aggregate increase in GNP in the OECD countries was not more than 1 percent over this same period.

We see in table 2–9 that the main effect of the oil- and energy-price rises of 1973–1974 was an "impact effect," although the world recession of 1974–1975 was also instrumental in pulling down energy use by depressing GNP on a worldwide basis. Since, in terms of percentage, the energy-price increases of 1979–1980 are much smaller than the earlier increases, and since no major oil-price increases are on the horizon, when the industrial world begins to recover from the present depression, the same type of recovery displayed here should be repeated. It must be admitted, however,

Table 2-9
World Energy Consumption, 1973–1978
(million tons oil equivalent)

	1973	1974	1975	1976	1977	1978
Developed countries[a]	3,280	3,217	3,214	3,321	3,349	3,369
Other noncentrally planned	462	490	514	559	605	637
Communist	1,520	1,585	1,673	1,752	1,856	1,950
Total	5,261	5,292	5,312	5,632	5,810	5,956
Composition						
Solid fuels	1,668	1,693	1,712	1,800	1,867	1,907
Liquid fuels	2,434	2,405	2,397	2,565	2,650	2,693
Natural gas	1,032	1,051	1,071	1,111	1,123	1,182
Hydro and nuclear	127	142	152	157	170	174

Source: United Nations, *World Energy Supplies 1973–1978.*
[a]OECD Countries, South Africa, and Israel.

that the recovery in consumption will not be as strong as before because there is more conservation as well as more energy-saving equipment in place now. In addition consumers were not as insulated from the last energy-price increases as they were from the 1973–1974 price rises, especially in the United States, which is a particularly heavy user of energy. I hope it is appreciated that the life-style of industrial countries is still based on a copious usage of energy; and the pleasing spectacle of smaller oil and energy intensities is being purchased at the cost of major inconveniences to the unemployed and their families, and increasing minor inconveniences to anyone not in the upper 5–15 percentiles of their country's income ladder.

Oil Price and the Economic Development of OPEC

Now we shall approach the matter of the availability of oil via the topic of economic development. In order for the reader to fully comprehend my analysis, however, the concept of *value added*—as opposed to profitability—must be understood. By way of clarifying these issues, a simple example in the form of a parable will be constructed.

Johan is a resident of Gaptown (Sydney), where he spends his time daydreaming, picking berries, and working on his book *The Step Six Model of the Demand for Sill at the Cafe Boulevard*. One day he discovers some oil bubbling up on his property, so he goes to see Mr. Dave, the banker. Mr. Dave likes the way this clean-cut young economist carries himself, so he gives (not lends) Johan $10,000, although the interest rate is 10 percent. Johan then buys a nondepreciable drilling and hauling truck that lifts and hauls enough oil to provide Johan with $2,000 a year after he has paid the operating expenses for the truck. Since Johan has no earning power—and needs none, since he lives on berries and the coffee he steals from the Gaptown College of Economic Knowledge—his profit is $2,000. This is also the value added (to the Australian economy) by his business activities. The profit rate (profitability) that is being realized, which will be defined here as the yield on invested capital, is $2,000/10,000 = 20$ percent.

Now Godfather Renfrew, whose taste for meddling is almost as well developed as his taste for moonshine, suggests to Johan that he expand his business to the extent that he can provide some work for his otherwise unemployable cousin Lenny. Johan then goes once more to Mr. Dave, who on this occasion is busy with some banking problems involving Manicura or Pago-Pago and, rather than take the time to draw up the papers for a loan, goes into his pocket and once more makes Johan a gift of $10,000. Johan then buys a minirefinery, which is operated by Lenny; Johan's truck now hauls refined rather than crude oil to market.

This operation brings him $4,500/year (net of fuel and maintenance costs). What about his expenses? Cousin Lenny draws wages of $1,000, which means that Johan earns a profit of $3,500 dollars. *Value added* (profit + wages) is $4,500, but the profit rate has fallen to (3,500/20,000) = 17.5 percent, where $20,000 is the total investment in the business (truck + refinery).

Johan, not being a generous person, is livid with rage, and goes to Godfather Renfrew with the following story: "Instead of hiring that worthless unteachable Lenny, I could have taken that second $10,000 and, by trading in my truck, bought a larger truck. Then I could have lifted and sold more oil, and probably continued to earn a 20-percent profit."

Godfather Renfrew, when he is sober, is even more selfish than Johan; but since he is now busy ordering wine for the Mutual Admiration Society Ball, he is intoxicated most of the time, so he delivers the following lecture on welfare economics:

> That's right, Joey. Your private profit would have been 20 percent, where now it is 17.5 percent. But that good-for-nothing Lenny is working, and that must be worth something; so in a sense it's possible to argue that the social profit is higher than 17.5 percent and might even be higher than 20 percent. Maybe that doesn't mean anything to you as a successful private businessman, but it does to me and everyone else who has to put up with Lenny idling around the college making a fool of himself. Besides, doesn't it make you proud to know that you are now an employer, your business is growing (you have a $4,500 turnover instead of $4,000), you helped reduce unemployment in our splendid community, you have set an example for other members of your lovely family, and you have turned Lenny's indolent hands and mind to purposeful activity instead of the nonsense he indulged in at the college. Then he smiles and returns to his cups.

Several years ago in Australia, most companies in the bauxite-alumina-aluminum industry made no secret of the fact that they preferred to invest money in the mining of bauxite and the production of alumina rather than in the processing of alumina into aluminum, although there is a very large increase in value added in the transformation of alumina into aluminum. (This was also true for some companies in North America, which preferred investments in mining and fabricating on the grounds that they were more profitable than those in smelting.)

For Australia, however, with an unemployment rate of more than 6 percent at that time, it was certainly possible to argue that more value added was needed, even though it meant a marginal lowering of the rate of profit for the firms engaged in primary production and basic processing.

The same thing is true in the OPEC countries today. Their oil industries are profitable, but profitability does not imply development. Regardless of how much oil these countries possess, it is still necessary for them to contemplate the conditions of their economies in a postpetroleum era, and/or the vulnerability of their physical and portfolio investments abroad in a world in which they lack political leverage because they no longer have oil, or because their energy resources have become less important in the world energy picture. If they are interested in useful models, it is only necessary to point to Taiwan, whose petrochemical industry is the twelfth largest in the world and, among other things, provides that economically successful country with one-third of its manufacturing jobs and one-third of its exports—mostly in the form of downstream products such as synthetic fibers, textiles, and plastics. Attention can also be paid to the fact that the Taiwanese petrochemical industry, among others, is in trouble because of the rise in price of their raw material and energy inputs.

The potential producers of chemicals in OPEC should also be aware that the largest business in Texas is not oil, but petrochemicals. If they cannot construct their future economic development on refining, petrochemicals, and perhaps a few other energy-intensive activities, they are not going to develop at all in the usual sense of the term. Another issue of subtle importance here should be mentioned in light of our underlying discussion. In order for the products of OPEC chemical plants to be assured of an entree to the international marketplace, it may be that the industrial world must feel a certain insecurity about the availability and price of its oil requirements. In a situation in which oil is inexpensive and abundant, and expectations are that it will remain so for a long period of time, the industrial output of the OPEC countries runs the risk of encountering serious trade barriers in Western countries.

At present the OPEC countries have about 3.5 percent of the world refining and petrochemical capacity. By 1985, or soon after, they intend to raise their slice of global refining to at least 8 percent. Just how much of this can be exported is uncertain, since there are a number of troublesome problems associated with the transportation of some refined products. On the other hand it may be well within the range of today's technology to develop supertankers for freighting refined products that display most of the efficiencies of the giant vessels carrying crude. Also, many crude-oil refineries in industrial countries are closing down or reducing the scale of their activities. Gulf Oil, for example, has decreased its worldwide refining capacity to 1.6 Mbbl/d from 2 Mbbl/d, closing two important refineries in 1980. This declining emphasis on refining and certain other activities is not the result of a worsening overall profit situation, however, since although Gulf-owned or -controlled oil production has fallen by 80 percent over the past ten years, its total assets have doubled and its profits have increased by 150 percent. These closures reflect the disadvantage of carrying on large-scale refining operations without having an assured access to adequate supplies of crude oil, and in particular relatively low-cost oil.

Before continuing, a brief summary of the petrochemical industry might be useful. As shown in figure 2-3, crude oil is converted in refineries to gasoline and a number of other products. Some of these other products, like naptha, are *feedstocks* that can be cracked into building blocks such as ethylene, which can then be further processed into *intermediate products*. These are then transformed to *final products* such as paint, synthetic fibers, plastics, additions for gasoline and lubricating oils, refrigerants, aerosol products, drugs, anesthetics, toiletries, detergents, and fertilizers. Building blocks such as ethylene can also be produced from natural gas. It is not inconceivable that OPEC will possess 6 million tons of ethylene capacity by 1985, which would put it in the same class as Japan where this product is concerned. Soon after, Saudi Arabia may be in position to supply at least 8 percent of the European bulk chemical market. Given that the OPEC countries will be producing feedstocks with the most inexpensive oil and gas in the world, they should be in position to launch a full-scale assault on the international petrochemical markets by 1990 (see figure 2-11).

In this connection it should be pointed out that in a declining world market a petrochemical boom has begun in Canada that is based almost exclusively on the relatively inexpensive gas that became available when the Canadian government imposed regulations on the price of gas to domestic buyers. In these circumstances Canadian chemicals are now capable of underselling those produced in the United States in both the United States and so-called third markets (such as Japan and Australasia). The natural gas of the Gulf is apparently to be made available to local industries at less than one-third of the Canadian price, however, and if necessary free. More important, since a major producer of chemicals, Japan, has started phasing out certain elements of its petrochemical industry in precisely the same manner as it has been reducing its aluminum-processing capacity, it seems reasonable to suppose that the Gulf may have the same chance to capture these operations as Australia enjoyed with the refining of aluminum. In addition, even before they are constructed, plants located in the Gulf may be able to arrange sales of both basic and intermediate products to Japan in deals that are associated with the supply of crude or refined products. To return to the aluminum analogy, Nippon Light Metals and Sumitomo Light Metals recently closed down 400,000 tonnes of aluminum-smelting capacity in Japan, and instead imported ingots (for further processing in Japan) from joint ventures that they were involved with in Indonesia and Australia.

One of the most important tasks facing the potential petrochemical producers of OPEC is holding down the construction and operating costs of the petrochemical installations they are building. One way this is being done is through a series of multibillion-dollar joint ventures with Western firms such as the Mobil Corporation, Mitsuibishi, and Exxon. These firms have already agreed to help with marketing products from the plants they build and, in some cases, are probably committed to operate technical schools of

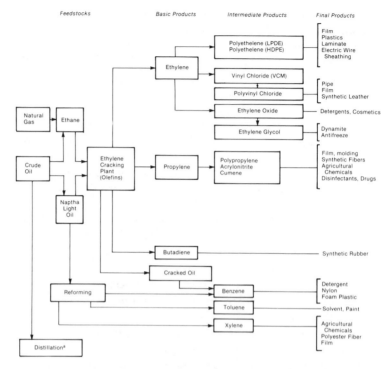

Feedstocks   Basic Products   Intermediate Products   Final Products

^aDistillation of crude oil into methane, with the basic product being liquid anhydrous ammonia and carbon black, and the final product fertilizer and tires.

Figure 2–11. The Petrochemical-Product Cycle

the type established at Kaduna Nigeria by Chiyoda (of Japan), which trained a thousand welders, pipefitters, boilermakers, electricians, and instrument fitters. In turn these firms will receive entitlements for the crude needed to feed their refineries in the United States and elsewhere.

What about the demand for petrochemicals? Because of their versatility, the increase in demand for the various products of this industry over the past few decades has been far larger (on a percentage basis) than the aggregate increase in world economic demand. Moreover, although the growth of per capita consumption of petrochemicals in developed countries is slowing down, it is growing in the LDCs; just as entitlements serve as the incentive that causes firms in developed countries to construct installations that may be in competition with their own facilities, so the lure of cheaper oil—as well as some ideological bonds—might give OPEC chemicals a strong competitive advantage in the Third World.

We can now attempt to relate the foregoing to the setting of the oil price. Obviously, if oil and gas prices rise substantially, it will make many refineries and petrochemical installations outside OPEC uncompetitive. Taiwan, as mentioned earlier, is a perfect example of a fairly large-scale producer that is in serious trouble because of the ascending price of its energy inputs. There are other countries in the same position. It can also be noted that with the present soft market in oil prices, a number of North American refiners of crude have been able to increase their capacity utilization by large amounts. Thus, as OPEC petrochemical and refinery capacity begins to come on stream, the owners of this OPEC capacity might be tempted to manipulate their market share via a systematic long-run policy predicated on a steady upward movement in the oil price.

Finally, I would like to respond to the argument that a country like Saudi Arabia is concerned with its oil being made obsolete someday by such things as synthetic fuel energy from photovoltaics. Remember that oil from the Gulf is inexpensive oil—the most inexpensive in the world by a factor of 10 or 20. In the event of some unforeseen technological breakthrough that deprives OPEC of its present advantages, the members will always have the option of signing long-term contracts with energy-poor countries like Japan and Germany, since any conceivable breakthrough will be a quantity rather than a price phenomenon.

A Parametric Programming Problem

In the foregoing discussion it was observed that in Canada at present a petrochemical boom is underway based on inexpensive natural gas that government regulation has kept below the world market price. This is a situation displaying external economies, where these economies are generated by the cheap gas. A similar arrangement could possibly come about in Australia with the refining of aluminum if Australian energy inputs were properly priced.

I will now set up a simple parametric program indicating the attractiveness of departing from world or opportunity-cost pricing when we have a produced input. Since there are some very important issues at stake here, however, a few introductory remarks are in order. Just now we see a tendency in many countries toward total deregulation and total reliance on market solutions. This tendency will not last for the simple reason that it represents an especially pernicious form of irrationality in a world of uncertainty, cartels, and quasi-state capitalism (of the type practiced in Japan). As Thornton Bradshaw, chairman of the Atlantic Richfield Oil Company, has pointed out, the only sensible course of action for most industrial countries is to forge an alliance between government and industry. For example, the Japanese government, which has supervised the greatest economic success story of the postwar world, has made it clear that it will not tolerate the enfeebling of Japanese industry in order to adhere to pseudoscientific canons of economic virtue of the type propounded by Frederick Hayek or Milton Friedman—although in principle they have nothing against their competitors being free to choose incompetence. (Other proponents of a government-business alliance are John Connally, the former governor of Texas, and secretary of the U.S. Treasury; and, at least implicitly, the brilliant New York investment banker Felix Rohatyn.)

Perhaps the best analytical apparatus for exploring this subject is found in the work of Chenery, who portrays the interdependence of investment decisions in an input-output setting. Moreover, Chenery's investigations suggest a simple linear-parametric programming exercise that can serve to conclude this note. Take a situation in which a commodity x_2 is produced using a primary input R and a produced input x_1. The price of this latter input to the industry producing x_2 is set at p_1'; however, this produced input can also be sold on another market at the fixed price p_1'', which could be viewed as a world price. One linear program describing this situation is:

$$Z = p_1''(x_1 - bx_2) + x_2(p_2 - p_1'b) = \text{Max.} \qquad (2.20a)$$

$$a_{11}x_1 + a_{12}x_2 \leqq R. \qquad (2.20b)$$

$$x_1 - bx_2 \geqq 0 \qquad x_1, x_2 \geqq 0. \qquad (2.20c)$$

Observe that in the objective function p_1'' is not multiplied by the output of x_1, but by the amount of x_1 in excess of that required by the activity for producing x_2. Also, when we simplify the objective function, and thereby obtain $Z = (p_2 - p_1'b - p_1''b)x_2 + p_1''x_1$, it becomes clear that if p_2 is high relative to p_1'', then, depending on the resource requirements of the processes, it might pay to adjust p_1' so that the production of x_2 can take place. Note also from equation 2.20c that if one unit of x_2 is produced, then b units of x_1 are required. As for the solution of this simple program, figure 2-12 should be instructive.

Two solutions are shown. In the situation depicted by $\bar{Z}$, p_2 is high enough and p_1' low enough to make the production of x_2 profitable, whereas with given p_2 and p_1'', Z^* shows an arrangement where there is no feasible value for p_1' that will make the production of x_2 profitable. The terminology being used here is also important: p_1' can be adjusted from a maximum value of p_1'' down to some value reflecting the cost of x_1. The parametric nature of the exercise calls for choosing a value of p_1' that will maximize Z;

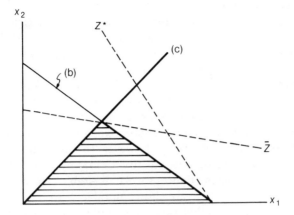

Figure 2-12. Convex Polygon for Parametric Programming Problem

in line with earlier observations, this value could be well under the world price p_1''

The Oil Price and the Financial Markets

We now come to some financial aspects of the oil price, of which the most important has to do with the recycling of a sizable portion of oil revenues to the financial institutions of the Western industrial world.

The point to emphasize here is that, as things now stand, the industrial world does not need these petrodollars. It needs OPEC's oil, but an important portion of this oil would not be forthcoming if the financial return on petrodollars were perceived as inadequate. Since the petroleum-exporting countries with major financial surplus (such as Saudi Arabia, Kuwait, and the UAE) comprehend that it would be impolitic—if not provocative—of them to purchase, or attempt to purchase, all the factories, skyscrapers, farmland, and so on in Europe and North America that they might find attractive, they hold a majority of their foreign assets in the form of financial assets that they hope will provide them with a satisfactory yield, and that are also secure from arbitrary confiscation or blocking (as happened with some financial properties of Iran). If they become convinced that their hopes were groundless, then they would have no choice but to keep their oil in the ground. This is an option that has very unpleasant ramifications for the energy-based economies of Europe and North America.

The case of Saudi Arabia is instructive here, particularly since a recent report in *Business Week* (27 July 1981) explained *why* the Saudis must keep their oil flowing. At the time this article was written, Saudi Arabia was lifting slightly more than 10 Mbbl/d. The revenue from selling this oil was used to finance imports of goods and services and to purchase assets abroad—mostly financial assets such as treasury bills, bank deposits, and other credit-market items. In 1980 imports of goods and services, plus foreign aid, came to $70 billion. With oil revenues slightly in excess of $103 billion, it appears that a petroleum output of approximately 7 Mbbl/d was needed to finance these imports; observing, however, that investment income was $12 billion, the necessary output falls to $(70/115) \times 10 = 6.026$ Mbbl/d.

Next we can examine the figures for Saudi Arabia's holding of financial assets, and their yields. These are shown in table 2-10.

With the average inflation rate in the industrial countries around 10 percent, the yield on these assets is not particularly impressive, although it is true that recently purchased credit market instruments carry a higher rate of return and that in some cases individual countries may be able to shop around among the suppliers of industrial goods in such a way as to avoid the higher-priced commodities. Generally, however, it is believed that many OPEC purchases (of such things as capital goods and engineering services) are made at exhorbitant prices. The observation of the *Business Week* article that these financial properties represent an important long-term diversification of assets given the uncertainties associated with the world oil market deserves some attention, however, because it has an interesting significance for the oil price—one that was apparently overlooked by the author of this article. If there was a movement toward normality on the world financial markets, it would imply a worldwide slide in interest rates and a much lower average earnings level for Saudi Arabian financial assets if the monetary authorities in that country intended to maintain a stock of assets as large as that held at present. On this point it might be useful to examine figure 2-13, which shows that during 1975-1978 the inflation rate in the OECD countries dropped from 12 percent to 6 percent, and interest rates declined accordingly; but interest rates declined faster than inflation rates. Accordingly, some OPEC countries curtailed their production of oil and financed some of their imports by selling off their bonds and shares. This could easily happen again, but this time it might mean a much larger decrease in the availability of oil since the OPEC countries are holding a much larger stock of assets, and these holdings are much more concentrated.

Now, regarding the other side of the market, the issue here is simple and basically reduces to the following: OPEC surplus revenues are channeled, either directly or indirectly, to the major international credit institutions who then lend to what they consider to be worthy borrowers. The

Table 2-10
Net Foreign Assets of Saudi Arabia, 1979-1981, and Their Yields

Year	Net Foreign Assets[a]	Foreign Investment Income[a]	Implied Average Yields (%)
1979	77	7.2	9.35
1980	110	12.0	10.90
1981[b]	160	14.0	8.75

Source: Morgan Guaranty Mongthy Bulletins: Euromoney (various issues).
[a]Estimated.
[b]Billions of dollars.

problem is that many of these borrowers are not, in the commercial sense, worthy, since they are borrowing money they cannot repay in the foreseeable future in order to finance activities that should not be undertaken, or continued, on economic grounds. In the past, however, it has been this unsound and/or unwarranted borrowing that has kept the international economic system in gear. Without this borrowing there would have been no place for OPEC surpluses to go, and sooner or later the OPEC countries would have eliminated the revenues generating these surpluses by cutting oil production drastically.

Since the beginning of 1982, many bankers are insisting that the era of cheap money is over. Bringing it to an end is the recent decline in OPEC financial surpluses as well as the growing unwillingness of some oil-producing countries to accept, or risk accepting, a negative rate of interest on their financial acquisitions: they want a positive real rate, although a large part of their money is financing projects with a negative return. Accordingly, in order for the mechanism cited in the previous paragraph to continue to function with at least a facsimile of its usual efficiency, economies displaying a pronounced lack of viability must go deeper into debt, only now they must pay more for this privilege.

Altogether 150 countries are about 550 billion dollars in debt, but Brazil and Mexico account for a fourth of this debt, and 13 other countries for about two-thirds. (Of these 13 countries, about 70 percent of their debt is now with private institutions, as compared to 34 percent in 1971.) In 1973 the debt of the nonoil-producing LDCs was 82 billion, they imported oil for 5.2 billion, and received development aid for 12 billion. In 1980 these figures were 293, 66.5, and 32 billion; while in 1981 they were 339, 77.5, and 35 billion. In addition, for these countries, the rate of interest on new loans has about doubled in the last few years and on the average is now 18 percent for a loan of 7 years. Thus, a very large and growing part of future borrow-

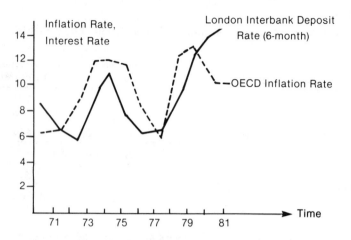

Figure 2-13. OECD Inflation and Interest Rates, 1970–1980

ing will be for the purpose of repaying debt and not for financing industrial or agricultural projects. Some countries in eastern Europe, particularly Poland and Rumania, are in a similar situation. For example, Poland's external debt doubled between 1976 and 1980.

What about the intermediaries in all this: the financial institutions, and in particular the private banks with tens of billions of dollars in dubious loans on their books. These institutions have neither the capital nor a sufficient amount of highly liquid investments in their portfolios to continue making the loans they are now making. This shortcoming is largely overlooked by the supposed experts, but for some years now the financial world has been holding its breath in expectation of a major bank failure, or at least a bank scandal that would severely weaken confidence in the financial system.

Thus, literally with the passage of every day, the banks and other financial institutions find their future lending constrained by the weight of non-bankable loans they have already granted; many LDCs are in the unenviable position of having to service a huge amount of debt with little or no relation to their legitimate needs. It therefore might be suggested that it is in the interest of both these parties to scale down their dealings. As emphasized previously, however, without these dealings there might be some problem in placing the OPEC surplus; and unless this surplus can be placed, the oil corresponding to this surplus would stay in the ground—thus contributing to a scarcity of oil that would be reflected in a higher price.

The key word here is might, because, as will be indicated, an intermediary in the form of an international institution could be the initial recipient of OPEC monies, granting OPEC lenders their real return, and also graciously absorbing any defaults by borrowers. Also, with the United States now running huge deficits, more petrodollars than ever before have a reputable sanctuary since naturally some of the OPEC surplus is being transformed in one way or another into U.S. debt.

The Oil-Transfer Problem

The transfer problem first assumed the status of a classic academic riddle in the course of some learned speculation on transferring the spoils of war from loser to winner. Specifically, a transfer of money first takes place, which, when spent by the winner, results in his being the recipient of a physical transfer of goods and services. The case most often cited in the literature of international economics is the post-World War I situation in which the Allies, following indemnity payments by Germany, bungled the physical transfer because of internal dissension and inept economic policies. Germany, on the other hand, had succeeded in raising production and restraining consumption to an extent that made it possible to meet the demands of the Allies as these were enumerated in the Treaty of Versailles. As a reward for their diligence they were later told that reparations had to be made in gold, not goods—which happened to be an economic impossibility at that time.

Academic economics in general has almost succeeded in presenting this and similar episodes exclusively in terms of similarities and dissimilarities in the taste patterns of the transferring and receiving countries. If the receiving country expands its consumption in exactly the same form as the paying country contracts its domestic purchases, then no global excess demand or supply is created at the initial prices for traded goods; thus relative prices do not need to change. The spending of the country receiving the transfer would exceed its production by precisely the amount of the transfer, and the spending of the country providing the transfer would be less than its production by the amount of the transfer. Most important, no price or exchange-rate changes would be required. In the simple aggregate models used to discuss this issue, it is the absence of these changes that indicates that a given transfer—in physical terms—actually matches the intended transfer, with no side effects (or secondary burden) that would increase the welfare of either party or decrease this welfare.

The oil-transfer problem contains some of these elements but is largely a different enigma. It is the change in the price of an internationally traded commodity that instigates the transfer. There is also another type of transfer from the countries that have gained from price changes—the oil-exporting countries (OEC)—to those that have lost—the oil-importing countries (OIC)—due to the inability of the OIC to raise their expenditures to the level of their income. More simply, this amounts to the OEC not drawing on the resources of the OIC by the maximum amount possible. Thus those issues that would have been regarded as subsidiary, had they been taken up in the context of the conventional transfer problem, here assume center stage.

Take, for instance, the recycling of petrodollars, which some economists feel is the nub of the oil-transfer problem. This involves the second kind of transfer mentioned in the previous paragraph, but with important ramifications. Here we have—at least until recently—a situation in which much of the recycling of petrodollars is not proceeding from the OEC to the main OIC, but to countries in the Second and Third World that could, at least in theory, do a large part of the spending that the OEC are unable to do out of their enlarged incomes, or choose not to do. As it happens, however, many of these Second and Third World countries are inefficient absorbers of capital; thus a substantial part of these petrodollars is being used to pay the interest on previous debts or to import consumer goods at ever increasing prices, rather than to finance productive investments.

Another way of approaching the foregoing turns on a subtle appreciation of the fact that the industrial countries need not petrodollars, but oil, and that this oil will not be forthcoming if the financial return on the OECs' foreign placements (oil revenues—expenditures) is inadequate. Adequate financial yields have been obtained by means of the risk premia put on the enormous amount of borrowing of petrodollars by less developed countries that could not make efficient use of this wherewithal, as well as a few industrial countries (like Sweden) that are capable of using it productively but for some reason choose to do otherwise. More important, however, has been the inflation premium superimposed on real interest rates, which, as explained in *The Political Economy of Oil,* gave OEC money managers the impression that they were earning more on their investments than was actually the case. As will be explained later, inflation was also useful in reducing the physical component of the first kind of transfer.

Before taking up this matter, however, an important point should be made with respect to the distinction between real and financial rates of return, since we may be on the brink of an era in which the supply of oil from the OEC will be formally linked to its real return (the actual amount of physical goods that can be purchased with oil incomes) through some sort of formula. At present the possibility definitely exists that a large part of the movement of petrodollars to various categories of borrowers via private financial institutions may have to be attenuated, and that instead OEC savings will be channeled directly to official institutions that can provide a return that, in either economic or political terms, is deemed satisfactory. The recent loan (of about $15 billion over a three-year period) by Saudi Arabia to the IMF at an effective interest rate in excess of 11 percent can be cited here, although it is unclear to me whether the real rate of return of the Saudi Arabian loan will exceed that which would have been realized if an equivalent amount of oil was kept in the ground. It is also true that the lending programs of the IMF are not above suspicion, even though the liabilities of the IMF are, in one sense or another, secured by the taxpayers of the countries belonging to the fund. What we have here is a transfer problem of the third kind—namely, the transfer of bad lending practices from the balance sheets of private institutions, where they would eventually be detected, to the bureaucratic routines of the IMF, where they can be concealed behind a particularly sophisticated variety of high-minded incompetence.

We can now go to the first type of oil-transfer problem referred to earlier. This focuses on the OEC obtaining goods and services from the OIC as a result of current purchases. The problem is not trivial because of the presence of the secondary burden, referred to earlier, which came into existence because of the high rate of inflation in the OIC and, as a result, reduced the purchasing power of oil. Another and perhaps more interesting complication is introduced by the fact that oil is sold for U.S. dollars. Thus there is a secondary burden on those OIC whose currencies depreciate with respect to the dollar, whereas countries whose currencies appreciate with respect to the dollar receive a secondary benefit. The latter category until recently included West Germany, Japan, and Switzerland. What about the OEC? At any point in time they could be experiencing a secondary burden or benefit in comparison with the previous day, week, or month; but the picture is clearer over a span of years: in the period between the initial oil price rises of 1973–1974, to the price rises of 1979, the OEC have definitely had a secondary burden imposed on them—although net real welfare gains have resulted from their receipt of the price-induced transfer. Whether these gains represented the maximum that could have been realized, however, cannot be discussed here.

This section can be rounded off with a little algebra. First we can look at the income term of the so-called Slutsky equation, which deals with the income effect of a change in price. Keeping the price of exports from the OIC constant, and considering only oil, the income loss of the OIC due to an increase in the price of oil is:

$$Y = -M \Delta p_m \qquad (2.16)$$

Here M is the amount of oil imported, and Δp_m is the change in the price of oil. Dividing both sides of this expression by Y, and multiplying the right-hand side by p_m/p_m, we get:

$$\frac{\Delta Y}{Y} = - \frac{p_m M}{Y}\left(\frac{\Delta p_m}{p_m}\right). \tag{2.17}$$

This is the transfer brought about by the change in the price of oil: if, for example, imports of oil by the OIC are 5 percent of their income, and import prices rise by 100 percent, then real incomes in the OIC fall by 5 percent. Note, however, that if substitution away from oil takes place, then by the Hicks-Allen term of the Slutsky expression, this loss will be decreased. The loss will also be smaller if the price of OIC exports rise. To see this, consider the income of the OIC to be expressed in units of their export good (such as machinery). Then we have:

$$\Delta Y = - M \Delta\left(\frac{p_m}{p_x}\right), \tag{2.18}$$

or

$$\Delta Y = - M\frac{p_m}{p_x}\left[\frac{\Delta p_m}{p_m} - \frac{\Delta p_x}{p_x}\right]. \tag{2.19}$$

As is clear from this expression, other things being equal, a high rate of increase in the price of OIC exports mitigates the transfer by reducing the purchasing power of OEC exports, as mentioned earlier.

Appendix 2A:
Oil and the Futures
Market

The purpose of this appendix is to clarify some of the mechanics of futures markets, given the growing importance of this kind of market for traders of oil and oil products. (Here, *traders* designates both buyers and sellers of oil. In the United States, for example, firms such as Gulf, Conoco, and ARCO are heavily involved in the futures markets.) As the reader will soon see, however, the issue is perfectly straightforward. It should be possible, just by reading this introduction and the third section of this appendix, to grasp the nub of the topic. At the same time, some theoretical material is presented on inventories and stock-flow models. As the textbooks of the future will certainly emphasize, anyone without at least a rudimentary insight into these matters will never have more than a superficial insight into why primary commodity prices behave so peculiarly.

First, it can be pointed out that at present more oil than ever is being sold on spot markets, which implies a deemphasis on long-term contracts. Since spot markets are capable of considerable fluctuation, a comprehensive futures market that functions in the habitual sense enables sellers and buyers of oil, oil products, and certain oil-intensive commodities to insure against price risk. This is an important point, and its significance should be clear before the termination of these introductory remarks; but a few preliminary observations seem justified before we continue.

In New York, where spot traders in the heating-oil market now account for about one-third of all buying and selling, there has been a general skepticism toward the futures market. Generally this was due to the erroneous assumption that oil prices tend only to rise, which would make this market uninteresting to the speculators who would have to provide the market with a large portion of its liquidity in the course of betting on the size and direction of price movements. Local supply and demand situations are often quite variable, however, (because of sharp changes in such things as the weather); thus there can be considerable price swings on regional spot markets even though the nominal world price of oil continues to rise. We see an indication of this situation in figure 2A-1, where spot and OPEC prices listed by the Rotterdam spot market are shown for the last six months of 1980.

Next we must make a clear distinction between the spot, forward, and futures markets, and the prices prevailing on these markets. The spot, or

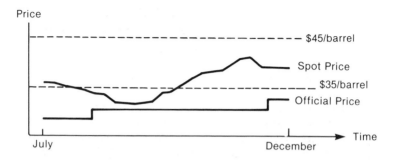

Figure 2A-1. The Spot and Official Price of Petroleum on the Rotterdam Spot Market, July–December 1980

cash, market need not be a market in the institutional sense of the word, but simply an arrangement between buyer and seller that calls for delivery—though perhaps not consumption—or a commodity in the immediate future. In the case of oil the Rotterdam Spot Market has literally become a household term; but although it is true that this district of Holland is blessed with extensive brokerage and oil-storage facilities, it bears only a passing resemblance to the conventional image of a market, since there is no physical locale and trading takes place by means of telecommunications. Similarly, the spot or cash price pertains to the immediate transfer of ownership of a commodity.

The forward market, on the other hand, involves forward sales—the sale of an item that will be delivered in the future at some mutually agreed-on price that is stipulated on a forward contract (or, in the case of some commodities, at a price related to the spot price of the commodity at or around the time of delivery). The forward market almost always involves physical delivery, and such things as the time and place of delivery are specified on a forward contract. The *long-term contract,* which is still the medium through which 80 percent of oil is sold today, is of course a forward contract, although its character seems to be changing. Of late these contracts often mature in a few months, compared with the five years or more that used to be typical. One of the main purposes of a forward contract is to eliminate uncertainty about price and other terms of delivery. As a result, large-scale consumers of oil and oil products have learned to limit their use of the spot market to covering unexpected shortages; selling excess supplies; and, in the same spirit, making marginal adjustments in their stocks.

As for the futures market, this is an arrangement that features paper transactions, with physical delivery of the commodity occurring in only a small minority of cases. Strictly speaking, a futures contract *is* a forward contract in that these contracts almost always refer to a particular month of delivery; at the same time, however, a futures market is so organized that sales or purchases can be offset, and the deliveries are unnecessary. In addition, futures contracts are bought and sold through an exchange, impersonally, with the validity of contracts guaranteed by the exchange. In order for this type of market to function smoothly, large numbers of traders are required—in particular *speculators,* who play a major role in generating the flow of contracts that permit other types of market actors (such as sellers and purchasers of physical commodities like oil) to avoid price risk. It should be made clear that these speculators, who are uninterested in the physical commodity, but have distinct ideas about the price of oil in the future, are buying or selling contracts with the intention of making an offsetting sale or purchase later. If, for example, the offsetting purchase price is less than that of the original sale, the speculator will make a profit.

To help us grasp this simple but crucial point, we can note the situation on the newly opened International Petroleum Exchange (IPE) in London on the first day of trading. The product being traded on this exchange is heating oil (or *gasoil*) in lots of 100 tonnes, and for delivery of up to nine months from the first delivery month, which is June. The June contract opened at $300/tonne and closed at $304.5/tonne, reaching at one point $305/tonne. Some speculators at the exchange that day believed that the price of oil was going to rise and that, as a result, a futures contract for the June delivery of heating oil would, at some point during or before June, be selling for considerably more than the aforementioned figures. These speculators therefore bought a futures contract for some multiple of 100 tonnes (the minimum lot), paying $300–$305/tonne, presumably in hopes that they would be able to make an offsetting sale of a contrct for the same quantity at some point before the delivery date, and for more than the purchase price. Similarly, there were speculators present who felt that the price would be much lower than the $300–$305/tonne range and who therefore sold a futures contract. These individuals thus intended to purchase, before the delivery date, an offsetting contract at a price lower than the one at which they made their original sale.

The third section of this appendix gives a detailed numerical example that deals with the insurance function of a futures market, but I would like to introduce this topic now. A heating-oil distributor on the east coast of the United States anticipated, in June of 1980, that his firm would have to make considerable purchases in the spot market during the coming autumn and winter if the firm's customers were to be satisfied. Given the possibility of sharp increases in the spot price of oil, this company bought futures contracts with delivery dates extending from September through February. As the spot price of oil increased, so did the price of futures contracts; and as this firm made offsetting *sales* of futures contracts prior to the delivery dates on these contracts, the profits from these paper transactions helped cover the rising cost of the spot oil they were buying. On the other hand, the competitors of this firm who had not used the futures market, but required more oil than they had previously purchased on the forward market, were forced to pay premium prices for their oil and were put in the distasteful position of attempting to pass at least a portion of these higher prices along to their customers.

In the foregoing example we say that the seller of the physical product, who at the same time purchases futures contracts, has *hedged* his position: he has opened a futures position opposite to the one held in *actuals* (the physical commodity). Also, the assumption has been made that the spot price of a commodity and the price of a futures contract for that commodity (or a related commodity) move in the same direction. The algebraic difference between these two prices is called the *basis,* or the *spread;* and the risk associated with hedging is termed the *basis risk,* since large changes in the basis could involve losses for the hedger. The statistical evidence indicates, however, that on balance a habitual user of the futures market for hedging will not regret this behavior. (To see this, suppose a trader buys some commodity at time t_0 for a price of p_0, expecting to sell this commodity at time t_i for $\bar{p}_1$. If he sells a futures contract now for q_0, and expects to make an offsetting purchase for $\bar{q}_1$, the expected gain from holding this commodity in inventory over the period $t_1 - t_0$ is obviously $G = (\bar{p}_1 - p_0) - (\bar{q}_1 - q_0) - c$, where c is the (marginal) carrying cost. This expression can now be rewritten to give $(q_0 - p_0) - (\bar{q}_1 - \bar{p}_1) - c$, where $q_0 - p_0 = b_0$ is the starting basis, and the maturity basis is $\bar{q}_1 - \bar{p}_1 = b_1$. It can be proved that the mathematical expectation of this expression, given that $\bar{q}_1$ and $\bar{p}_1$ are stochastic, is equal to $-c$. Over the long run a trader making a large number of hedges should expect to pay only the carrying and purchasing cost of his inventories, not to experience an aggregate windfall gain or loss.

In the foregoing example, the seller of the physical commodity (heating oil) turned his price risk over to the seller of the futures contract, who, if he was a speculator, expected that the price of the commodity would decrease. If this condition were fulfilled, the price of the futures contract would also normally decrease; thus it would be possible to buy an offsetting contract at a price low enough to make a profit. The reader should now observe carefully what would have happened had the spot price of heating oil fallen. The speculator, as pointed out, would have been happy; but what about the distributor of heating oil? His situation could be better, since he must sell his futures contracts at a loss (because a fall in the price of the physical commodity will usually be accompanied by a fall in the price of these contracts). At the same time, however, he is getting his heating oil at a lower price.

Thus the insurance or hedging function of a futures market becomes clear: a loss on the paper transaction was balanced by a gain in the actuals market, just as in the preceding paragraph a gain on futures was neutralized by the heating-oil distributor having to pay a higher price for spot oil.

A general rule has now begun to take shape: anyone buying a physical commodity under circumstances that make it necessary to guard against a price increase should also buy a futures contract. Later, about the time of assuming ownership of the commodity, an offsetting futures contract (for the same quantity) is sold. Conversely, if one is selling a physical commodity at an unknown price, the proper procedure is to begin by selling a futures contract. Then—again assuming that the price of the futures contract moves in the same direction as the price of the physical commodity—a fall in the price of the physical commodity would be counterbalanced by the ability to buy an offsetting futures contract at a lower price that that for which the original contract was sold.

I will now interrupt this exposition to introduce the important topic of inventories; but in order to tie together certain loose ends, I offer the following observations:

1. The foregoing examples dealing with the futures market referred to heating oil rather than crude oil. This is because, with their impersonal arrangements for carrying out transactions, futures markets must deal in very homogeneous commodities in the event delivery is taken. Crude oil, with its widely varying chemical properties, does not fit this description; heating oil, however, which is about 30 percent of the refined barrel, does, and is also widely tradable. As will be explained in the next section, however, since the price of heating oil tends to move up and down in phase with the price of crude oil, the trader in crude oil can guard against price risk by dealing in futures for heating oil, or for that matter some other oil product such as benzine or naptha (for which futures market contracts may soon be introduced). Still, there was once a broad futures market for crude oil in the United States that was extremely active until the mid-1930s. At that time, however, the volume of transactions on this market dwindled considerably with the introduction in that country of a system of prorationing that greatly reduced fluctuations in crude-oil prices. With almost stable prices guaranteed over long periods on all markets, dealing in futures was judged superfluous by many traders; and this particular medium faded away.

In *The Political Economy of Oil* I make the point that introducing a contract for crude oil does not present an insurmountable problem. I have received several queries from people who think that it does. It seems clear, however, that a contract could be designed on which there are a number of deliverable grades of crude oil, with one grade (the basic grade) deliverable at par, and the others at premia or discounts, with these varying in accordance with the price differences prevailing the in the actuals market.

2. Although only a small proportion of all futures contracts are settled by delivery, this arrangement does take place from time to time. As a result, the contracts do specify delivery points, which in the case of the IPE have been specified as Amsterdam, Rotterdam, and Antwerp. This also seems to be a favorite location for the London Metal Exchange (LME); in the case of the LME, however, it has been suggested that the utility of the exchange could be widened if points of delivery were established in many different parts of the world.

Finally, I will close this introduction by reminding the reader that regardless of the neatness of the futures market, a great deal of speculation takes place in physical items. A producer of some product who puts it in inventory without having a commitment from a buyer, and does not take steps to eliminate price risk (which in this case means a fall in price) is speculating—regardless of the strength of his belief in the imminent appreciation of the price of the item. The same is obviously true of someone buying a commodity on the spot market and holding it unhedged with the intention of selling it later on the spot market, presumably at a higher price. This behavior is called being *long* in the commodity. It is also possible to speculate by being *short* in the commodity. This entails contracting to sell something in the future for a price that the speculator believes is higher than the price at which the commodity could be purchased on the spot market at the time the commodity is supposed to be delivered to the purchaser. The term *short* simply means selling something that one does not own.

When a commodity is held in inventory in an uncertain world, someone must be speculating because there is a price risk that must be accepted. A futures market introduces an arrangement whereby this risk devolves on persons (speculators) who are not averse to accepting it. Thus it may well be true that the reapportioning of risk that comes about through these markets represents a clear welfare gain for the community as a whole in the same sense that insurance markets provide a clear social gain.

It is also possible to apply the expressions *long* and *short* to transactions in future contracts. A market participant is short in futures if selling these contracts, and long in futures if buying them.

Inventories: An Introductory Analysis

As already noted, one of the most important functions of a futures market involves the shifting of risk—including the risk inherent in holding stocks. Another major function has to do with providing information that facilitates making rational decisions about inventories. Obviously the most valuable piece of information required here has to do with the price of the commodity in the future; in the absence of faith in their own forecasting capacity, many market participants rely on prices quoted on the various futures exchanges.

Before going deeper into these matters, let us make a clear distinction between long- and short-run prices. In the case of conventional petroleum, the ultimate determinant of supply is the finite amount of this commodity within the crust of the earth's surface, as well as the intertemporal price-maximizing decisions of its owners. By way of contrast, demand is a function of such things as economic activity in the main industrial countries and, to an increasing extent, the Third World countries; global population and its rate of growth; conservation; and the technical and economic possibilities of alternatives to oil. The ineluctable long-run trend of the oil price appears to be upward, although from time to time various economists may say otherwise.

On the other hand, short-term prices tend to be characterized by considerable oscillation, regardless of the underlying long-run trend. An illustration of this situation can be found in figure 2A–1, where, even though the nominal trend price as mirrored in the official quotations is clearly rising, the spot price nevertheless moves in a cyclic fashion. The key to these cyclic movements lies in speculatives tides of bullishness and bearishness fueled by fantasy, naiveté, or irrationality. Thus a relatively minor lapse in demand might cause the spot price of oil to make a sharp deviation downward, which in turn might convince an influential group of market analysts and economic journalists that world markets will soon be awash with oil. Exactly this type of situation prevailed toward the end of 1978, when it was claimed that so much oil was reaching Europe that it could not be stored, and so-called reliable sources were forecasting a production of 20 million barrels of oil per day by Saudi Arabia before 1990.

The only possibility, however, for this kind of logic to prevail over a long period is for a substantial number of market participants to conclude that the underlying economic picture is one of excess supply. If this is not the case, then an upward pressure on prices may be only a step away. Conjecture like that of Professor Peter Odell about supergiant fields of inexpensive oil on the verge of discovery is either ignored or put into a proper time perspective; and a general reassessment of the overall supply-demand situation is initiated, with a premium now placed on facts rather than wishful thinking.

Speculators begin to sense a rapid upswing in prices and increase their purchases of futures contracts, confident that they can sell them later on for much higher prices. As a result of this activity, the price of these contracts will tend to increase; and since some market participants and analysts regard the price of futures contracts as a rough forecast of the spot price of a commodity *in the future,* the feeling that a bull market is at hand is reinforced. In the same vein, markets for commodities such as tin, copper, and zinc perform a forward-pricing function in the sense that they are used as a basis for forward contract pricing or for tendering for such contracts; this may well be the situation with oil at some point in the not too distant future. Last but not least, if it is true that the price of futures contracts serves as an estimate of prices in the future, then inventory holders will increase their demand for the commodity in the belief that it will be to their advantage to buy now and store, instead of relying on the spot market at a time when world demand might be escalating. Needless to say, this kind of behavior places an additional upward pressure on the spot price.

We have now arrived at the main topic for this section. First, note that if producer or consumers' inventories are low, then each extra unit held in stock reduces the possibility that deliveries or production will have to be scaled down because of some unforeseen event such as the absence of an input. Remember that both producers and purchasers of industrial raw materials are bound by contractual obligations to their customers; inventories must be held as long as uncertainty exists as to whether an essential input or promised output will be physically available during the period when it is required. By extension, even if the expected money yield from holding and later selling a commodity does not cover such things as its storage cost, this negative aspect is counterbalanced by a positive *convenience yield* when the size of inventories is small relative to the amount of the commodity being used as a current input in the production process. In this type of situation an effective price system must function in such a way as to ration existing stocks among the demanders of inventories. This often calls for a departure from normality in the form of the spot price ending up at a premium to the (expected) future price. This phenomenon is called *backwardation*.

The same type of reasoning makes it clear that if existing inventories are large in relation to the amount of a commodity being used as a current input, then there is little incentive to hold more. In these circumstances convenience yields are small, and stockholders require that the expected future price of the commodity be sufficient to cover such things as storage, handling, insurance, and other charges—unless, as sometimes happens, futures contracts can be sold at a price that, in the opinion of the stockholder, provides security against a fall in the price of his inventory and covers carrying costs. Otherwise these stocks are put on the market, driving down spot prices and, with an unchanged expected price, widening the gap between present (spot) prices and expected prices to the extent that holding the existing stock is justified. Figure 2A-2 illustrates some of these concepts for a typical storable commodity.

In figure 2A-2 p_s is the spot price and p_f the *expected* future price of the commodity. As pointed out on several occasions, a proxy for p_f might be the quoted price of the commodity on a futures (or forward) contract. At Z we observe the shift from backwardation at low levels of the inventory/con-

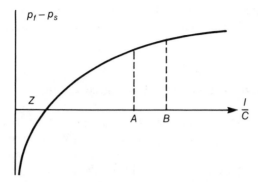

Figure 2A-2. Equilibrium Values of the Inventory/Consumption Ratio
with Given Spot and Expected Future Prices

sumption (I/C) ratio for this commodity to the normal condition called
contango, where the expected future price exceeds the spot price. Until
recently, even though there has been a light surplus of petroleum on the
world market, the spot price has shown a tendency to exceed many
estimates of the expected price. It should also be made clear that a curve of
this type is capable of shifting up or downward from time to time. For instance, this type of curve for petroleum may be shifting upward, which
would mean that with unchanged price expectations, some attempt will be
made to reduce inventories that are being held in private ownership. This in
turn would mean that, other things being equal, there will be a noticeable
downward pressure on the spot price of oil that could, under certain circumstances, influence the official price.

The diagram in figure 2A-2 resembles the supply-of-storage curve
developed by Michael Brennan (1958), although Brennan had inventories
instead of the inventory/consumption ratio on the horizontal axis, and
complemented his analysis by a demand for storage curve. As far as I am
concerned, Brennan's exposition, as it stands, is incapable of providing an
insight into commodity markets in general or the oil market in particular, if
only because the level of inventories (independent of consumption) cannot
possibly have any explanatory value except in the special case in which consumption is constant. Given the many disequilibrium situations in which the
adjustment process includes changes in consumption, however, this case is
too special. In these disequilibrium situations changes in, for example, expected prices cause a reallocation of the commodity between inventories
and current supply, thereby affecting the spot price, which in turn influences present consumption. As for the reason that I instead of I/C was

used, I would have to point to the ease with which *I* could be fitted into a mathematical analysis. Relevance, apparently, was not taken into consideration.

The Futures Markets: Some Practical Considerations

This section will enumerate a few of the more mundane characteristics and possibilities of futures markets. We can begin by mentioning cross-hedging, which has already been alluded to earlier where it was pointed out that at present, futures markets for oil are primarily for heating oil. If there is a reasonably strong correlation between the price movements for heating oil and those for crude oil, however, then the futures market for heating oil can be used to hedge crude oil and, for that matter, other oil products or energy-intensive industrial or agricultural inputs such as fertilizer. The principal problem here is generating enough activity on an exchange that handles only heating-oil contracts to satisfy the huge potential volume of hedgers that could, under certain circumstances, come onto the market.

Early in the first section of this appendix an example was given of a heating-oil distributor who bought futures contracts in order to hedge the price risk associated with the probable purchase of a great deal of heating oil on the spot market at some point in the future. A situation that is sometimes called *forward-commitment protection* involves a trader who has agreed to sell some physical item (such as oil) in the future but does not wish to hold the item in inventory or to make a commitment for it now. This person is therefore short in the commodity, but the financial exposure involved can be reduced by taking a long position in futures (that is, buying futures). In all but the details, these two operations amount to the same thing.

Before making some remarks about speculators, it might be useful to give a final example, along the lines of the one mentioned in the previous paragraph. Suppose that a firm has signed a contract to supply a customer with a certain product that will be manufactured and delivered at some future date, and, when calculating the price that was entered on the contract, estimated that they would have to pay $300/tonne for the oil needed to manufacture the item. Suppose the contract was signed in April, and the product is scheduled for delivery in September. It is true that the price of oil might be $300/tonne in September; by the same token, however, there could be a world economic upswing about that time that would cause a sharp increase in the demand for oil. As a result, the price might turn out to be $400/tonne. If this firm were forced to buy oil at $400/tonne instead of $300, it could suffer a loss on the sale.

One way to guard against such a misfortune might be to order the oil in advance at a fixed price. If this is possible, then it is the supplier of the oil who accepts the price risk. For the purpose of this example, however, assume that it is impossible to find a supplier willing to make this sacrifice. Another arrangement might then be to order the x tonnes desired but to agree to pay a price that is somewhere in the vicinity of the spot price prevailing on some major oil market (such as the Rotterdam Spot Market) at or around the time that the oil is purchased. Assume that this is the scheme that is finally adopted—or, what is approximately the same thing, the oil is simply purchased on a spot market at the time of manufacturing the product.

Should the buyer of this oil decide to hedge his price risk on the LPE, he could buy a futures contract for a quantity of oil equal to that being purchased on the spot market. Let us say that the firm is buying 1,000 tonnes of oil in September; thus in the futures market ten 100-tonne lots will be bought for forward delivery in September or later (although, as emphasized earlier, no delivery need take place). Let us say that the price of oil on the futures contract is $305/tonne, where the contango or forward premium existing here reflects such things as storage costs, insurance, and so on. A 100-tonne lot would thus cost $30,500. On futures markets, however, a great deal of buying takes place on *margin*—which for the IPE means having to put down only 5–10 percent of a purchase. (Similar *leveraging* is available on the New York Merchantile Exchange.) In addition, a small fee must be paid the member of the exchange who handles the order. On the IPE this may be as small as 0.15 percent.

Next, let us assume that the price of oil does rise, and in September spot oil sells for $350/tonne. As explained earlier, however, the price of futures contracts should also rise; the assumption is that in September the offsetting sale of a futures contract that closes out the hedger's position can be made at $356/tonne. The hedger's bookkeeping per 100-tonne lot then takes on the following appearance:

− $30,500	Purchase of a futures contract for 100 tonnes of oil
+ $35,600	Offsetting sale of a futures contract for 100 tonnes of oil
− $35,000	Purchase of 100 tonnes of oil
$29,000	Net payment for 100 tonnes of oil

To this last figure a small brokerage fee must be added, and some economists would argue that interest income on the margin that was paid in April should also be added.

Several things are important here. First, in this example the hedger does not pay $30,000 for a 100-tonne lot, but $29,900. This is due to a change in the basis or spread. The starting basis was $305 − $300 = $5/tonne, but the basis at the time of the sale was $356 − $350 = $6/tonne. It had changed, as is often the case—and on this occasion to the advantage of the hedger.

Obviously, this basis change could have gone the other way, in which case the hedger would have paid slightly more than the $30,000 he had budgeted for this oil; over the long run, however, there should be no losses or gains due to changes in the basis. The only people who can be certain of making money from the exchange are the floor members charging a commission for their services—and the management. Note also that had no hedging taken place, the loss on the total purchase of 1,000 tonnes would have been $50,000.

The matter of delivery must also be mentioned. If the owner of the futures contract insists on delivery, he can be accommodated. In fact, if the seller of the contract produce the physical commodity, the IPE is backed by the International Commodities Clearing House, whose function is to arrange for delivery if necessary. Here it should be appreciated that delivery can only take place into an approved storage area, which for the IPE means designated installations within the Amsterdam-Rotterdam-Antwerp area. This type of arrangement would normally be of only limited interest to a firm located in, say, Liverpool or Belfast. On the other hand in the United States deliveries are becoming more common. Since its introduction about two years ago, the contract for number 2 heating oil has resulted in the delivery of 8 million barrels of this commodity; in December 1980 deliveries jumped by 100 percent over the usual delivery volume, with some of the major oil companies in the United States on the selling end.

I will conclude with a few observations on speculation. In recent years there has hardly been a futures market anywhere in the world whose management has not taken great pains to emphasize that the purpose of these institutions is primarily to enable traders to insure themselves against losses, and that under no circumstances are they intended to provide speculators with a new type of casino. In addition, there has been considerable conjecture about the possible destabilizing effects of futures trading: certain categories of traders, a number of economists, and even some official bodies have claimed that the speculation found on most futures markets has led to the violent price oscillations that are occasionally observed on spot markets, and that as a result severe losses have been imposed on both producers and consumers. Many people feel that it would be better if speculators were to turn to some other line of work; and the chairman of the IPE has even found it necessary to assure the media (in this case the *London Financial Times,* 20 February 1981) that speculators will account for less than 30 percent of the clientele of his establishment. The opinion here, however, is that if Mr. Chairman finds his job congenial, and would like to continue in its present capacity, then—although he may turn speculators out of his heart—he would do well not to turn them away from his door.

Some say that futures-market speculators are uninformed amateurs whose principal distinction is the damage they cause third parties, such as their relatives and creditors. Others say that relatively well informed speculators eventually come to master the art of predicting prices, insofar as this is possible, and thus provide other market participants with a great deal of valuable information at a fairly low cost. This information is also provided to individuals who do not use the futures market, thereby lowering transaction costs for the community as a whole.

This may be so; but if large numbers of uniformed speculators are willing to pay for the opportunity to make profits from trading, and if these novices keep flooding the market—replacing persistent losers who cannot sustain losses and must find some other ways to keep themselves occupied—then the information that is being provided to the market can hardly be of the highest quality. I have no doubt that some professional speculators do have the knack of forecasting commodity prices and making themselves rich in the bargain, but I am not so sure it is their wisdom that we see reflected in various market quotations around the world.

A Theoretical Excursus

This section will examine briefly several theoretical considerations. On the basis of the preceding discussion, it should be clear that futures contracts can be both demanded and supplied by hedgers as well as speculators. If we were to summarize the behavior of a typical hedger who was hedging against a fall in the price of the physical commodity, we might postulate that the higher the price of futures contracts, the greater the supply of these contracts. Why? The answer is that not all traders facing price risk hedge, and even those who do hedge may not hedge all their risk; but the higher the price this category of trader can get for the sale of a futures contract in the presence of a probability that the price of the commodity might fall, the more tempting it is to enter into a hedging operation.

The assumption here is that there is some value of the price of a futures contract, say p_f' below which a typical trader who is a candidate for hedging against a price fall would not be interested in selling a futures contract. The supply schedule for this type of trader is shown in figure 2A-3. Similarly, for traders who are in a position where price risk involves a price rise (for example, traders who are going to buy a commodity in the future at an unknown price), the lower the price of a contract, the more appealing the purchase of a futures contract becomes. The assumption here is that there is a price p_f'' above which this type of trader would not be interested in buying a contract. The demand schedule for this type of trader is shown in figure 2A-3. Now, if we aggregate over all hedgers, it is conceivable that we get the schedule D^h—S^h, shown in figure 2A-3c, where p_f^* distinguishes the

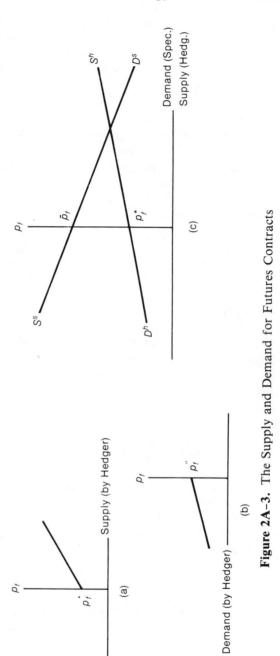

Figure 2A–3. The Supply and Demand for Futures Contracts

point at which a net supply of futures contracts by hedgers passes over into a net demand, or vice versa.

What about speculators? If we assume that speculators are equally prepared to buy or sell futures contracts, then the key thing for our analysis is the estimate of the value of these contracts in the future. Let us say that for speculators as a group, the expected future price of these contracts is $\bar{p}_f$. Then if the existing price of a contract is greater than $\bar{p}_f$, they will sell contracts, expecting to buy them back later at a profit; if the existing price is less than $\bar{p}_f$, they will buy contracts (expecting that later on they can make offsetting sales). This kind of behavior will generate the schedule S^s—D^s shown in figure 2A–3c.

It must be noted that the S^s—D^s schedule is an aggregate: all speculators will not have $\bar{p}_f$ as their expected future price of contracts. Then, too, although the intersection of the D^h—S^h and S^s—D^s curves takes place in the right-hand quadrant of figure 2A–3c, where hedgers are net suppliers of futures contracts and speculators are net demanders, this intersection could have taken place in the left-hand quadrant, where the opposite situation would have prevailed. On the London Metal Exchange, however, it appears that on a distinct majority of trading dates, hedgers are net suppliers of futures contracts. This makes sense because hedgers are largely traders in the physical commodity and, to an overwhelming extent, are interested in protecting themselves against a fall in prices.

Finally, it should be recognized that the schedules D^h—S^h and S^s—D^s are idealized constructions in that it was expedient to specify a unique p_f^* and $\bar{p}_f$: conceivably, in the aggregation of individual schedules (particularly for hedgers) we might have ended up with curves that were discontinuous at the vertical axis. As far as I can tell, however, this changes nothing; on the basis of the information we have from exchanges such as the LME and the New York Commodity Exchange (COMEX), the situation in the vicinity of the intersection of the schedules is realistic; thus the situation at the vertical axis is irrelevant.

Next it should be recognized that in deciding how much stock should be hedged, and how much left unhedged, we are dealing with a *portfolio* problem. The simplest possible exposition of this problem would assume that by hedging we eliminate all stochastic factors, and ensure that at the end of a given period, when we sell the stock acquired at the beginning of the period, we obtain P as our net per-unit revenue. What about unhedged stock U, which is the total amount of stock T minus hedged stock H? The selling price of this unhedged stock is a stochastic variable $\bar{P}$, and thus we can write as our end-of-period revenue from the sale of hedged and unhedged stock:

$$R = (T - H)\bar{P} + HP.$$

Since $\bar{P}$ is stochastic, we have for the *expected* value of R, $E(R)$:

$$E(R) = (T - H)E(\bar{P}) + HP.$$

Here $E(\bar{P})$ is the expected value of the price of the stock at the period's end, and in theory could be obtained from a frequency distribution of possible prices (for the end of the period) conjured up by the stockholder. Now, the variance of R is:

$$\text{Var } R = E[R - E(R)]^2 = E\{(T-H)\bar{P} + HP - [(T - H)E(\bar{P}) + HP]\}^2.$$

This can be simplified to give:

$$\text{Var } R = E[(T - H)(\bar{P} - E(\bar{P}))]^2 = (T - H)^2 E[\bar{P} - E(\bar{P})]^2$$
$$= (T - H)^2 \text{ Var } \bar{P}.$$

Since standard deviation is the square root of variance, we get immediately:

$$\sigma_R = (T - H)\sigma_{\bar{P}}.$$

Observe that $\sigma_{\bar{P}}$ is also derived from the stockholders' estimate of the likely values of the market price at the end of the period, along with their probabilities. Substituting H from the last equation given into the previous relation gives:

$$E(R) = PT + \left[\frac{E(\bar{P}) - P}{\sigma_{\bar{P}}}\right]\sigma_R$$

Assuming that $E(\bar{P}) > P$ (which means that the expected future return on a unit of unhedged stock exceeds that of hedged), the foregoing expression becomes a linear opportunity locus of alternative combinations of the expected revenue from stockholding, $E(R)$, and the risk, σ_R, which slopes up to the right. If we can define a convex indifference curve involving return and risk, then the arrangement shown in figure 2A–4 could be expected to prevail.

Had the tangency in figure 2A–4 taken place at A, all the stock would have been hedged. On the other hand, had the slope of the indifference curves decreased as σ_R increased, the optimum point for a utility-maximizing stockholder might have been point B, where $U = T$ and none of the stock is hedged. Thus the image of the point F on the horizontal axis, F', indicates the percentage of the total stock that the inventory holder left unhedged. In figure 2A–4a the proportion OF'/OB' of total stocks T is unhedged (whereas $F'B'/OB'$ is the percentage hedged).

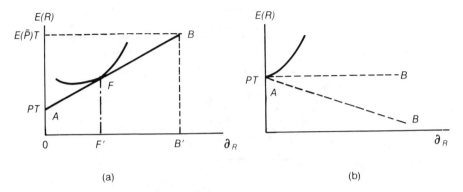

Figure 2A-4. Determining the Amount of Unhedged Stock

What about the case where $E(\bar{P}) \leq P$! In this case the opportunity locus would either be flat or slope downward, as shown in figure 2A-4b. In these circumstances a rational stockholder with a convex utility curve would choose to hold all stock hedged. In the figure this is shown by a *tangency* at A—in other words, a corner solution.

One final point should probably be made in this section. The inadequacy of the simple flow model featured in the basic economic textbooks cannot, for our purposes, be overemphasized; as a tool for discussing short-run price behavior in most markets for industrial raw materials, it is not only inadequate but also, taken at face value, misleading. Basically, the situation we are interested in is as shown in figure 2A-5. An extensive discussion of this type of model is found in Banks (1977, 1979b); but the essential thing to understand in this construction is that a full equilibrium must include both stock and flow equilibria. Letting $I(\ )$ be the demand for stocks, where here the arguments of I are left unspecified, a full equilibrium at time t would call for:

$$I(\ ) - \int_0^t [s(p) - d(p)] dt = 0.$$

and:

$$x = d(p) - s(p) = 0.$$

In other words, the demand for stocks is equal to the supply of stocks; and excess flow demand is equal to zero. However it is possible to have a *market* or *temporary* equilibrium when

$$d(p) - s(p) = k[I(\ ) - \bar{I}].$$

That is, the amount of stocks desired by inventory holders during a given period—which is a fraction $0 \leq k \leq 1$ of the *total* change in stocks that is

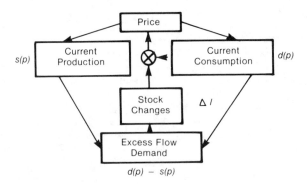

Figure 2A-5. A Simple Stock-Flow Model

to take place over a number of periods—is equal to the actual change during that period (which is excess flow supply). Naturally, these two concepts of equilibrium are not the same. (See the discussion beginning on page 179 for a thorough, but elementary, review of this topic.)

Conclusions

I have attempted in this appendix to provide the reader with a basic insight into the functioning of futures markets in general, and a futures market for oil in particular. At present the largest exchange on which oil futures are sold—the New York Mercantile Exchange (NYMEX)—is experiencing a period of explosive growth. The International Petroleum Exchange (IPE) in London has similar aspirations.

Although I would like to say that the existence of futures markets enhances the aggregate efficiency of a modern economy, no real evidence exists that would make possible a categorical declaration to this effect; by the same token, however, it certainly must be possible to argue that the gambling (speculation) that takes place on futures markets is no more harmful, intrinsically, than that taking place at the Randwick Race Track in Sydney or the New York Stock Exchange, and I am unaware of any serious threats to either the mode of operation or existence of these establishments. In fact, since the operation of futures markets also enables a large number of individuals and firms in the productive sector or the economy to relieve themselves of undesirable price risk, these markets could in some sense be said to provide a service to the community that is no less valuable than that provided by insurance companies. If it is true, as I believe it is, that intuition is more important in economics than in any other intellectual discipline, then my intuition tells me that futures markets are—at least in practice—mostly harmless—and, on balance, have an important service to render the world economy.

In the examples presented in this chapter, the futures exchange seems to be the ideal instrument for sparing buyers and sellers from those embarrassments that can result from not being able to forecast the price of oil correctly; but it has yet to be proved that it will be possible for any futures market, or combination of futures markets, to generate enough liquidity to insure even a small fraction of the total world purchasers of oil against precipitate rises in the price of oil, or sellers of oil against precipitate falls in the price of oil. I suspect that one reason for the great success of many futures markets is the large amount of speculation they have attracted relative to the amount of hedging that takes place on these markets. It seems likely, however, that a ceiling must exist on the amount of speculation that can be mobilized.

Futures markets are traditionally the object of a great deal of attention by the regulatory authority of various governments, and the management and members of several exchanges accused of indulging in practices that are not in the public interest. I believe that much of this harassment is unjustified, and that the individuals engaged in it should find more constructive things to do with their time; but I see no reason to be optimistic on this score. Futures trading has existed for at least two hundred years, and it has seldom been free of criticism by individuals and legislatures with an overdeveloped sense of public righteousness, but a deficient education in basic economics. Indeed, it could probably be argued that monetarist economics, applied in a somewhat diluted version, has done more damage to the economy of the United Kingdom and the welfare of its citizens than all the sharp practices and manipulations practiced by all the futures exchanges in the world, since—to use the suggestive teminology of John Cary—the first bottle of brandy was "disposed of" on the Rotterdam exchange in the late seventeenth century.

3 The Economics of Natural Gas

For the most part natural gas is found in an environment similar to that of crude oil. Indeed, natural gas has on occasion been termed *gaseous petroleum*. The hydrocarbons of natural gas are, however, lighter and less complex than those of crude oil; and natural gas occasionally contains water and gases that are not hydrocarbons. Although it is often believed that gas and oil are found in reservoirs or huge underground caverns (and this terminology is in general use), the truth is that they originate in water-coated pore spaces in rocks—for the most part sedimentary rock that, for geologic purposes, can be classified as organic shale. This shale, originally the remains of prehistoric plants and animals, was transformed into oil and gas by heat, by the pressure of the earth acting over millions of years, and by various chemical reactions.

Some accumulations contain oil but no gas, whereas others—where the "cooking" process continued until the hydrocarbons were reduced from liquid oil to molecules of gas—contain gas but no oil. It is believed that gas can be found at a much greater depth than oil; and a new drilling technology now under development may be capable of reaching gas at depths of 30,000 feet or more, which represents the frontier of current technology. If and when this technology permits downward probes of 30,000–50,000 feet, world reserves of gas should show a dramatic increase.

For many years oilmen treated gas as a pest. Through its exsolving and expanding, it provided the pressure that drove a great deal of oil to the surface; by the same token, however, it could cause blowouts that were costly in both lives and money. Since the 1860s trillions of cubic feet of associated gas have been flared and it was not until the 1930s that the first major U.S. pipeline was constructed to carry gas as a by-product of oil. Still, in 1978 about 200 billion (10^9) cubic meters of oil field gas were flared, out of a total world production of 1,500 billion cubic meters (1,500 Gm^3), with another 100 Gm^3 reinjected for the purpose of raising the pressure of various oil deposits and thereby increasing the amount of oil that can eventually be removed from these deposits. Since one billion cubic meters of natural gas (1 Gm^3) has the energy content of 0.86 million tons of oil (0.86 Mtoe), and one barrel of oil per day (1 Bbl/d) is the equivalent of 50 tons of oil per year, then 200 Gm^3 of gas is the equivalent of 3.44 million barrels of oil per day ($200 \times 0.86 \times 10^6/50$). This is more than the total petroleum output of the North Sea. At least two-thirds of the gas flared in the world belonged to the OPEC countries.

Today in the United States laws exist that restrict flaring. There are some plausible economic explanations for the flaring of gas, however, although, from the social (as opposed to the private-profit) point of view, it could be argued that it might be better for all concerned if as much flared gas as possible is saved and used in the future. Still, at present very few countries are in the enviable position of being able to refrain from wasting gas if it means having to accept a lower level of oil production. At the same time it is useful to appreciate that Saudi Arabia and Kuwait may fall into this category in the not too distant future, since if their petrochemical investments begin to pay off, it will raise the value of their gas—because of the possibility of using this gas as a basic input in the chemical industry— and thus will tend to reduce the flaring of gas. Insofar as this gas is associated with oil, the output of oil could also decrease. (Saudi Arabia, for example, has a gas-gathering scheme that should reduce flared gas significantly by 1990.)

In most industrial countries the combination of a low price, cleanliness, and ease of handling has led to a rapid expansion of gas distribution and storage systems. Because of the extensive investments required, as well as some doubts about the ultimate reserves of gas (and the reliability of suppliers) there has often been a great deal of reluctance to introduce gas; but this reluctance has almost always been dispelled when purchasers get a chance to appraise the utility of this energy source under operational conditions. The popularity of gas may also have been enhanced by the tendency of governments to regulate the price of gas.

Since gas is usually sold by just a few large firms, and gas purchasers must make very large investments that lock them into the use of gas, it has traditionally been considered appropriate by the governments of large gas-producing and -consuming countries (such as the United States and Australia) to regulate the price of gas. A great deal of controversy is inevitably associated with the topic of regulation, particularly since most deregulators and their economic advisers avoid the issue that where monopoly or oligopoly exists on both the buying and selling sides of the marketplace, along with huge "lumpy" investments and market failure caused by the uncertainty that cannot be avoided in geological undertakings, free markets will often result in an inefficient allocation of resources. It must be admitted, however, that regulation in itself does not guarantee an improvement and, in some cases, can make things much worse.

In the markets for natural gas, regulated or not, it is customary for buyers to attempt to protect themselves from debilitating ad hoc price rises by the use of long-term contracts. Ideally, the period of validity of these contracts would match the longevity of the equipment using the gas, but this ambition is seldom realized. It is also useful to remember that the seller of gas has problems with ensuring that once he has made the investments necessary to supply his customers, these customers will continue to use his gas, not the low-cost supplies of some alternative supplier should these become available. Thus he too is amenable to using long-term contracts. In these circumstances, with bargaining between buyers and sellers who have only a vague idea of the future availability and price of this commodity, a sizable portion of the price theory taught in elementary courses in economics becomes quite irrelevant for determining prices and contract lengths.

Gas Types and Uses

Natural gas from a well consists mainly of methane (85 percent); heavier hydrocarbons collectively known as natural-gas liquids (composed of ethane, propane, butane, pentane, and some heavier fractions); water, carbon dioxide; and nitrogen and some other nonhydrocarbons. Before dry natural gas can be distributed to consumers, some undesirable components must be removed and, by decreasing the share of heavier hydrocarbons, a uniform quality attained.

The last-mentioned operation takes place either at the gas well itself or in special installations. It is at this point that the natural-gas liquids (NGL) can be separated out. (NGL should not be confused with liquified natural gas—or LNG—which to a considerable extent consists of methane and ethane.) It is becoming commonplace that in those regions where the availability of natural gas is greater than the absorption capacity of local markets—and for one reason or another it is not possible to transport the gas to foreign markets in its original form—the processing of natural gas into gas liquids is a valuable economic activity. The most important constituents of NGL are butane and propane; in liquid form these are called LPG. In many countries LPG is sold under the name *gasol* or bottled gas and is often delivered to larger bulk tanks. LPG is free of sulphur and most other contaminants and can be used in place of natural gas and as a replacement for such things as naptha and gasoline. Figure 3-1 summarizes the above discussion.

Although it was not taken up in chapter 2, the distillation of crude oil in a refinery will also yield methane, ethane, propane, and butane; but the introduction of new technology for the production and transportation of natural gas, together with the progressive increase in the price of crude oil, has resulted in the weakening of the position of refineries as the principal suppliers of LPG. This could have a pronounced significance for the price of oil because, should important existing and potential producers of natural-gas liquids in OPEC agree to decrease the price of petroleum, they could

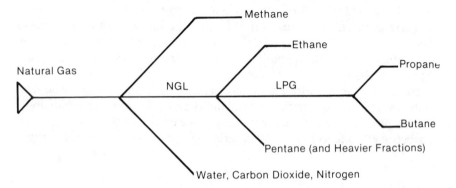

Natural Gas: Methane + NGL + (Water, Nitrogen, and so on)

NGL: Ethane + LPG + (Pentane and Heavier Fractions)

LPG: Propane + Butane + (Mixtures of Propane and Butane)

Figure 3–1. The Principal Ingredients of Natural Gas

be undercutting their future ability to gain a larger share of both crude-oil refining and the market for LPG. Higher crude-oil prices today will mean a lower level of investment in refineries by present-day refinery owners in the industrial countries, and, perhaps, a change in the output mix of refineries in such a way that OPEC countries can greatly increase their sales of certain products.

An example can be cited here. The LPG plant established by the Bahrain National Gas Company (BANAGAS) has been described by its chairman, Assan Fakhro, as the best decision ever made by that country. The plant, which cost $95 million, is owned by the Bahrain National Oil Company (75 percent), Caltex (12.5 percent), and the Organization of Arab Petroleum Exporting Countries (12.5 percent). It opened in 1979 and by 1 July 1981 had produced 87,700 tonnes of propane and 74,100 tonnes of butane—exporting 55,000 and 53,946 tonnes, respectively. The feedstock for this plant is provided by associated gas from Bahrain installations, which until recently was flared. The lesson here is clear: given the success of this particular plant, the availability and the cheapness of gas throughout the Middle East, and the influence of Middle Eastern countries on the world price of oil and gas, this region has the potential to provide itself with an impressive advantage in the production of petrochemicals using propane and butane as starter materials.

Before discussing the pricing of natural gas, let us note the uses of natural gas. Natural gas has been widely used as a primary source of energy for power generation, especially in Japan. It has also been important for heating purposes, both in industry and in the household sector. Table 3–1 gives some idea of the uses of natural gas within the European Economic Community (EEC) in 1970 and 1975. Despite the impression that may be gained from these figures, the progress of natural gas in the aforementioned uses is highly uncertain over the immediate future. Holland, the most important producer in Europe, is going to experience a substantial decline in the output of its gas fields; the long-run status of Norwegian and British production and of imports from the USSR, Algeria, and elsewhere is uncertain. This issue will be taken up later in this chapter.

The Pricing of Natural Gas

The next topic is the price of natural gas, but first some things must be said about units. Thermally, or in terms of heating values, 1,000 cubic feet of natural gas is equivalent to 0.178 barrels of crude oil. In addition, 1,000 cubic feet is equal to 28.3 cubic meters (1 cubic meter = 35 cubic feet). Thus, 1,000 cubic meters of natural gas is equal to $(1,000/28.3) \times 0.178 = 6.3$ barrels of crude oil. We should also know that one cubic foot of natural gas has a heating value of 1,035 British thermal units (2.61×10^5 calories = 1.055 million joules), or 1,000 cubic feet ($1,000 \text{ ft}^3$) has a heating value of approximately one million British thermal units (Btu).

As mentioned earlier in this chapter, the price of natural gas has often been regulated in the main producing countries (especially the United States). This has meant that, relatively speaking, gas has been one of the cheapest sources of energy. In most countries, including the United States, regulation is becoming less popular; therefore, a closing of the gap between petroleum and natural-gas prices is to be expected. (This would mean that in

Table 3–1
The Uses of Natural Gas in the European Community

	Generation of Electricity	Use in Industry	Household Uses	Use Outside the Energy Sector
1970	17,969	19,811	13,237	4,197
1975	31,645	43,825	46,512	5,645
Percentage increase	76	121	251	34

Source: Economic Commission for Europe.
Note: In 100 tons of oil equivalent.

terms of heating values, or thermal equivalents, the price of gas would tend to rise toward that of oil, which is about $5.75 per million Btu at present.) Even so, given the extensive investments required to increase the consumption of gas, its buyers will definitely require some regulatory protection. The expectation is that they will receive this protection.

The theory advanced by antiregulatory politicians and economists that freeing the market for natural gas will lead to sizable increases in supply, but not in price, is observably false—as the executives of a number of energy firms have not hesitated to point out. Confusing the issue further is the willingness of certain well-known economic theorists to supply so-called scientific proof of the availability of natural gas in the event of a rise in its price. Some years ago Professor Paul MacAvoy published some econometric evidence that if the U.S. government decontrolled the wellhead price of interstate gas, allowing new gas to rise in price from $0.54/1,000 ft^3 to $1.00/1,000 ft^3, then domestic discoveries would climb to 33×10^{12} ft^3/year by 1980. In 1980, however, with the price of new gas at $2.00/1,000 ft^3, discoveries were still at the 10×10^{12} ft^3/year level they have averaged over the past decade. Unfortunately, however, some politicians believed this pseudoscientific econometric claim, and still believe it even though it has been proved false.

At present the price of domestic natural gas in the United States averages slightly more than $2/million Btu (MBtu), but it is creeping up slowly because of deregulation. This price will also be affected by the price of imported gas. For example, U.S. and Algerian energy officials have been negotiating an agreement that calls for the importing of considerable amounts of Algerian gas into the United States. The minimum expected price of this gas, FOB, is somewhere between $4/MBtu and $4.5/MBtu, where FOB signifies free-on-board and is the price of the gas at an Algerian port. The delivered price would then be higher by the amount equal to the transportation cost. It should be noted that this price ($4.0–$4.5/MBtu) is approximately double the rate Algeria was receiving when it abruptly reduced it supplies to the United States in the spring of 1980—apparently for the purpose of strengthening its negotiating position. Sonatrach, the Algerian state oil and gas company, wanted between $6/MBtu and $7/MBtu, where the higher figure meant that the price of Algerian gas would have been on a par with Algerian oil (which at that time was priced at $40/barrel). With a price of $6/MBtu, plus transportation and distribution charges, the delivered price of the gas in the United States would average more than $8/MBtu. This is very high compared with the price of domestically produced gas; on the other hand, the amount of gas imported would not be large enough to cause an appreciable rise in the average price of gas in the United States.

Algeria has also been a major supplier of LNG to France, where its

FOB price for the recently agreed upon deliveries was $5.1 MBtu. Similarly, Britain has agreed to pay $4.5–$5.0/MBtu for Algerian gas, which is considerably above the $1–$1.5/MBtu that it was paying on a previous fifteen-year contract, but still lower than the $7/MBtu desired by Algeria. Here it should be noted that according to recent estimates by the *Petroleum Economist,* production and gathering costs in Algeria are somewhat in excess of 25 cents/MBtu, and that pipeline operations and other domestic transport add another 25 cents to the cost. The cost of liquefaction is probably about 0.75 cents/MBtu, although it could be considerably higher because of the rapidly rising price of liquefaction equipment.

The reader should understand that there are very expensive machines and structures involved in the export of natural gas, particularly when liquefied natural gas is involved. This high capital intensity often leads to tremendous financing problems. In these circumstances, not only must gas prices be high, but also, in order to be certain of a market over the long term, it has become customary to sell gas on the basis of contracts with a duration of up to twenty years. Given the likely upward trend in energy prices over time horizons of this length, prices of $6–$7/MBtu FOB might not be too high—unless these prices are indexed. Here it is interesting to note that in 1977 U.S. officials declined to purchase Mexican gas at a price that was well under $3/MBtu, although apparently this position has been reconsidered. As would be expected, however, Mexican officials now have another vision of the worth of their gas—one that is more in line with that being advanced by Algeria. (And gas prices usually *are* indexed.)

The asking price for Norwegian and British North Sea oil is also around $4.5–$5/MBtu. British gas buyers are interested in purchasing some Norwegian gas in order to top up its standby reserves, and have bid $3.75/MBtu for Statfjord gas and $4.40–$5.00/MBtu for Heimdal gas (excluding transport costs). On the other hand, a continental consortium (consisting of Gas Unie of the Netherlands, Distrigaz of Belgium, Gaz de France, and Ruhrgas of West Germany) has put in counterbids and expects to win this gas. With the gas transported through a pipeline, the cost to Germany at an Emden terminal should be about $5.50/MBtu, whereas the French will have to pay about $6.00/MBtu. Basically, however, Norway's position on gas pricing is similar to Algeria's: gas should be priced at the thermal parity.

The situation for Holland is also of interest. Holland has been the major producer and exporter of gas in western Europe (and until recently the world export leader); its Groningen gas field is one of the most important assets of this kind anywhere, considering the energy requirements of the country in which it is located. Unless the situation has changed recently, however, Holland is importing foreign gas at a price that is higher than that being paid for Dutch gas by such affluent gas importers as West Germany and Switzerland. The reason for this unusual situation is that in the 1960s the directors of the company exploiting the Groningen field—acting on information available at that time—were concerned about not being able to sell their gas in the long run in competition with such items as oil and nuclear power. Long-term contracts were therefore entered into with many buyers, some of which called for deliveries ranging over a twenty-five-year period. Thus, after the oil price rises of 1973–1974, a large amount of Dutch gas—because of contractual obligations—was still being sold at prices well under the international price for this and other energy materials; as a consequence, a precious national energy resource was being depleted at an unacceptable rate. Present-day textbooks unfortunately have little to say about this kind of situation, preferring to deal with idealized scenarios in which market participants have all the information they need about the present and future in the form of market prices. The Dutch government had to resort to a drastic improvisation: it threatened to cut off supplies to the purchasers of Groningen gas unless these buyers declared themselves willing to renegotiate existing contracts.

In 1980 about 7 trillion cubic feet (Tft^3) of gas (equivalent to approximately 3.3 Mb/d of oil) was exported, with the Netherlands and the USSR the largest exporters, followed by Norway and OPEC. The Shell Corporation estimates that by 1990 this export figure could reach 14–16 Tft^3, which at that time would represent 22 percent of world consumption. Of these exports the USSR should be responsible for 25 percent. As for the price of Soviet gas, this is expected to be very competitive with other sources; at present it appears that the gas that will be sold to western Europe will cost approximately \$4.75/MBtu at some point close to the German border.

Just now it appears that the price of Soviet gas to other buyers may be creeping up. In 1979 Finland bought 993 million cubic meters of Soviet gas for \$61 million; in 1980 imports dropped to 925 million cubic meters, but the total price was \$111 million. In these circumstances Finnish authorities have begun to make plans for switching from gas to coal and peat, although they have not stopped trying to negotiate price decreases with the Soviets. One of the arguments they presented is that if the price of gas comes down, it may convince the Swedish national gas company (Swedegas) to extend the Soviet natural-gas pipeline across the Gulf of Finland, thus giving the Soviets access to the Swedish energy market.

Observe the calculation here. 925 Mm^3 for \$111 million gives a price of \$0.12 per cubic meter of gas. Multiplying this by 28.3 cubic meters (per thousand cubic feet) gives a price of \$3.396/1,000 ft^3, or approximately \$3.396 per million Btu. Also, since 1,000 ft^3 of natural gas is the equivalent of 0.178 barrels of crude oil, the equivalent oil price is then 3.396/0.178 = \$19/barrel. The new price thus comes close to doubling the old price, but it is still well under thermal parity.

As is often the case in economics, one of the conditions for a country to engage in a considerable amount of international trade in a particular commodity is that it is itself a major producer *and* consumer of the commodity. To a large extent this is the case with natural gas, where the strength of the domestic market in countries like Holland and the USSR has provided a stimulus and base for a high rate of production that rapidly led to an exportable surplus. The United States has had a somewhat different history, with domestic supply expanding rather closely in phase with domestic consumption; but it could hardly have been otherwise. Given the per-unit cost of interocean transportation for gas, which in some cases is almost five times the cost for an equivalent amount of energy in the form of petroleum, it is very easy to understand that not much scope existed for the United States to become a major exporter. Both Holland and the USSR do almost all their exporting through pipelines, and in fact the United States hopes to import a great deal of gas via this medium from Mexico and Canada.

The Supply of Natural Gas and Some Final Comments on Regulation

At present the total amount of ultimately recoverable natural gas is estimated at between 6,000 Tcf and 12,000 Tcf—between 115 and 235 times the present annual level of world consumption. Proved reserves of natural gas are estimated at 2,250 Tcf or 64 Tcm, which in turn is equivalent to 60 billion tonnes or 450 billion barrels of petroleum (compared with remaining crude-oil reserves of 600 billion barrels). As pointed out by Nordine Ait-Laoussine, these figures imply that the prospects for proving up additional reserves of gas are brighter than those for oil, because the ratio of proved to potential reserves is lower for gas than for oil (31 percent against 60 percent).

Next we shall look at the natural-gas situation in the United States. Conventional proved reserves in the United States amount to almost 200 Tcf (8 percent of world reserves), and potential reserves are 500 Tcf. There is also a potential supply of high-cost conventional gas (about 200 Tcf) available in deep-basin wells. These reservoirs, which are 15,000–30,000 feet below the surface of the earth, require equipment that is much more expensive and technically complicated for their exploitation than do ordinary wells. The average gas well costs between $275,000 and $350,000 at present, whereas a deep well can cost anywhere from $3.5 million to $15 million, since deep drilling rigs must be much heavier and more powerful than ordinary rigs in order to handle more than a million pounds of pipe extending five miles down into the ground. Still, with the average price of natural gas approaching $6/MBtu on world markets, deep drilling makes sense in some localities. A few years ago Standard Oil of California (Socal) was producing gas from a deep well in the Tuscaloosa Trend and selling it on the uncontrolled interstate market for $2.10/MBtu. According to Socal officials, the company made a profit on these transactions.

Unconventional gas—both in the United States and, to a certain extent, in the rest of the world—consists largely of gas in coal seams, Devonian shales, tight sands, and geopressure zones. The estimated potential resources from the first of these is 300 Tcf. Chemically, this gas consists largely of methane seeping from coal beds, especially in the Appalachian region. (Note that the gas in the Groningen field originates from high-grade coal deposits under the North Sea.)

By way of contrast, Devonian shales are shale-rock formations close to the earth's surface that contain large deposits of slow-flowing gas. The present estimate of maximum potential resources is about 500 Tcf, mostly in the northeastern United States. The recovery problems associated with the Western Tight Sands are similar to those for Devonian shales, and in the United States most of these resources seem to be located in the Rocky Mountain region. Current estimates of the potential resource base here seem to center around 600 Tcf, and it could be that as much as one-quarter of this is fairly low-cost gas. Finally, there are the geopressured zones along such water boundaries as the U.S. Gulf coast, where reservoirs of salt water under high pressure contain dissolved methane. The estimate of the ultimate resource base here is astronomical: apparently, gas technologists are thinking in terms of 2,000 Tcf as a reasonable approximation. Theories are also being advanced that beneath the Arctic oceans and the permafrost of Siberia lie reservoirs of methane hydrates (solutions of gas and water) that might contain up to 35 million trillion cubic feet. This is only a theory, however.

It should be noted that in the United States the proving of new reserves has not kept up with the consumption of natural gas. As a result the reserve/consumption ratio has fallen from 17 in 1966 to less than 10 in 1980. In addition, production has been falling on an absolute basis since 1973. Some people claim that this is due to regulation of the gas price; between 1955 and 1973, however, both production and exploration expenditures increased at a steady pace, and since 1973–1974 natural-gas prices have risen very rapidly. The actual, as opposed to the imagined, dilemma is that a large part of the low-cost conventional natural gas in the United States has already been discovered; from now on more emphasis must be placed on higher-cost offshore supplies and, perhaps, the unconventional gas mentioned earlier, assuming that the technology for obtaining large amounts of this gas will be available in the not too distant future. As for completely deregulating the gas sector and permitting an uninhibited rise in the price of gas in order to finance the location and exploitation of high-cost gas, the profit outlook has seldom been rosier for energy companies than it is now, at least with respect to the situation for the entire U.S. industrial sector. A complaint levied against many of these companies is that, instead of using their huge profits to increase the supplies of energy materials, they are investing progressively larger amounts of money in other activities. The mining sector, for example, is regarded with great favor by many energy-company executives. Table 3–2 presents some recent figures for the output and reserves of natural gas.

It is ironic that a large part of the gas-regulation apparatus in the United States was set up during the presidency of Dwight D. Eisenhower, probably the most conservative U.S. president of the postwar period. The theory advanced by his administration was that since gas is a necessity for millions of moderate-income households, if its price were permitted to soar without restriction, then there could be serious repercussions on the cost-of-living index and hence on such things as wages and the rate of inflation. There were also long-run political and ethical considerations involved in energy pricing, or so reasoned Associate U.S. Justice Sherman Minton when he invoked the general-welfare clause that enabled the Federal Power Commission to regulate the price of gas. The reader should also be aware that low-priced gas directly helped create tens of thousands of jobs in the United States, and probably hundreds of thousands indirectly. This contradicts the assertions of many liberal economists that U.S. industry enjoyed a free ride between 1954 and 1974. The employment situation in Great Britain today also gives us some insights into this matter: in part because the Thatcher government has not allowed British industry to benefit from that country's energy self-sufficiency, and this is one of the reasons why unemployment in Great Britain will soon reach the levels attained during the Great Depression. Even more discouraging is the effect that high energy prices have on investment and, through investment, on productivity. Unfortunately, the failure to give British industry a free ride will be felt long after Prime Minister Thatcher has been returned to private life.

The Energy Sector of the USSR

By virtually any criterion, the USSR is the most energy-rich country in the world. It possesses not only major supplies of all the oil, gas, and coal it requires for energy independence, but also a form of government that permits the unprotested use of such fuels as uranium and thorium in nuclear reactors. The USSR's only disadvantage is geographic: the center of gravity of the Soviet energy sector is rapidly moving eastward, toward the immense and largely unexploited resources of Siberia. Because the cost of transport, infrastructure, and labor in that part of the USSR are well above the national average, there could be considerable difficulty in getting at these

Table 3–2
The Reserves, Production, and Exports of Natural Gas

Country	Reserves, 1980 Tcf	Gross Production, 1980 (Gcm)	Net Production, 1980 (Gcm)
USSR	1160	336[c]	—
Iran	535	20.0	8.0
United States	198	538.0[c]	—
Saudi Arabia	100	53.0	15.0
Canada	90.5	100.5[c]	—
Algeria	79.5	43.0	19.0
Holland	60.9	96.6[c]	—
Qatar	53.0	6.4	5.2
Nigeria	36.0	24.5	1.0
Australia[c]	32.3	6.8	—
Kuwait	30.0	8.7	6.9
United Kingdom	28.8	39.4[c]	—
China	24.4	17.0[c]	—
Iraq	24.0	11.3	1.7
Libya	21.0	20.3	5.3
Indonesia	21.0	29.0	18.0
U.A.E.	18.4	14.0	7.2
Norway	16.0[a]	4.4[c]	—
Mexico	50.5	—	—
Pakistan	15.5[c]	—	—
Malaysia	15.0[c]	—	—

World Reserves of Gas (end of 1981), Tcf		Gas Exporters, 1980 and 1990 (Estimated)			
		1980: 7 Tcf		1990: 14.3 Tcf	
United States	198	Pipeline gas	(83%)	Pipeline gas	(62%)
Canada	90.5	USSR	26%	USSR	26%
Latin America	176.3	Netherlands	24%	Netherlands	7%
Western Europe	152.7	Norway	14%	Norway	7%
Middle East	761.9	Canada	12%	Algeria-Iran	7%
Africa	211.7	OPEC	2%	Canada-Mexico	13%
USSR	1160.0	Others	5%	Other	2%
Eastern Europe	8.8				
China	24.4				
Other Eastern Hemisphere	127.7				
		Lng	(17%)	Lng	(38%)
		OPEC	12%	Algeria	9%
		Others	5%	Indonesia	7%
				Other OPEC	9%
				Non-OPEC	13%

Source: The OPEC Annual Report, 1981; Shell Briefing Service (various issues); M. Folie and G. McColl (1978).

[a]In fields on stream and under development. Total estimated reserves at present are 50.7 Tcf, but this figure is expected to increase.

[b]Net Production = Gross Production − Amount Flared − Amount Reinjected.

[c]Production Figures from 1976.

riches. Still, the Soviet government has no choice. The world price of energy is steadily increasing, which discourages any thought of imports; and the export of energy materials is the surest and most economical way to earn the currencies needed to purchase outside technology. In 1977 Soviet energy exports to the West amounted to $3.5 billion, and they are probably about four times that amount today. By 1986–1987 the Soviets could be exporting energy materials for $20–$25 billion, mostly to western Europe. Although oil is now the major energy export, natural gas is termed the export of the future. More important, natural gas is the only major Soviet energy material that has expanded at the pace stipulated in the most recent Soviet Five-Year Plan (1976–1980). The output of natural gas rose from 10.2 Tcf (288 Gcm) in 1975 to 14.4 Tcf in 1979, and exceeded 15 Tcf in 1980 (of which about 1.9 Tcf was exported with 0.945 Tcf going to western Europe). This amounts to a 50-percent increase in half a decade, compared with a 22-percent increase in the production of oil and only a 3-percent increase in the production of coal, although 15 percent was anticipated.

No less than 23 percent of world coal, 35 percent of world natural gas, and 9 percent of global oil reserves are located within the boundaries of the USSR, or directly offshore. Insofar as production is concerned, the shift is from oil to coal and gas, but the USSR is still the largest producer of oil in the world and seems likely to remain so. In 1980 oil production averaged slightly more than 12 Mbbl/d, or 603 million tonnes for the year (of which 90 million tonnes were exported to eastern Europe and 60 million tonnes elsewhere). Although output is decelerating, however—and may rise by only 1 percent per year in the near future—yearly output of crude petroleum should reach at least 620 million tonnes in 1985, and perhaps as much as 645 million tonnes. At present only 1.2 percent of Soviet energy production is nuclear based, and known Soviet reserves of uranium are not impressive with regard to quantity or grade; but the Soviets undoubtedly have more than enough to support their current nuclear program, and show no great urgency about searching out more. In addition, the USSR is believed to be extremely rich in thorium, the third-largest energy resource in the world (after coal and oil). Like uranium, thorium is a fissionable substance; it is at least as naturally abundant as uranium, and the next generation of reactors is expected to be especially suited to operating on thorium. Last but not least, because of its enormous size and the natural drainage of the main rivers, the USSR probably has the largest hydroelectric potential of any country in the world. This potential is being exploited rapidly, but the most exploitable hydropower resources are, again, in the eastern part of the country, far from major population and industrial centers.

Table 3-3 gives an overview of Soviet production of oil, natural (and associated) gas, and coal, and shows the share of each of these fuels in the Soviet energy supply. Although there is a belief in certain quarters that Soviet oil production is due to peak in the near future, the rate of expansion revealed in this table could only have taken place if it were based on very

Table 3–3
The Production of Oil, Gas, and Coal in the USSR, 1960–1980

	Oil[a]	Share[b]	Gas[c]	Share[b]	Coal[d]	Share[b]
1960	147.9	30.5	45.3	7.9	509.6	53.9
1961	166.1	32.4	59.0	9.7	506.4	50.5
1962	186.2	34.2	73.5	10.9	517.4	48.8
1963	206.1	32.8	89.8	12.4	531.7	45.9
1964	223.6	35.1	108.6	13.9	554.0	44.2
1965	242.9	35.9	127.7	15.6	577.7	42.9
1966	265.1	36.7	143.0	16.5	585.6	40.7
1967	288.1	37.8	157.4	17.2	595.2	39.4
1968	309.2	39.2	169.1	17.9	594.2	38.0
1969	328.4	39.9	181.1	18.3	607.8	37.3
1970	353.0	41.1	197.9	19.1	624.1	35.4
1971	377.1	41.8	212.4	19.5	640.9	34.6
1972	400.4	42.3	221.4	19.5	655.2	34.0
1973	429.0	43.2	236.3	19.9	667.6	33.0
1974	458.8	43.8	260.6	20.8	684.0	32.1
1975	490.8	44.7	289.3	21.8	701.3	30.0
1976	519.7	45.0	321.0	23.1	711.0	29.0
1977	546.0	45.2	346.0	23.7	722.0	28.0
1978	571.4		372.0		724.0	
1979	593.0		406.0		773.0	
1980	603.0		435.0		716.0	

[a]Including condensate, in millions of tons.
[b]In percentages.
[c]Natural and associated gas, in billions of cubic meters.
[d]In millions of tons.

large reserves, or if the Soviet energy managers overlooked or ignored the criteria of optimal economic extraction (as explained in chapter 2).

Were this table drawn up ten years from now, it would reveal a very different trend for Soviet energy usage. Nuclear energy is scheduled to account for a major share of the increase in electrical generating capacity in European Russia until the end of the century, and if possible coal production is to be revitalized. There are some uncertainties associated with the rate of increase in oil production; and, as pointed out by Stern (1980) in his important and comprehensive book, any slack in Soviet oil production undoubtedly will first affect Soviet clients in eastern Europe. For a number of years now the USSR has been attempting to dampen the increase in its oil exports to eastern Europe; in the face of the prospect of competing for costly Middle Eastern oil, the annual rate of growth of oil consumption in these countries has declined significantly—from 12.2 percent in 1970–1973 to 6.3 percent in 1974–1978. As pointed out by their Soviet suppliers, however, these countries will have to do better than that. Specifically, the significant reexporting of imported Soviet oil (which was obtained at less than the world price) must cease, and a more rapid substitution of coal and nuclear energy for oil must take place. In their latest Five-Year Plan the USSR has provided for a 20-percent increase in total fuel and energy deliveries to eastern Europe, but it was not specified whether this would apply equally to all kinds of fuel. In June 1980 Moscow announced that oil deliveries to Comecon (the eastern European trading bloc) would remain frozen at current levels, though implying that additional deliveries might be possible if they were paid for in hard currencies.

Soviet Natural Gas

The Soviet trade in natural gas will depend to a considerable extent on domestic consumption: if indigenous requirements are high, there will be less to export. Given some of the uncertainties associated with Soviet oil, in conjunction with that country's hard-currency requirements, a larger percentage of Soviet gas definitely will have to find its way into the export market. To help bring this about, Soviet planners are calling for higher domestic gas prices, hoping to provoke a shift by domestic users of gas to brown coal, hydro, and nuclear energy. Increases in domestic gas consumption would be limited, where possible, to industrial users; and the household and municipal sector—which uses only about 12 percent of total gas production—would be squeezed as much as possible. Notice has also been served on Soviet industry by Mr. Tikhonov, the prime minister, that some 165 million tonnes of fuel will have to be saved through various conservation programs; as a result, restrictions and quotas almost certainly will be imposed on all sectors of the Soviet economy. In theory these measures imply that managers who are unable to keep up their output because of an inability to economize on or substitute for present inputs will have to find some other line of work.

Where trade with countries outside eastern Europe is concerned, it was only in 1974 that Italy, Finland, and the Federal Republic of Germany began to receive Soviet gas in any quantity, although Austria was the first western European country to import Soviet gas (in 1968). This trade took place through extensions to the pipeline network that had been constructed to link Soviet gas fields to eastern European purchasers. A major terminal of this network is at Uzhgorod on the Soviet-Czech frontier, and there is a ma· branch going into Bratislava, where it is extended in two directions:

toward East Germany on the one hand, and Austria and Italy on the other, with the end station being in the vicinity of Milan. West Germany now receives its Soviet gas from a branch of the East German line, which proceeds all the way to Paris.

Today Soviet revenues from sales of natural gas are in excess of $3 billion, with probably more than half of this being received in hard currencies. Because the price of natural gas is steadily rising all over the world as attempts are made to bring its price (in dollars per heating unit) up to the level of petroleum, a situation has come about whereby Soviet revenues could increase faster than physical exports as contracts are renegotiated. If the Urengoy pipeline taking gas to western Europe goes ahead as planned, increased gas exports in addition to the rise in energy prices could increase the USSR's revenues from gas sales by up to $7 billion, and eventually as much as $10 billion—at least in the long run.

An important advantage possessed by Soviet gas is that it can be delivered to its main markets by pipeline. Intercontinental pipelines are an extremely economical means of delivery, yielding rapidly decreasing costs with increasing quantities over a given distance; and in western Europe the building of piplines and storage and distribution facilities would not encounter the technological difficulties that exist in Russia. Moreover, where the Urengoy pipeline is concerned, the Soviets are considering the use of two pipes running parallel instead of a single conductor operating under extremely high pressure. In some respects this is an ideal arrangement for gas purchasers in that only one of the lines needs to be installed initially; thus potential buyers of Soviet gas can scrutinize the project in more detail before making or widening their commitments. Obviously the main barrier to this deal is political: allowing the USSR to become a major supplier of a key western European industrial input, and in the bargain obtaining cash to finance military or political activities elsewhere in the world.

No answer to this dilemma can be supplied here, although it should be evident that the problem will become much less critical when other important sources of natural gas become capable of delivering larger amounts, and when LNG technology is better developed. In these circumstances a stoppage in the delivery of Soviet gas could be largely offset by introducing supplies from elsewhere. Moreover, the Soviets would have reason to be extremely careful about their behavior, since the natural-gas market is characterized by very long term contracts; once customers are lost or alienated, they are out of the picture indefinitely. In the light of its crucial need for Western technology and, quite often, agricultural products, it is hard to visualize any international *political* situation, short of World War III, that would cause the USSR voluntarily to interrupt its sales of this energy resource to the capitalist world. Still, the U.S. government has recently attempted to convince the West Germans that it would not be in their interest to increase their purchases of gas from the Soviets by the amount now planned. At the same time President Reagan's diplomats have tried to persuade the Norwegian government that it would be a nice display of generosity if the present Norwegian energy policy could be altered in such a way as to permit larger deliveries of gas to western Europe. It has been said, for example, that a more intensive utilization of the huge gas field in the Norwegian North Sea known as 31-2 would make it possible for Norway to deliver as much gas to West Germany as that country intends to buy from the USSR; but even a cursory scrutiny of the relative reserves possessed by these two countries makes it clear that if gas is to be purchased over a very long time horizon, then the USSR is a more suitable supplier. In addition, if the USSR proves to be a reliable supplier of gas, then it may be possible at some point in the future for western European countries to cooperate with that country in the exploitation of other natural resources. The opinion also exists in West Germany and elsewhere that if no resort is made to Soviet gas, increased bidding will have to take place for OPEC oil. This is not viewed as a desirable outcome.

The Federal Republic of Germany is undoubtedly the most important potential customer of the USSR. Provided with a steady supply of energy materials and other natural resources, that country is capable of remaining, at least qualitatively, the premier industrial power of Europe. Thus the natural-gas deals serve as important forerunners of similar arrangements for iron ore, aluminum-bearing materials, and so on. In addition, the manufacturing of equipment needed by the Soviets to exploit their natural resources has an important spinoff effect on the German (and European) steel industry, which now is not particularly optimistic about its future. In the case of some gas purchases (by Ruhrgas) that took place not too long ago, payment was made in the form of steel pipes manufactured by Mannesmann, with Mannesmann receiving credits from the German government via a consortium of financial institutions.

The aforementioned deal was a kind of warm-up for the transaction now in the intermediate stage of planning. Soviet gas (from the giant Urengoy field) would be transmitted to six western European countries. France and West Germany are, according to original intentions, to get between 353 and 424 billion cubic feet each per year; Italy is to get 247 billion cubic feet per year; and Belgium, Holland, and Austria were originally scheduled to obtain 177 billion cubic feet each per year. The total cost of this project will be around $15 billion (U.S.). Most of the financing will be handled by a syndicate of European banks, with a very large West German representation led by the Deutsche Bank; and the interest rate on the Soviet loan will be 10–11 percent. The price for this gas, expressed in dollars, appears to be about $4.75/MBtu. Consequently, this could turn out to be a very expensive undertaking for the USSR, particularly since the unit of account for the agreement is the West German mark, and it is not unlikely that the mark will appreciate relative to the dollar over the medium term. Once the gas starts flowing, however, the USSR should be in a good position to repay this loan in a fairly short time. For Germany this project means that the Soviets raise their share of German gas supplies to 28–30 percent; despite some security problems, however, the Germans are fundamentally in favor of getting more Soviet gas, pointing out that they will be obtaining only 5 or 6 percent of their total energy supplies from Russia. Sweden, Switzerland, and Spain have also been named as potential customers for Soviet gas, particularly since French enthusiasm waxes and wanes in tune with the possibility of obtaining gas from other sources—for example, by underwater pipeline from Algeria. But now France is definitely set to receive Soviet gas (as well as Algerian gas).

As mentioned earlier, Austria was the first western European country to import gas from the USSR, and conditions exist that indicate that it is only a matter of time until Austria purchases a great deal more—perhaps as much as 70 percent of its entire gas consumption. Since the completion of the Western Austrian Gas Line (WAG), increased supplies from the USSR appear increasingly attractive, even though a trans-Mediterranean pipeline from Algeria to Italy could easily be extended into Austria. As for Italy, the Soviets have been paying for steel pipe delivered from Finsider with coal, iron ore, and gas; but Stern, among others, believes that Algerian gas delivered via Tunisia and Sicily is generally a more appealing proposition. Italy also produces some gas from its installations in the Po Valley; but the output of this source is static, and no new onshore supplies are expected to be found.

Ruhrgas and Gaz de France are extending the Prague-Waidhaus-Mannheim-Mendelsheim pipeline to Paris, since as explained above the gas that could be delivered to France from Algeria is more expensive. In addition, Franco-Algerian relations could deteriorate because of North African or OPEC politics, or even the question of Algerian immigrants in France. Finland, on the other hand, has no likely source of supply for gas other than the USSR right now, and would like greater access to the Leningrad terminal of the Northern Lights pipeline. As for Sweden, the problem has been a reluctance to make large investments in pipelines and various accessories in the face of what have been regarded as uncertain supplies. The opinion here, however, is that, given their comparatively modest requirements of energy materials in relation to the total energy exports of the USSR, the Swedes should be thinking about a very large expansion of trade with that country. Table 3–4 gives some indication of the present (1980) and projected (1986) dependence on Soviet gas of some important clients of the USSR.

Until recently the USSR was also a fairly large importer of gas, from

Table 3-4
The Gas Dependency of Some Western European Countries

	1980	1986
Western Europe	9	25
Austria	50–55	70
France	7	30
West Germany	16	30
Italy[a]	15–20	30
Belgium	—	170 Gcm
Holland	—	170 Gcm
Switzerland	—	20

Note: For Belgium and Holland the figures represent gas imports in billions of cubic meters.
[a]Estimated.

Iran and Afganistan. The arrangement here is similar to the coal trade between the United States and Canada. The consuming district of one country is proximate to the producing district of another, and at the same time far from the producing areas within its own borders. Thus it imports foreign supplies. Soviet imports of gas were almost 10 Gcm in 1978 and were payment for such things as a steel plant and a gas pipeline constructed for Iran, and mechanical equipment and technical aid provided Afghanistan. At one time the possibility existed for the Soviets eventually to obtain at least 25 Gcm/year of Iranian gas, but the revolution in Iran has completely altered trade prospects between these two countries. In the early days of the revolution the Iranian Gas Trunkline (IGAT) was sabotaged, apparently out of pique at the low price the Soviets were paying for their supplies. Later, after the Soviets agreed to pay a higher price, flow was partially increased; but in general the Iranians are showing reluctance in expanding trade with the USSR, and the presence of a large amount of Soviet manufactured armor and artillery in the service of the Iraqi army has not increased the desire of the Iranians to do more business with the USSR.

Soviets Offshore

Like most of the other major oil and gas producers, the USSR has begun to prospect its offshore areas for energy materials. Everything indicates that this activity will increase drastically in the coming years, assuming that the Soviets can buy and then utilize properly the techniques required to master underwater exploration and exploitation.

Preliminary calculations indicate that 4.5 million square kilometers of Soviet continental shelf and inland seas contain oil and gas, but some question exists as to just how the Soviets are going to use this information. Such hunting grounds as the Cara Sea, off the northwestern coast of Siberia, rank among the most inhospitable environments in the world; as luck would have it, this is the most geologically promising area.

Another locale of some interest is the Soviet Pacific coast, particularly the northern part, where the USSR and Japan have a joint venture in operation. Some oil was found there in 1977, and some observers have claimed that this area is as rich in oil and gas as the North Sea.

Conclusion

In *The Political Economy of Oil* I made a number of predictions about the Soviet energy sector. These were controversial at the time, particularly since they contradicted the well-publicized beliefs of the U.S. Central Intelligence Agency (CIA).

Since that time the CIA has changed its time. It would have done so eventually anyway, since the Soviet energy statistics for 1981 have just been published, and a few details about the recent geological survey of western Siberia are now available. As things stand, the energy sector has become one of the brightest stars on the entire Soviet economic scene—some say the only bright star. The oil and gas sectors have fulfilled—and to a certain extent overfulfilled—their 1981 plan targets. Reserves of gas and oil are in fact even larger than earlier estimates—at least if Soviet figures are to be believed. What appears to have happened is that the Soviets have begun to explore those parts of western Siberia that they earlier overlooked, and they are also drilling deeper. Among other things, it appears that gas deposits at Urengoy—the source of a large portion of the gas destined for western Europe—now amount to 70 Tcf, a great deal more than previously announced.

Soviet drilling is also apparently becoming more efficient. In 1981 the gas industry output showed a 6.7-percent increase over 1980; while the output of the oil industry increased by 1.1-percent increase over the previous year. The figure for oil, however, is subject to change when the final statistics become available. The important thing here is that western Siberia may continue to be a major producer of oil for years to come, despite the usual assumption that this region is on its last legs. The only energy-producing activity that did not register impressive gains was coal, but nobody expected it to. The story here will probably go as follows: when a large market for coal appears outside the USSR (which will probably take a decade or more) and when the Soviets can afford to inject a large amount of Western tech-

Table 3–5
Some Recent Soviet Energy Statistics, Projections for 1985, and Information about Commodity Trade

	1965	1970[a]		1975[a]		1980[a]		1985[a]	(Plan)
Petroleum (million metric tons)	242.9	353.0	(7.8)	490.8	(6.8)	603.0	(4.2)	632.5	(1.0)
Gas (billion cubic meters)	127.7	197.9	(9.2)	289.3	(7.9)	435.0	(8.5)	620.0	(7.3)
Coal (million metric tons)	577.7	624.1	(1.6)	701.3	(2.4)	716.0	(0.4)	785.0	(1.75)
All mineral fuels[b] (million metric tons)	908.7	1,168.7	(5.2)	1,516.5	(5.3)	1,859.8	(4.2)	2,167.5	(3.1)
Electric power (billion kilowatt-hours)	506.7	741.0	(7.9)	1,038.7	(7.0)	1,295.0	(4.5)	1,575.0	(4.0)

Source: English language summaries of various Soviet publications, particularly *The National Economy of the USSR*.
[a]Average annual increase, in percentages over years, shown in parentheses.
[b]Excludes minor fuels. Expressed employing the following transformation: 1 ton equals 7,000 kilocalories.

Table 3-5 continued

SOVIET COMMODITY TRADE, 1981

Exports	Amount	Value
Gold	280 tonnes	$4,600 Million
Platinum	439,000 ounces	$200 Million
Diamonds	326,000 carats	$175 Million
Petroleum	1.2 Mbbl/day	$12,000 Million
Natural Gas	22.65 Gcm	$2,500 Million
Imports		
Grains	45 Million Tonnes	$7,500 Million
Sugar	4.2 Million Tonnes	$1,500 Million

Note: Diamond exports consist of polished stones into Antwerp; Oil and Gas are sales to the West only; and the sugar imports represent raw value.

nology into the coal industry, then that activity should react in the same manner as the gas industry. The latest figures for the Soviet energy sector are shown in table 3-5.

Since many economists and others would hardly take any publication seriously unless it contained some unflattering references to the ambitions of the Soviet government, I will end this chapter on a negative note. Soviet oil production still falls well under the goals set under the tenth Five-Year Plan, and it seems clear that the targets of future plans will have to adjust to the realities of Soviet technology and personal motivation. Put more simply, the Soviet government is going to have to lower its sights considerably, and this may be true throughout the energy sector. Even with natural gas, where genuine success has been attained, the rate of growth of output is lower than it was a decade ago.

Some researchers see the Soviet problem as being one of diminishing returns to drilling (and mineral exploitation in general); others are of the opinion that the failure to raise the tempo of expansion in certain sectors (such as energy and agriculture) will severely constrain the Soviet economy as a whole. The matter of agriculture cannot be taken up here; but in my opinion, although diminishing returns might show up for some—or even many—mining and drilling projects, this phenomenon does not have any ominous significance for the Soviet mineral sector as a whole—at least in the medium run. This sector still has enormous amounts of undiscovered resources, and the increasing ability of the Soviets to employ foreign technology successfully in what they consider crucial activities should eventually enable them to exploit these resources at a satisfactory pace, and at a reasonable cost. (An example that can be cited here is the 1980 Olympic games, which even some hostile observers claim to have been the best organized of all the modern Olympics). The constraint on Soviet energy production (and indeed on all Soviet output) continues to be the immovable presence of a rigid, unimaginative bureaucracy intent only on preserving and expanding its own privileges. Still, as has been clearly shown in the building of Soviet military power, miracles are sometimes possible in countries without stock exchanges or Playboy Clubs. The question is, however, whether there will be enough of them.

For an extension of these topics the reader is referred to my paper "Soviet Natural Gas and the Western European Energy Crisis: The Solution is the Problem," which is published in the transactions of the 1982 meeting of the International Association of Energy Economists (Churchill College, Cambridge University, June 1982). Among other things, this paper explains that because of the presence of oil, the exploitation of huge Norwegian gas fields such as 31-2 will almost certainly not take place until the 1990s.

Appendix 3A:
Some Elementary
Capital and Investment
Theory

Among other things, capital theory involves determining the amount of a durable capital goods that will come into existence—goods that give rise to a stream of products or services over time. It is also concerned with how an asset such as a mine or an oil or gas deposit shall be depleted over time. The key expression here obviously is *time,* but it might equally well be uncertainty, since the art of reading the future has not progressed very much beyond the point to which it was carried by crystal gazers during the Dark Ages.

In order to keep the following analysis simple, uncertainty will be assumed away; and the mysteries of the future will be taken as common knowledge—free to all those with a curiosity about them. This, of course, implies a perfect capital market wherein unlimited amounts can be borrowed at the given rate of interest, and such things as default are unknown because the borrower—in addition to being honest—possesses a complete knowledge of the outcome of his investment. Naturally, this is a very artificial assumption; but introducing imperfect capital markets leads to even more artificiality in the effort to duplicate reality. Finally, although the wording here indicates a concern for oil, the same discussion applies to gas if our units are, for example, thousands of cubic feet of gas instead of barrels of oil.

Our approach to this problem will be via a project of the type indicated figure 3A–1a. Here, beginning at time t_0, consumption is $\bar{C}$. In the period from t_1 to t_2 it is reduced to $\bar{C} - x$ in order to realize $\bar{C} + y$ in the period from t_2 to t_3. According to the usual tenets of capital theory (and common sense), we can define the rate of return ρ on this project in the following manner:

$$\frac{\Delta C_1}{\Delta C_0} = \frac{y}{x} = \frac{\text{return}}{\text{cost}} = \frac{\text{return} - \text{cost}}{\text{cost}} + 1 = \rho + 1. \qquad (3A.1)$$

$$\rho = \left| \frac{y}{x} \right| - 1. \qquad (3A.2)$$

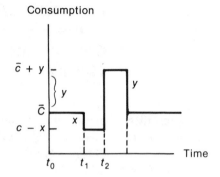

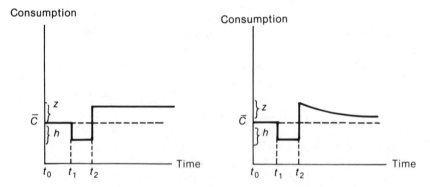

Source: Reprinted by permission of the publisher from Ferdinand E. Banks, *The Political Economy of Oil* (Lexington, Mass.: Lexington Books, D.C. Heath and Company, 1980), p. 61.

Figure 3A–1. Investment and Subsequent Patterns of Consumption

Generally $y > x$, since otherwise there would be no point in abstaining from x, although, admittedly, there are situations in which this is not true.

In the first situation shown in figure 3A–1b, h is given up over the interval t_1 to t_2, giving rise to a return z over an infinite period. (This is equivalent to purchasing a bond known as a perpetuity.) Now, with p the price of this commodity, and r a constant discount rate:

$$ph = \int_0^\infty pze^{-rt}dt = \frac{pz}{r} \qquad (3A.3)$$

Thus

$$z = rh \qquad (3A.4)$$

It should also be clear that an analogy to equation 3A.2 could have been obtained by using simple discounting on the arrangement in figure 3A-1a. Thus:

$$px = \frac{py}{1 + r}$$

or

$$r = \frac{y}{x} - 1$$

The problem with this exposition is that it does not take into consideration the very real problem of depreciation. If we have machinery or structures, they can be expected to depreciate over time. If this depreciation takes place exponentially (or radioactively), then the yield of these facilities is progressively diminished. Thus we would have, as in figure 3A-1b:

$$ph = \int_0^\infty pze^{-at}e^{-rt}dt \qquad a > 0 \qquad (3A.5)$$

Observe here that radioactive depreciation has the effect of driving the return z asymptotically toward zero. Integrating this, we get

$$r = \frac{z}{h} - a \qquad \text{or} \qquad h = \frac{z}{r + a} \qquad (3A.6)$$

As mentioned, the effective yield has been reduced by the depression. Another interpretation of these results is that in the presence of depreciation, the price of the facilities must be reduced [from z/r to $z/(r + a)$].

Now we can go to an oil field. Here recoverable reserves R can be defined as cumulative output. Thus

$$R = \int_0^b q_0 e^{-at}dt \qquad (3A.7)$$

where q_0 is the current output rate and a is the rate of decline for the field. If $b \to \infty$, then the $R = q_0/a$. Next we can recall that as with any asset, if an oil field is to be developed, the discounted value of its net receipts must equal the cost of investment. Then we get, with I being the cost of investment and p the income per barrel

$$I = \int_0^b pq(t)e^{-rt}dt \qquad (3A.8)$$

With $q(t) = q_0 e^{-at}$ we get

$$I = \int_0^b pq_0 e^{-at} e^{-rt} dt \qquad \text{(3A.9)}$$

or

$$P_{min} = \frac{I(a + r)}{q_0(1 - e^{-(a + r)b})}$$

(Notice that if $b \to \infty$, we get a direct analogy of our previous results). The income per barrel required to bring about 1-unit increase in output is p_{min}, with the cost of investment given and the pattern of output known.

4 The World Coal Industry

Coal is formed from the remains of trees that have been preserved for millions of years under a special nonoxidizing condition, for the most part in swamps where, after falling, they either did not rot or rotted very slowly. The conversion process itself took place by photosynthesis. In general coal seams lie on so-called *underclays* that are associated with coastal swamps; their quality is largely, but not entirely, a function of their age. Top-grade coal requires a gestation period of a few hundred million years, and scientists have calculated that the average time required to accumulate enough vegetable matter to form one meter of coal eventually is about 1.6 million years. Similarly, a coal seam 1 meter thick would have been compacted originally from a 120-meter layer of plant remains.

It is possible to distinguish a spectrum of coals, ranging from peat through anthracite. *Peat,* which is brown, porous, and often contains visible plant remains, is the lowest class of coal, with an average energy content of 8.4 GJ/ton. (Here G signifies *giga,* which is a billion, and J signifies the basic energy unit *joule.* Thus 1 GJ = 1,000,000,000 joules. The matter of energy units and equivalents will be taken up next.) Peat also has a very high moisture content. Next we come to *lignite,* which can be regarded as the transition link to *hard coal* (bituminous plus anthracite coal). Lignite also contains a great deal of water (on the average 41 percent of its composition is moisture), and its average heat content is 14.7 GJ/ton. *Bituminous* coals, on the other hand, are characterized by a low moisture content, and the moisture content of a typical anthracite coal is only 3 percent. Where energy values are concerned we distinguish between *subbituminous* coal, with an energy value of 25.1 GJ/ton, and *bituminous* coal, with an average energy value of 29.3 GJ/ton. *Anthracite* coal, which is jet black and difficult to ignite, has an average energy value of 33.5 GJ/ton.

In some countries it is common to categorize coal as soft coal or hard coal. Soft coal consists of brown coals and lignite, whereas hard coal includes bituminous coal and anthracite. In this system peat is regarded as a fuel type in itself, though one that is markedly inferior. In countries like Sweden and Finland, however, it may still be an important energy resource. Still another system divides coal into two classifications: brown coal and black coal. Brown coal is geologically young, high in water content, and has been subject to little pressure; black coal has been subjected to more intense pressure, is considerably lower in water content, and contains much more carbon. Black coal ranges from subbituminous coal (which is usually dull black and waxy in appearance) to anthracite, and is divided into two general categories: coking or metallurgical coal, and thermal or steaming coal. East Germany is the world's largest producer of brown coal, with 250 million tons/year, followed by the USSR with 180 million tons/year. Among the noncommunist countries West Germany is the largest producer of brown coal and lignite; but Australia's resources of brown coal are probably much larger, and the thickest seam of brown coal found anywhere in the world is at Loy Yang in Australia.

Several units are used to measure energy. Physicists like joules, while engineers seem to prefer British thermal units (Btu) or kilowatt hours. Others use the calorie or the kilocalorie (1,000 calories). The transformation between them is easily obtained: 1 calorie $= 4.184$ joules; and 1 Btu $= 1.055 \times 10^3$ joules $= 252$ calories $= 0.252$ kcal. On the average we also have one metric ton of anthracite coal $= 2.80 \times 10^7$ Btu $= 7.06 \times 10^7$ calories; and one metric ton of bituminous coal $= 2.89 \times 10^7$ Btu $= 7.28 \times 10^9$ calories. With this information it is easy to show that, on the average, one metric ton of hard coal equals 4.9 barrels of crude oil in energy value. Thus 50 million tons of oil per year, or one million barrels per day of oil, is the energy equivalent of 71.4 million metric tons of coal per year.

In addition to its energy content, coal can be graded with respect to the amount of waste materials it contains—primarily sulphur and ash—and also its gas content. Table 4–1 lists some characteristic of coals with respect to these criteria.

Coal finds its principal uses in the generation of electricity and the production of heat (through direct burning), in metallurgical processes, and as a feedstock in the production of coal gas and synthetic oil and fuels. In the

Table 4–1
Some Physical Characteristics of Coal

	Carbon Content[a]	Volatile Matter[a]	Calorific Value[b]	Moisture Content[a]
Peat	60	53	16,800	75
Brown coal or lignite	60–71	53–49	23,000	35
Subbituminous coal	71–77	49–42	29,300	25–10
Bituminous coal	77–87	42–29	36,250	8
Anthracite	81–91 +	29–8	36,250	8

[a]Average percentage.
[b]Kilojoules/kilogram (KJ/Kg).

United States in 1972, 57 percent of all coal mined was burned in electricity-generating plants; 12 percent was used for the production of process heat; 15 percent was used as coking coal, and went to the making of *coke* (a strong, porous block) in the coking ovens of the steel industry; and 10 percent was exported. A very small amount was used in the chemical industry (for example, to make synthetic dye). As for the production of synthetic fuel, oil, and gas from coal, this takes place only on a pilot scale in most of the industrial world; but the Republic of South Africa already has a commercial installation for this purpose operating at Sasolburg, and the El Paso Power and Light Company has leased the South African process with the intention of perfecting it and producing synthetic gas and liquids in the United States.

On a worldwide basis the major portion of coal production is of hard coals (70 percent), with most of this output earmarked for thermal purposes. A small proportion of hard coal (about 25 percent, and this seems to be declining) is of the metallurgical variety. The best coking coal is bituminous coal with a low or medium gas content and a low sulphur content, since large amounts of gas and sulphur contaminate the charge of ore and limestone that coking coal is supposed to support. Almost all brown coal is used for power generation, but since in general this type of coal cannot be stored or transported because of the danger of spontaneous combustion, it is usually fed into power stations very close to the mining site. Another possibility is to turn it into briquettes by compressing the coal and shaping it into blocks. In this form brown coal can travel without being doused with water constantly, and its energy content is also increased with reference to its weight: a tonne of brown coal from the Latrobe Valley in Australia has an energy content of 9.7 GJ; when briquetted, however, a tonne has an energy content of 22.3 GJ. Table 4–2 outlines the market for coal in three of the main consuming areas.

Because of the expense and/or the lack of facilities for transporting coal, the tendency of late has been to locate as many coal-based power stations as possible close to coal fields. Transport expenses also explain why

Table 4–2
The Uses of Coal in Three Important Consuming Areas

	North America	Western Europe	Japan
Electricity generation	77	57	18
Iron and steel	10	20	66
Industry	11	9	5
Residential and commercial	2	14	11

most of the nonmetallurgical coal entering into international trade is anthracite coal, since shipping coal with a low energy content is patently unprofitable. Still, a huge amount of coal must be moved over large distances every day—for the most part by train, but also in pipelines (as a slurry); one of the reasons for the relatively slow expansion of the world coal industry is the difficulty in financing the investments in coal-transportation facilities that would be required if coal were to find a more widespread use in countries without a domestic coal industry. This problem is aggravated in very large countries like the United States and, to a much greater extent, the USSR.

We conclude this section by saying something about the three methods for mining coal. The first is *underground mining,* whereby a vertical or inclined shaft from the surface intersects a coal bed of satisfactory thickness and quality, and miners and their equipment enter this shaft, drive lateral openings from it into the coal face, and extract the coal. In a coalfield recoverability depends on the accessibility of the coal (which in turn is influenced by things like seam thickness, depth, and type of terrain); and in this type of mine it varies from 25 to 70 percent. *Strip mines,* on the other hand, are developed by removing the overburden covering a minable deposit of coal, then breaking up the deposit by digging and/or blasting, and finally removing the broken coal. In general strip mining (which is sometimes called *open-pit* or *open-cast* mining) is much less costly per ton of coal produced than underground mining, even when complete reclamation of the land disturbed by stripping takes place. Recovery rates are also higher: from 85 to 95 percent of the coal in place.

The third method, which is applicable to only a few coal deposits, is called *auger mining.* Large horizontal augers bore into coal beds that are exposed on the side of hills, and push out broken coal as they rotate. Obviously, this kind of technique can be used only in special situations. At present, in the United States, more than half the production of coal results from strip mining; this rate will increase as the center of gravity of U.S. coal mining shifts westward toward areas where strip mining is the rule. An interesting and important feature of strip mining is its capital intensity. The expansion of the Australian coal-mining sector has been accompanied by a large-scale shift to open-cast mining; as a result there has been a significant increase in the productivity of labor as the amount of capital per unit of labor has gone up, as well as a reduction in the number of employees on the newer mining sites.

Reserves, Resources, and Production

Reserves are essentially the known inventory of a natural resource that can be extracted economically, although this definition is generally extended to include unknown materials that, when found, will be no more expensive to exploit than those currently being won (such as any oil that will be found in Saudi Arabia in the near future) *and* known materials that are slightly sub-marginal economically but will probably be promoted to the category of exploitable properties as a result of technological progress or a rise in their price. Everything else can be labeled *resources.* The U.S. Geological Survey has compiled estimates of fossil-fuel resources on the basis of such things as the anticipated frequency of these resources in certain types of rock formations. Resources also include known accumulations of a material whose grade is too low to warrant exploitation at any time in the forseeable future, but that might be exploitable at some distant point in the future because of the radically different application of a known technology.

We can now look at table 4-3, which shows some of the most recent estimates of reserves, resources, and production. Among other things we see that over 60 percent of reserves are concentrated in just four countries, and the same is true of 90 percent of known resources—although coal deposits can be found in more than eighty countries. Where resources are concerned, these four countries are the USSR (45 percent), the United States (24 percent), China (13 percent), and Australia (6 percent). Of the less developed countries, India is perhaps the only one with a major coal-production and -exporting capability, although some geologists believe that Columbia, Indonesia, and Botswana could become significant coal export-ers. The same is true of Canada—a country with small coal reserves but relatively large resources. As things now stand, only a few countries have the capability to generate a large enough surplus of coal over domestic requirements to become major exporters between now and the end of the century. One of these, Canada, has already been mentioned; the others are the United States, Australia, and South Africa.

Of the countries named in table 4-3, the United States must rank as the number one coal power, regardless of the patently larger amount of resources possessed by the USSR. In the United States, not only are there huge reserves, but they exist proximate to a highly efficient transportation network. In addition, with the gradual shift in coal-mining operations from underground installations in the East to open-pit mining in the western United States, unit mining costs could easily fall since labor productivity in open-cast mining is often several times that of underground mining. (Still, it should be recognized that at present 70 percent of U.S. reserves can be reached only by underground mining.) It seems almost certain, however, that there will be appreciably higher transport costs in the West, and also higher costs due to reclamation requirements and the expense involved in recruiting and keeping the additional personnel that will be required if there is to be a substantial increase in production.

According to the Energy Modeling Forum at Stanford University, by the year 2000 the share of the states west of the Mississippi River in U.S.

Table 4-3
The Reserves, Resources, and Production of Coal in 1977; and Some Forecasts for 1985 and 2000

Country	Reserves[a]	Resources[a]	Production[b]	Production Forecasts[c]			
				1985[d]	2000[d]	1985[e]	2000[e]
USSR	110	4,860	510	851	1,100	—	—
United States	167	2,570	560	842	1,340	837	1,181
China	99	1,438	373	725	1,200	—	—
Australia	33	600	76	150	300	109	285
Canada	4	323	23	35	115	40	71
West Germany	34	247	120	129	145	124	125
United Kingdom	45	190	108	137	173	111	120
Poland	60	140	167	258	300	—	—
India	12	81	72	135	235	—	—
Republic of South Africa	43	72	73	119	233	—	—
Others	56	229	368				

Note: The expression *million* (or *billion*) *tonnes of coal equivalent* (Mtce or Btce) refers to standardized tonnes with heating values of 12,600 Btus/pound.

[a]Gtce.

[b]1977: hard coal only (Mtce). In 1977 910 Mt of lignite and brown coal produced.

[c]Mtce.

[d]World Energy Congress Forecasts.

[e]International Energy Agency.

coal output should be up to at least 60 percent, compared with a current figure of slightly less than 20 percent. Looking exclusively at supply constraints, production in the United States is capable of expanding by a factor of four by the beginning of the next century; but there are doubts about whether there will be a demand for this much coal. The demand for coal is heavily dependent on the requirements of power stations, and at least in the near future the deceleration in the world economy will reflect on the amount of electricity required. Of course, the sharp increase in the price of oil has stopped the substitution of oil for coal that was taking place in all countries with a substantial electricity-generating capacity; but the conversion of oil-fueled installations to coal has proceeded much more slowly than originally expected. In the United States, for example, domestic oil prices did not increase in step with world market prices after the first oil-price shock, whereas at the same time coal prices increased. Thus it was decided that many installations would continue to use gas and oil as their sources of energy.

As for the USSR, coal production during the 1970s increased by about 2 percent per year. It is expected that great efforts will be made to raise this figure, but some doubts exist as to whether the Soviet transport system and the competence of the Soviet bureaucracy are equal to the task. At present eastern Europe and Japan—the world's largest importer of coal—are important buyers of Soviet coal; predictions are that Japan will eventually be interested in a much larger share of Russian output. With this customer, however, the USSR faces increasing competition from both China and Australia, and possibly from Canada, since Japanese economic policy is distinguished by a consistent and conscientious effort to diversify the origins of its natural-resource imports.

The Australian situation will be taken up in detail later in this chapter. Because of its political reliability, stability, and relatively small needs, Australia is a major potential source of energy materials for the entire industrial world. China, in contrast, has enormous resources; but no one knows just when or how these will be developed, or what will happen to them afterward. The largest coal fields in China are in the north of the country, although other regions probably contain sizable deposits. Chinese sales of coal to Japan may also be increasing, but a major constraint on these dealings is the limited capacity of the Chinese rail system, as well as the lack of facilities at ports for handling larger amounts of coal.

Some Projections of Coal Consumption

The purpose of this section and the next is to review some of the results of the coal-forecasting exercises carried out over the past few years. Production is taken up here, and the next section looks at trade.

At one time coal was the most important energy resource in the world. Later it lost ground to other fuels, particularly oil. This decline seems to have been arrested, however, and the pendulum may be swinging back in the other direction. When we consider the high probability that world oil production will peak and begin to decline before the year 2010, then it becomes essential to take a closer look at the capabilities of coal, both quantitative and qualitative. The 750–850 billion tons of recoverable coal reserves, which adjusts to about 650 billion tons of hard-coal equivalent when the inferior heating value of soft coals is considered, is sufficient to cover two hundred years of consumption at present use rates. In terms of an oil equivalency, this comes to 3,000 billion barrels of oil, which compares quite favorably with the roughly 600 billion barrels of known oil reserves. By the same token, if 25 percent—2,500 billion tons—of the 11,500 billion tons of total known world coal resources is recoverable, this is the energy equivalent of approximately 12,000 billion barrels of oil. With the estimated quantity of remaining ultimately recoverable oil at 1,600 billion barrels, these coal resources represent a healthy energy reserve.

The latest and perhaps the most authoritative of the coal-forecasting exercises is the World Coal Study (WOCOL), which involved more than eighty experts under the supervision of Dr. Carroll Wilson of the Massachusetts Institute of Technology (MIT). This exercise also drew on other energy studies and worked in terms of scenarios rather than a theoretical model of the world coal economy.

Two reference cases were postulated. Case A assumed a 1.75-percent annual energy growth rate up to the year 2000; case B assumed an energy growth rate of 2.5 percent per year. I will not comment on these assumptions except to say that they are much lower than the energy growth rates postulated in earlier studies, and they imply very low aggregate (macroeconomic) growth rates.

In case B world coal production should move to 6,780 Mtce in the year 2000 (from 2,450 Mtce in 1977). The theory is that the OECD (which consumes 85 percent of the energy used outside the centrally planned economies) will display a demand for coal that is likely to increase at a rate of 3–4 percent a year until 1985, and then accelerate as policies designed to encourage the shift from oil to coal take hold. Under case A, OECD coal requirements should at least double, and will constitute 37 percent of the OECD's energy requirements in the year 2000. If oil supplies should be limited, however, this last figure will rise to 55 percent; if oil supplies are limited *and* there are delays in building out nuclear capacity, it could approach 67 percent. All this should be contrasted with recent energy-consumption trends in the OECD. In the past twenty years the share of coal in the rising OECD energy requirements has been constant, whereas 67 percent of the increase in OECD energy needs was met with oil. My opinion is that breaking this pattern will be much more difficult than many of the experts of WOCOL believe; in fact, were it not for the growing opposition to nuclear power in Europe, it would probably be impossible.

The major use of coal between now and the year 2000 should continue to be for electricity generation. Coal-fired capacity should at least double and perhaps increase by a factor of three. By that time the coal share of electricity generation in the OECD might have reached 40 percent (compared with 32 percent in 1977). In Australia this figure could reach 85 percent, in North America 50 percent, in Europe 35 percent, and in Japan 15 percent. These figures, of course, reflect the domestic availability of coal. As for metallurgical applications, the share of these in total coal use could fall from today's 25 percent to 15 percent.

The use of coal in industry is also expected to expand. At present this application requires only about 9 percent of the OECD's total coal consumption, but WOCOL projections indicate that industrial demand might increase by 5–7 percent annually after 1985. The industries in which coal use should be the most expansive include cement, petroleum refining, and paper; and the countries for which the largest increases are projected are Canada, the United States, France, and the United Kingdom. In addition, the use of coal as feedstocks for synthetic gas, fuel, and liquids installations should be accelerating in the 1990s in both Europe and the United States. Only recently plans were announced to build a coal-liquefaction plant in Germany to produce about one million metric tons of gasoline a year—5 percent of West Germany's requirements. The construction of this plant is supposed to begin in 1984, and it will take five or six years to build. Present estimates are that the price of the gasoline would be 50–60 pfennigs more per liter than current gasoline prices of 1.50 DM. Producing synthetic gas and liquids might be an attractive alternative to importing energy resources for countries with ample coal supplies, and would help take the pressure off the world oil price. Owners of oil- and gas-fired equipment would also be spared the cost of converting to coal; and the investments in new energy media could be carried out by producers and processors of energy instead of by consumers, which could eventually result in important scale economies, and reduce pollution by centralizing the utilization of coal. Even so, this is a costly way of obtaining energy and is justified at present only because the prices of gas and oil seems likely to continue rising.

International Trade

Today's international trade in coal is very moderate and appears to be expanding very slowly. Currently less than 10 percent of world coal production is sold outside the country where it is mined, but it is believed that

Europe and Japan will eventually raise their consumption of coal by a large amount. If there is another oil-price shock or two, of course, the demand for coal will accelerate everywhere.

It is difficult to forecast exports and imports for any commodity; for one with as many variables attached as coal, no forecasting exercise can be expected to provide unassailable results. Interesting attempts to project the international trade in coal have been made by both the IEA and WOCOL, however; both approach this matter by attempting to quantify the obvious potential for such things as converting from oil-fired electricity generation to coal (and nuclear) power plants, and then attempting to piece together some kind of schedule for realizing that potential in terms of the growing availability of coal around the world.

The IEA study predicts that by 1990 the total import requirements of coal will rise to about 300 million tonnes of coal equivalent (Mtce), which will be divided evenly between steaming and coking coal. This can be compared with a present level of trade of 100 Mtce of coking coal and 30 Mtce of steaming coal. Two opposing forces are at work in this projection. The first is the relatively slow growth of the world steel industry, which would tend to slow down the growth in coking-coal trade; the second is a more widespread application of coal to base-load electricity generation in the OECD. It should be remembered though that if there is a revival in enthusiasm for nuclear power, then there could be a spectacular fall in the projected increase in coal-fired generation. The same thing is true if it turns out that the environmental problems associated with coal cannot be mastered. (For more on this topic, see the paper on Soviet natural gas referred to at the end of the previous chapter.)

WOCOL expects Australia to be the most dynamic exporter of coal in the medium term, with a projected increase in steaming-coal exports to 36 million tonnes from a current level of 7 million tonnes. By the turn of the century Australian exports are put at 120 million tonnes, but the Joint Coal Board of Australia disagrees with these estimates, regarding them as completely unrealizable if present trends are maintained. Another point of interest is the failure of western Europe's two most important coal producers, Britain and West Germany, to take a greater interest in exporting coal, although West Germany could become a modest exporter of coking coal (while still importing some steaming coal). Britain's attitude toward coal exports might change, however, should world consumption of coal begin to accelerate and revenues from its North Sea oil and gas assets begin to shrink. In contrast, the coal industries of France and Belgium are on the decline, not even showing signs of recovery as the price of coal increases. Similarly, the coal-mining industry in the Netherlands appears to belong to history; but it should be remembered that some of the best coal in Europe—perhaps in the world—lies just under the floor of the North Sea in the vicinity of the Groningen gas deposits; if these deposits are large enough, an attempt will eventually be made to exploit them. The preceding discussion is summarized in table 4–4, which shows the trade projections of the International Energy Agency. Also given in some instances are the projections of WOCOL, which have been put in parenthesis.

Lately, considerable attention is being paid the cost of an international coal-supply chain involving, for example, the movement of coal from Australia to northern Europe. Most of the investment required for such a chain is for user facilities such as power plants or synthetic fuel facilities: generally these are at least three times as expensive as the remainder of the coal-supply infrastructure, coming to about $500 per annual ton of coal. Elsewhere in the chain, investments in mines, transport, and loading ports tend to be greater than those for receiving ports and domestic transportation. According to the Shell Corporation, an annual ton of coal equivalent in 1980 cost $158, of which $59 represented investment in sea transport facilities.

According to the Workshop on Alternative Energy Strategies, in order for the United States to become a major coal exporter, massive investments in production and transport facilities would be necessary. Increasing U.S. production capacity to 2,000 Mtce by the year 2000 would require a thirteenfold increase in western coal production and a doubling of eastern underground mining: nine new slurry pipelines and expanded rail facilities involving 1,400 new coal trains and 3,200 new conventional trains. Slurry pipelines require a ton of water for each ton of coal they carry, and in addition pose the severe environmental problem of disposal of the water at the pipeline's terminal. Also needed would be 500 new coal barges and 9,400 new coal trucks, plus a major expansion in port facilities in order to permit much larger coal carriers to be loaded and serviced. In 1975 dollars the investment required came to $118 billion, and this would be larger today.

Coal Prices and Electricity Prices

The structure of prices is considerably more complicated for coal than for oil. Among other things, physical residues (such as ash) and sulphur-dioxide content are extremely important in the commercial classification of coal; and considerable complications result from the type of long-term contracts used, since these can run up to twenty years and different prices are associated with different contract lengths.

At present coking coal is the dominant coal product on the international market. During most of the 1970s there was an excess demand for this commodity, and prices increased steadily until 1979, when they stagnated somewhat. It is also true, however, that the general rise in energy prices has

Table 4-4
Some Expected Trade Patterns for Coal in 1985, 1990, and 2000

	Importers			Exporters			
	1985	1990	2000	1985	1990	2000	
Coking coal							
Western Europe	31	41	57	70	78	93 (152–247)	North America
Japan	83	90	104	43	54	75 (160)	Australia
Less developed countries	16	19	26	20	23	25	Centrally planned economies (net)
Other	5	5	6				
Subtotal	134 (163–177)	155	193 (253–302)	134	155	193	
Thermal coal							
Western Europe	63	112	252	4	10	50	North America
Japan	14	33	77	14	38	120	Australia
Other	3	4	6	23	26	41	Centrally planned economies (net)
				5	17	36	Less developed countries
				34	60	90	South Africa
Subtotal	80 (101–148)	149	337 (295–673)	80	149	337	
Total	214	304	530	214	304	530	

Source: International Energy Agency, 1978.
Note: Figures in parentheses are estimates of WOCOL. All figures rounded.

pulled up the price level of coal. Twenty years ago in western Europe the price of fuel oil and coal (calculated in terms of energy value) were very close, but then the price of oil began to fall and coal lost a great deal of its attraction. The oil-price rises have begun to reverse this situation. The EEC's statistics show that whereas the average price of steaming-coal imports in 1977 was $30.4/ton, this figure had reached $51.70/ton in 1980, an increase of 70 percent. This is not as large a price increase as that experienced by oil in the same period, however. The price of steaming coal has in general been about one-half that of coking coal, but comparatively large variations in the former are common because of significant differences in energy and sulphur content.

Shipping costs are also an important determinant of the price of coal. As table 4-5 shows, these vary considerably depending on the origin of shipments and their destination. Average prices could conceivably be reduced by increasing the size of coal transports, but the use of very large coal carriers makes sense only if suitable port facilities are available. At present this

Table 4-5
Some Transportation Costs for Coal

	FOB: Mine	CIF: Port of Delivery	Dollars/Btu (CIF)
To Northwest Europe from:			
United States			
East, underground	20–35	39–59	1.85
West, surface	8–18	31–50	2.19
Canada: West, surface	15–20	36–50	1.92
Australia: Underground	15–25	34–43	1.63
Surface	12–20	32–43	1.52
South Africa: Underground	10–15	26–35	1.41
Poland: Underground			
(FOB port)	23–31	31–39	1.46
To Japan from:			
United States			
East, underground	20–35	44–64	2.05
West, surface	8–18	31–50	2.00
Canada:		35–45	2.00
Australia: Underground	15–25	29–36	1.38
Surface	12–20	27–36	1.33
South Africa: Underground	10–15	26–33	1.36
Poland: Underground			
(FOB port)	23–31	36–44	1.67

Source: U.S. Department of Energy.
Note: All figures 1979 dollars/tonne.

is not the case in many parts of the world, and since these facilities are expensive, and on the average require two and a half or three years to construct, their absence has had and will continue to have an important influence on potential users of coal.

As pointed out repeatedly in this chapter, the major use of coal should its production be greatly expanded is in the generation of electricity; but there is considerable doubt whether, given its very noticeable effect on the environment, coal is preferable in all situations to other energy forms. Many coal enthusiasts enjoy making the point that (with the exception of water power) coal is the most inexpensive way to obtain electric power, but this is not universally true. According to figures published in April 1981, there is a wide spectrum of electricity prices throughout the industrial world (see table 4-6). The low price of electricity in Australia and South Africa can be explained by inexpensive coal. Among the large industrial countries, however, the situation in Sweden and France is due to the large share of nuclear energy in total electricity consumption, just as the high cost of electrical energy in Belgium and Britain can be traced to their extreme reliance on oil (and the fact that in Britain the present government insists on handicapping British industry by making it pay market rates for power).

The capital costs of nuclear power plants are higher than those of coalfired plants, which in turn are higher than those of oil-fired plants; fuel costs are lowest for nuclear plants and highest for oil-using installations. With the present range of fuel prices, nuclear energy has turned out to be the most inexpensive source of electricity in many countries. There is a very powerful antinuclear lobby operating in most countries in the industrial world, however, which chooses to ignore such things as money costs. As far as I can tell, this lobby is attempting to say something about its lack of faith in the capacity of politicians to set up adequate safeguards against nuclear accidents or to guarantee the safe storage of used nuclear fuel. Given the

Table 4-6
Some Comparative Electricity Costs, 1980–1981, in Index Form

Canada	(55.1)	United States	(100.0)
South Africa	(67.6)	Italy	(113.9)
Australia	(71.3)	West Germany	(116.0)
Sweden	(78.9)	Ireland	(120.7)
France	(88.8)	Britain	(133.5)
Holland	(93.0)	Belgium	(139.0)

Source: various publications.
Note: The cost of electricity in the United States during the period April 1980–April 1981 was, on the average, 5.31 mills per kilowatt hour, where a mill is one-tenth of a cent.

intellectual—and in some cases moral—shortcomings of many politicians, I find this attitude understandable.

Australian Coal

Not long after the colonization of Australia, coal was discovered near a settlement that is now called Newcastle, which has become one of the most important coal-mining centers in Australia. Several years later, in 1979, a coal outcrop was sighted near Coalcliff, south of Sydney and in the vicinity of Wollongong, which is also one of the most important coal-mining localities in that country; the following month Lieutenant Shortland located a coal deposit at the mouth of the Hunter River. It has now been officially decided that the first coal mined in Australia came from outcrops on a cliff face of the Hunter River estuary. This was in 1799.

In 1801 about 150 tonnes of Hunter River coal was exported to India, and later the same year coal was shipped to the Cape of Good Hope. The coal-mining district in Australia gradually began to spread, and mines worked by convict labor were soon established at Newcastle and elsewhere in the vicinity of Sydney. As for Queenland, coal was first located near Ipswitch in 1827, but not until 1870 and the discovery of the Aberdare seam did the Queensland coal-mining industry became important.

There is also coal in the states of Victoria, South Australia, Tasmania, and Western Australia—although the deposits in these latter three states are relatively unimportant, and the brown coal of Victoria is not exported. The main coal-producing *basins* or deposits are Sydney Basin; Bowen Basin (Queensland); Clarence-Moreton Basin (located largely in Queensland, but stretching into New South Wales); Galilee Basin (Queensland); and the brown-coal deposits of the Latrobe Valley in Victoria.

The growth of Australian production was hindered for a long time by the resistance of mine workers to a comprehensive mechanization of the mines. The big issue was fear of unemployment, although the primitive machinery employed during the nineteenth century also jeopardized the safety of the miners. When mechanization did arrive in Australia, it arrived with a rush; today, output per man-shift is among the highest in the world—some say the highest. Another factor contributing to the jump in productivity in Australia was the shift to open-pit (or open-cut) mining that began in the 1930s. Eventually the coal-mining sector was able to become extremely capital intensive and, as a result of the nature of open-pit mining, safer rather than less safe. Tremendous efforts were also made to avoid the environmental damage associated with open-pit mining. For instance, Queensland and New South Wales companies are required to lodge a bond of $123 (Australian) per hectare as a pledge against restoring land worth $2 per hectare.

Both Australian mine owners and the Australian government would like to see a much larger export of Australian coal. In fact, if the demand is forthcoming, the intention is to raise coal exports by a factor of four or five in the coming twenty years. In 1978–1979 Australian coal exports came to more than 38 million tons, which represented 53 percent of Australia's salable output. Japan bought 66 percent of this amount; Europe received 23 percent, Korea 5 percent, and Taiwan 3.5 percent. Plans exist for exporting 200 million tonnes by the year 2000 if the market exists and if Australia possesses the port and transport infrastructure needed to meet a growing world demand. At present the coal-mining industry is one of the largest contributors to government revenue of any industrial sector, and every dollar earned from the sale of coal results in tax revenues of 36 cents. Of this amount 21 cents went to state governments as royalties (since resources in the ground normally belong to the local community—that is, the state), land tax, payroll tax, shipping fees, rail freight, and other service charges. It is interesting to note that tentative plans have been drawn up for the large-scale use of pipelines to transport coal to Australian markets or overseas terminals. In the United States coal particles of medium size, suspended in water, have been transported along a 174-kilometer pipeline at a speed of 136 tonnes per hour.

A major problem that will have to be met with a rapid expansion of the coal industry is the growing cost of both production facilities and infrastructure. The cost of infrastructure has risen so rapidly that it has become more profitable to exploit coal deposits by underground mining in or near the built-up areas of New South Wales, than to develop open-cut mines in some of the more remote areas of the country (assuming there is no difference in the richness of the coal deposit). In New South Wales the development of an underground mine designed to produce one million tons of coal per annum can be expected to cost $25–$100 million and to take from three to five years. By the same token, to develop an open-cut mine producing 3 million tonnes of coal per year in Central Queensland will normally cost about $400 million, but could cost much more. For instance, a 45-cubic-meter dragline used to remove soil and rock lying above the coal seam can cost as much as $20 million.

The Australian coal industry is privately owned. The big names in this industry are Utah Development Company, CRA, and Shell Australia. A considerable amount of the coal industry is foreign owned or controlled, and in percentage terms foreign owners may control more than for the Australian minerals industry as a whole (where the figure is about 52 percent, with firms from the United States and United Kingdom holding about 80 percent of this). There is a considerable amount of antagonism in Australia toward the foreign ownership of mineral resources, and therefore it seems unlikely that foreigners will be encouraged to acquire more mineral properties in that country. At the same time, expectations are that the Australian energy sector will expand more than any other sector; and as things now stand this is patently impossible without more overseas investment—and presumably more overseas ownership, at least in this sector.

Regardless of the ownership issue, however, it is generally recognized throughout the country that the mining industry has made an important contribution to the welfare of the average Australian, particularly for those individuals who are employed in this industry. Wages in coal mining are much higher than the average Australian wage; and the total tax burden (which in the coal industry is augmented by certain levies) is higher for producers than in the United States, Canada, and South Africa. (The Australian company income tax is 40 percent of taxable income, with taxable income defined as sales revenue minus operating costs, interest charges, depreciation allowances, and special deductions such as investment allowances). It could be reasoned, however, that the larger tax burden (of which a reasonable part falls on foreigners) is a payment for the kind of political stability that does not always exist in other parts of the mineral-producing world.

If we consider the experiences of the largest importer of Australian coal—Japan—it is difficult to conclude that the Japanese have come to regret their association with the Australian coal industry. In the past Japan has mostly purchased coking coal from Australia, but in the coming years the trade in steaming coal will undoubtedly be more important—as is even more the case for European importers of Australian coal. Coal exports to Japan have generally taken place under long-term contracts having a duration of up to fifteen years; but although quantities are specified, plus-minus options of as much as 10 percent are allowable, subject to prior notification. At one time prices were set in terms of a base price that could be escalated for certain costs; but present arrangements often permit a price adjustment that is a function of market conditions. Coking-coal contracts with Japan generally carry one of the following stipulations:

1. Regular price reviews can take place outside contract provisions.
2. Contracts provide for regular reviews, although price is expressed in terms of a base price and escalatable components, with definite limits being placed on these escalations.
3. Contracts may provide for the price to be set by annual or biannual reviews, with a no-price, no-contract clause. Obviously, this type of arrangement is virtually equivalent to a short-term contract.

Because of some bad feeling that has surfaced in Australia concerning the alleged abuse of coal and iron-ore contracts by their overseas customers, it has been suggested that new contracts contain the following features:

Table 4-7
International Hard-Coal Trade Matrix, 1979
(in millions of tonnes)

	Benelux	France	Germany (FR)	Italy	U.K.	Canada	Japan	CMEA[a]	Others	Total
European communities	5	8	1	2	—	—	1	—	1	19
United States	4	3	2	4	1	17	13	—	12	60
Canada	—	—	1	—	—	—	11	1	1	14
Australia	1	2	1	1	2	—	27	4	1	40
South Africa	2	8	1	2	—	—	3	—	5	23
USSR	—	1	—	1	—	—	2	15	3	24
Poland	2	4	2	3	1	—	—	15	6	41
Others	1	1	1	—	—	—	1	2	3	8
Total	15	27	8	13	4	17	59	36	32	229

Note: Some totals do not add because of rounding; — signifies less than one million tonnes.
[a]Centrally planned countries of eastern Europe (excluding the USSR).

1. Pricing reviews should occur at intervals of not less than two years.
2. If buyers and sellers cannot agree on a pricing formula, then it should be the option of sellers to determine whether the contract should be terminated with some notice or whether the existing pricing arrangements should continue.
3. There should be a limited scope for adjustment of contract tonnages.
4. If possible, prices should be denominated in a basket of currencies to take account of currency instability.
5. Most contracts today are on an FOB basis. More contracts on a CIF basis should be resorted to.

The trade matrix shown in table 4–7 makes clear the position of Australia in the world trade picture. The consumption of coal in Australia has increased during the postwar period by 2.5 percent a year on the average, although in the last decade the rate of increase has been about 3 percent. Where exports are concerned, the intention is to sell as much coal as possible, subject to the constraints mentioned earlier. The Australian Bureau of Mineral Resources has provided forecasts for coal exports, as shown in table 4–8. These figures should be regarded as the potential supply in the face of expected price and cost developments.

The brown coal of Victoria is another story. Since its heating value is only one-quarter to one-third that of black coal, there is no possibility of its being exported. For many years it appeared that brown coal would not even have a market in Australia, since it could not compete with domestic gas. In a world of rising energy prices, however, the situation has changed; among other things, brown coal has become very important for the industries of Victoria—in particular aluminum refining. The thickest seam of brown coal in the world is found at Loy Yang, and the electricity generated from this coal may be some of the most inexpensive in the world. Some Australian

Table 4–8
Some Export Forecasts for Australian Coal

Year	Forecast Exports[a]
1978	37,500
1980	46,500
1985	86,000
1990	120,000
1995	150,000
2000	185,000

Source: Australian Bureau of Mineral Resources.
[a]In thousands of tonnes.

economists have argued that this cheap electricity should not be used to subsidize the Australian aluminum industry. In my opinion, however, this is exactly what it should be used for.

5 Uranium

The precursor of today's market for uranium began with the production of the first atomic bomb during World War II, and continued after the war as a market in which sellers of uranium supplied the raw material needed to increase the world's arsenal of nuclear explosives. In North America the demand side of the market was dominated by the Atomic Energy Commission (AEC), which also completely regulated the price. This price was immobile for long periods of time and in fact remained almost constant during the entire period 1951–1962; but with the passage of the Cold War and the rapid evolution of electricity-generating equipment employing nuclear fuels, a worldwide civilian market for uranium was eventually created. During the 1960s the AEC gradually abandoned its desire to supervise the disposition of all uranium produced outside the centrally planned economies. (Here it can be mentioned that in the past forty years about 600,000 tonnes of uranium has been produced.)

The production, use, and disposal of nuclear fuel has now been reduced to a relatively simple routine, assuming that it is managed by conscientious and intelligent men and women with attitudes and professionalism roughly similar to that of airline pilots or ground-control personnel. For the light-water reactor (LWR), which today comprises about 90 percent of the stock of nuclear generating equipment, the nuclear fuel cycle takes on the following appearance.

Uranium is mined in either underground or open-pit installations from ores having a concentration that averages about 0.25 percent. This implies that 4,000 tonnes of ore must be removed to obtain one tonne of natural uranium, and in addition it may be necessary to remove overburden having a weight of ten to twenty times the amount of ore that will be mined. Some elementary processing (called *milling*) transforms the ore into a substance known as *yellowcake,* of which about 80–85 percent is uranium oxide (U_3O_8). Yellowcake is then purified by chemical means to get uranium oxide, which is *converted* to uranium hexafloride (UF_6), which in turn is used as a feedstock for the enrichment process that increases the proportion of U-235 in natural uranium from 0.7 percent to about 3 percent. The enriched fuel is then fabricated into the pellets or fuel rods used in nuclear reactors. These activities can be easily located in figure 5–1

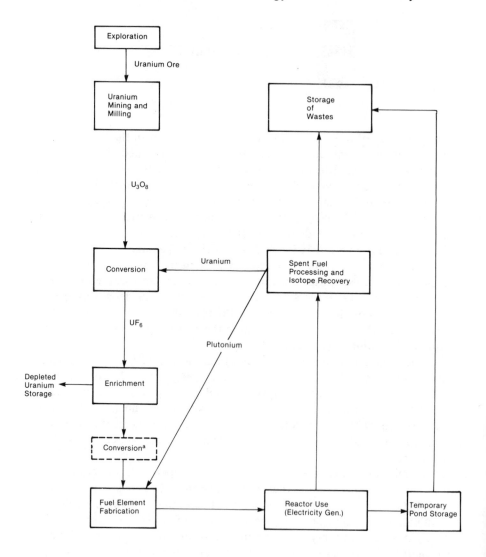

[a]In, for example, pressurized water reactors the enriched uranium hexafluoride from the enrichment plant is converted to uranium dioxide (UO_2). After some processing these are fabricated into the fuel crates for insertion into reactor cores.

Figure 5–1. The Nuclear Fuel Cycle

The rest of the fuel cycle consists of irradiation of the fabricated fuel in the reactor to produce steam for power generation; temporary on-site storage of the spent nuclear fuel until the most dangerous forms of radiation are dissipated; reprocessing of this fuel; and storage of radioactive wastes for a long period. On the average slightly more than 45 percent of the cost of producing a kilowatt-hour of electricity by nuclear means can be attributed to obtaining uranium oxide; and about the same amount is required for conversion to UF_6, enrichment, and fabrication. The remainder is for waste disposal. For reasons having to do with transportation economics, the milling of uranium into yellowcake almost always takes place in close proximity to the mine; as a result, total transport costs in the fuel cycle are extremely low.

The richness of the uranium ore that is mined today compares unfavorably with that of the ore that was being extracted from the Canadian mines that supplied the major portion of the uranium used by the United States in the 1940s. At the same time, however, it can be noted that rich and easily mined deposits may be on the verge of large-scale exploitation in Australia and Canada. For example, around Midwest Lake, Saskatchewan, there is said to be almost 40,000 tonnes U of reserves with an average grade of 3.5 percent. (The designation *U of reserves* signifies the pure uranium metal. On the average 1 metric ton U corresponds to 1.3 short tons of U_3O_8; remembering that 1 tonne = 1 metric ton = 2,204 pounds, whereas 1 short ton = 2,000 pounds, we see that U_3O_8 contains 84.8 percent U.) By way of comparison, Swedish ores have a uranium content of only 0.02–0.03 percent; even with large-scale exploitation, Swedish uranium would cost a great deal relative to the average world cost. Since fuel costs are only a relatively small percentage of the cost of electricity generated from uranium, however, this fact by itself does not indicate that it is uneconomical to mine Swedish ore.

The general abundance of uranium is about 4 parts per million (ppm) in the earth's crust, and 3 parts per billion (ppb) in seawater (where ppm signifies 10^{-6} parts of uranium and ppb is 10^{-9} parts of uranium). It has been estimated that about 8×10^{13} tonnes of uranium can be found in common rocks; with known and postulated techniques, 4×10^{11} tonnes of uranium can be obtained from onshore deposits. Similarly, 4×10^9 tonnes of uranium are to be found in the oceans, and it has been said that as much as 2×10^9 tonnes might eventually be extracted.

Table 5–1 provides information on the production of uranium in 1976 and gives some idea of the pattern of ownership in this industry. The most important change that could take place here is a more rapid expansion in Australian output, although it should not be forgotten that considerable pressure is being exerted in Australia against expanding production of this material. For example, trade unions recently refused to load nuclear materials for shipment abroad.

It was in the United States that production climbed most rapidly during the period 1945–1963, reaching 10.9 thousand tonnes of uranium in 1963. From 1963 to 1967 production fell to 7.4 thousand tonnes of uranium; but as the private market became more important, production climbed again, reaching 10.2 thousand tonnes in 1973. Stagnation then set in, but by 1978 (for reasons given later) production had reached 14.4 thousand tonnes. Ura-

Table 5-1

The Production of Uranium (1979), in Tonnes, and Some Important Producers

Country	Production (1979)	Principal Domestic Producers	Main Foreign Companies Involved
United States	14,400	1. Kerr McGee 2. United Nuclear 3. Anaconda 4. General Electric	
Canada	6,600	1. Denison Mines Ltd. 2. Rio Algom 3. Eldorado	a) Uranerz (Fed. Repub. Ger.)
South Africa and Namibia	7,200	1. Vaal Reefs 2. Buffelsfontein 3. Rossing Uranium	a) Rio Tinto Zinc (U.K.) b) Rio Algom (U.K.)
Australia	500	1. Queensland Mines 2. Peko-Wallsend 3. Western Mining 4. Mary Kathleen Uranium	a) Getty Oil (U.S.) b) Western Nuclear (Canada)
France	2,300	1. CEA 2. Rothschild 3. PUK	
Niger	2,200	1. ONAREM	a) COGEMA (France) b) OURD (Japan)
Gabon	1,000	1. Cie des Mines d'Uranium de Franceville	a) Imetel (France)
Others	800		

nium mining in the United States is for the most part concentrated in four states (New Mexico, Colorado, Wyoming, and Utah); and one of the most important features of this activity in the United States is the progressive decline in ore grade. There is some evidence that production costs in the United States are higher than for any other of the major uranium producers.

Canada and South Africa are also major producers of uranium. In Canada the major part of production and concentration has occurred around Elliot Lake in Ontario, with a large percentage of production being exported to the United States and Britain. The average ore grade has also fallen rapidly in Canada, although this situation could be reversed as production increases in Saskatchewan. As for South Africa, uranium is a by-product with gold; thus the supply of uranium is determined by the supply of gold, which in turn is determined by a number of things, including price and the policies of the South African government. South African and other foreign companies have a considerable interest in the Rossing uranium mine in Namibia, which according to present estimates contains reserves of over 100,000 tonnes; but rather than being the prize that some people make it out to be, Namibian uranium appears to be very high-cost uranium.

In the countries discussed earlier, the constraints placed on the uranium market by governments, and the political and psychological climates in which governments operate have generally been such that price has had a relatively minor role in determining output. It is said that this situation is changing at present, but strategic considerations still carry some weight in determining the distribution of U.S. uranium, whereas countries like Canada have begun placing more emphasis on energy independence and thus incline toward increased export restriction. The French government, though an aggressive champion of nuclear power, possesses only limited uranium resources in mainland France and therefore has obtained important ownership interests in both Gabon and Niger. Most of the uranium produced in France is controlled by COGEMA—Cie Generale des Matieres Nucleaires—and this organization is also very active in Africa. Niger, in particular, is expected to become a very important producer of uranium in the near future; but it seems likely that the bulk of its exports will be directed toward France.

Electricity and Nuclear Energy

The most important buyers of uranium are the electricity-generating companies or *utilities,* which not only must secure a supply of fuel that will last the lifetime of their reactors, but also guarantee the efficient disposal of the spent fuel. If we look at the rate of growth of electricity demand, we see that between 1940 and 1974 it often outstripped even the growth in aggregate energy demand, which in turn was higher than the aggregate rate of economic growth in the world economy. On the average the rate of growth of electricity demand was slightly in excess of 7 percent; but from 1974 onward, as energy demand fell, electricity demand also declined. Most predictions now are for a growth in the consumption of electricity in the 2.5-3.5-percent range for the rest of the 1980s, and perhaps longer.

The expected fall in the growth of electricity demand is incontestably related to the decline in the aggregate rate of world economic growth, and in particular the deceleration of output in the most energy-using sectors of the major industrial countries. Electricity prices have also risen, though not as rapidly as the escalation in price of most primary fuels; this has not only helped reduce the demand for electricity but has also had a deleterious effect on economic growth. For instance, the increasing price of electricity has led to much higher costs for industries such as aluminum refining and the automobile industry; and for various reasons (such as declining real incomes and the inertia caused by government regulation) these costs could not be passed on to final consumers. As a result, much of this type of industry is in a turmoil that has resulted in the global relocation of some capacity and the phasing out of some other. In addition, the electric-power industry in some countries did not respond rapidly enough to the fall in demand for its products, and thus now and in the foreseeable future must contend with an unwanted margin of excess capacity. (Note here that a fall in demand for electric power does not lead, in the short run, to a fall in the price of this power, since in almost all countries this price is set by public authorities on the basis of average—not marginal—costs.) In this connection I would like to mention again my belief that there is a causal link between higher oil (and energy) prices and decelerating growth in the world economy. Thus higher oil prices do not automatically have the salutary effect for such industries as nuclear and coal that was once taken for granted, since with lower growth less energy is required on the whole, and this tends to slow down the ordering of new coal and nuclear facilities. Also it is difficult to finance new equipment in these circumstances.

In 1978, 220 reactors worldwide were supplying about 2 percent of world energy consumption, the equivalent of 3 million barrels of oil per day. In terms of electricity, however, the contribution of nuclear energy is much more impressive, generally running to more than 10 percent of energy consumption in most industrial countries. Table 5–2 provides a survey of world nuclear generating capacity and some guesses concerning nuclear capacity about the year 1990.

In examining the last column of table 5–2, it must not be forgotten that in the long and sorry history of the forecasting arts, few vistas have been as resistant to accurate prognoses as has nuclear energy. In 1974 the AEC (the predecessor of the U.S. Department of Energy) predicted that by 1985 the noncentrally planned countries would possess an installed nuclear generating capacity of 640 GW_e, although December 1981 estimates center about 200–250 GW_e. (Here the subscript e signifies electrical energy.) This expected capacity represents 5.8–7 Mbbl/d of oil equivalent; by the same token, the probable range of capacity at the end of this century (850–1,200 GW_e) represents 24–34 Mbbl/doe.

Nuclear Generating Capacity

The demand for uranium is derived from the demand for nuclear equipment, and in the nature of things this equipment will not be demanded if the follow-through components of the fuel cycle (such as enrichment capacity) are unavailable. If we work with a once-through cycle, a light-water reactor requires about 5,000 tons of U_3O_8 per GW_e for a thirty-year life. This input,

Table 5-2
Nuclear Power Plants: Existing, under Construction, and on Order as of January 1979

Country	Installed		Under Construction or on Order	
	Number	Capacity (MW)	Number	Capacity (MW)
Argentina	1	319	1	600
Bangladesh	—	—	1	200
Belgium	4	1,660	4	3,800
Brazil	—	—	3	3,116
Bulgaria	2	840	2	840
Egypt	—	—	1	622
Philippines	—	—	2	1,240
Finland	2	1,080	2	1,080
France	16	8,321	32	32,195
Germany (FR)	15	8,857	15	16,436
India	3	580	5	1,080
Iran	—	—	6	7,192
Italy	4	1,447	5	3,959
Japan	23	14,523	8	6,857
Yugoslavia	—	—	1	632
Canada	11	5,516	15	10,056
Korea (South)	1	564	4	3,134
Cuba	—	—	1	420
Luxemburg	—	—	1	1,247
Mexico	—	—	2	1,308
Netherlands	2	500	—	—
Pakistan	1	125	—	—
Poland	—	—	—	—
Rumania	—	—	2	1,020
Switzerland	4	1,926	3	3,007
USSR	28	9,820	32	30,685
Spain	3	1,073	14	13,158
United Kingdom	18	8,118	6	3,750
Sweden	6	3,740	6	5,720
Republic of South Africa	—	—	2	1,844
Taiwan	2	1,220	4	3,714
Czechoslovakia	2	530	3	1,260
Hungary	—	—	4	1,680
United States	72	52,477	133	148,065
Austria	—	—	1	692
East Germany	4	1,340	7	3,780
Total	224	124,586	332	315,259

Source: Australian Atomic Energy Commission, 1978.

considered over an entire stock of reactors whose total capacity might reach 1,000 GW_e by the end of the century, does not imply an early exhaustion of present known resources of uranium. It also appears that design improvements may be capable of reducing uranium input per KW_e by 25–30 percent—though not, some experts feel, before the next century. It is true, however, that reactor production capacity (at present about 50 GW_e/year) will have to be fully utilized if a generating capacity of 1,000 GW_e is to be achieved by the year 2000. At the same time mining output will have to reach the vicinity of 150,000 tonnes/year from its present capacity of 40,000 tonnes/year, and both conversion and enrichment capacity will have to expand in phase with the increase in generating capacity. In addition, as pointed out later, some rearrangement will have to take place in the location of conversion and enrichment facilities, since at present there is the possibility that extensive political constraints could be placed on the use of nuclear installations as a consequence of the concentration of conversion and enrichment capacity in only a few countries.

Something must be said here about the matter of reprocessing spent fuel, as well as the tails assay—a measure of the amount of U-235 left in the waste material following enrichment. Where the first topic is concerned, spent-fuel elements are removed from the reactor (typically about 30 tonnes of spent fuel per year for each 1,000 megawatts (MW_e) of generating capacity) and after storage to reduce the radiation level, are reprocessed in such a way as to separate the uranium in the spent fuel, any plutonium that has been produced, and the radioactive fission products. The uranium can be disposed of (at considerable cost) or reconstructed to UF_6 and fed back into an enrichment plant, where it is reconverted into fuel.

disposed of (at considerable cost) or reconverted to UF_6 and fed back into an enrichment plant, where is it transformed into fuel.

nium input needed to produce a given amount of enriched uranium. For example, increasing the tails assay to 0.3 percent from 0.2 percent requires a 20-percent increase in the input of natural uranium, but at the same time there is a saving in enrichment capacity of about the same amount. Thus any forecast of uranium requirements must turn on the assumptions made about the tails assay.

As mentioned earlier, it may happen that 1,000 GW_e will be generated by nuclear equipment in the year 2000 in the noncentrally planned world, but no one can know for sure. Something can, however, be said about expected generating capacity up to the year 1986, and consequently about uranium demand. This information is shown in table 5-3. It is also clear that the cost of nuclear capacity is increasing by almost 20 percent a year, much faster than the increase of the general price level in the industrial world. This is due to longer construction cycles (which have now reached more than ten years, compared with five or six years in 1970), growing materials intensity, and a more complicated licensing process. In fact, some

Table 5–3
Estimated Nuclear Generating Capacity and Uranium Demand, 1980–1985

Year	GW$_e$ Capacity	Demand (10^3 Tonnes U)	National Stockpiles[c] (10^3 Tonnes U)
1980	118	28.4	101.2
1981	137	30.7	103.5[a]
1982	156	33.5	103.0[b]
1983	184	36.5	102.2[b]
1984	207	40.2	101.2[b]
1985	232	43.8	—

[a]Estimate.
[b]Projection.
[c]United States plus western Europe.

of the major suppliers of nuclear equipment are in serious financial difficulties; and several of them are leaving the industry (such as AEG-Telefunken of Germany, which departed with known losses of about 1.7 billion DM). One of the consequences of these departures may turn out to be that the figures for generating capacity and uranium demand referred to in table 5–3 may turn out to be too high.

Before leaving this topic let us consider uranium stocks, since in recent years there has been a rapid increase in the rate at which inventories have been expanded around the world. Given present expectations about the future availability of uranium, however, present inventory levels are probably too high, and will almost certainly be decreased in the near future. Some estimates of future inventory levels can also be found in table 5–3.

Inventories have been discussed at some length in chapter 2, but because uranium stocks are somewhat special, a little more should be said about them here. As with other commodities, uranium stocks are important because of the lack of substitutes for uranium and because, given the characteristics of the market and the growth in nuclear capacity, supply interruptions are possible at any time. Moreover, uranium is relatively easy to stockpile and is attractive from the point of view of the energy content of each cubic foot of space devoted to inventories. Here it should be appreciated that despite the slowdown in the growth of nuclear capacity, many buyers consider it preferable to continue stocking low-cost uranium purchased on long-term contracts than to cancel these contracts and risk having to buy high-cost materials at some point in the future. In addition, in the United States, reneging on commitments made with enrichment plants can involve heavy penalties. The size of these penalties probably keeps stocks of uranium well in excess of levels required to ensure against supply disruptions or unforeseen escalations in price.

The Supply of Uranium and Nuclear Energy

As indicated earlier, there should be no problem with supplying uranium to the growing stock of nuclear equipment unless there is a sharp acceleration in the pace of reactor construction, which now seems unlikely. The total amount of uranium in the crust of the earth amounts to 10^{13} tons, enough to run existing reactors for several million years if all this uranium could be won. As it happens, however, only a small portion of this 10^{13} tons occurs in usable concentrations; and in the traditional deposits, particularly in the United States, ore grades are falling rapidly. As with many minerals, it has been suggested that a tenfold decline in the grade of uranium ore would mean that quantities of recoverable uranium would increase by a factor of 300 *if this thinner grade could actually be exploited*. As explained in Banks (1979b), this is a variant of Lasky's law. Since the classification of uranium resources is, as always, in a state of flux, I will only present a summary table here showing low- and high-cost *reserves* (see table 5–4). It could be argued though that the most accessible categories of so-called uranium resources should be added to these figures.

The low-cost uranium listed in this table signifies U_3O_8 costing less than \$30/pound; at present there appears to be enough of this available to last at least through 1990. Extending our horizon forward another ten years changes the picture, however. If the rate of depletion of reserves during the 1990s attains and holds a level of 0.3 million short tons of U_3O_8 per year, then it seems almost certain that the price of uranium will have to rise if higher-cost ores are to be exploited at a suitable rate. It should not be forgotten, however, that technological progress involving existing types of reactors, as well as the introduction of new varieties of equipment, could result in an appreciable lowering of uranium requirements per unit of generated electricity—nor can it be forgotten that a more dramatic version of the Three Mile Island incident in Pennsylvania would result in a drastic scaling down of nuclear expansion.

Another factor that might slow down the construction of nuclear facilities is the situation alluded to earlier concerning conversion and enrichment capacity. Just now most of these facilities in the noncommunist world are located in the United States. This has important consequences both for the cost of operating nuclear equipment and for the incentive to invest in this equipment: enrichment contracts include financial penalties for reduction in or cancellation of contracted amounts, and as a result some utilities have had to carry larger inventories then they consider economical. If present plans are carried out, however, world conversion and enrichment capacity is scheduled to expand quite considerably in the coming three or four years; and, everything else remaining the same, the costs associated with this phase of the fuel cycle should fall. France, the Netherlands, and the United Kingdom are especially active in the building of centrifuge enrichment plants.

Table 5–4
Estimated Uranium Reserves and Resources, 1977
(in 10³ tonnes U)

	Uranium Reserves		Uranium Resources	
	U_3O_8 Price $30/Pound	U_3O_8 Price $30–50/Pound	U_3O_8 Price $30/Pound	U_3O_8 Price $30–50/Pound
United States	523	120	838	215
South Africa and Namibia	306	42	34	38
Australia	289	7	44	5
Canada	167	15	392	264
Niger	160	—	53	—
France	37	15	24	20
India	30	—	24	—
Algeria	28	—	50	—
Gabon	20	—	5	5
Brazil	18	—	8	—
Argentina	18	24	—	—
Sweden	1	300	3	—
Others	50	17	35	38
Total	1,657	540	1,510	590

Source: OECD documents.

Capital Costs

Capital costs are an important item in the production of almost all goods and services. Unfortunately, however, most textbooks in economics do not give their readers a satisfactory introduction to the topic. Thus most people, when they hear that the production of nuclear energy involves low operating costs but high capital costs, do not quite know what to make of the latter term.

I intend to clarify this matter with a simple numerical example. I would strongly advise readers to work through this example until they understand it perfectly, since it contains some extremely important economic concepts. First, let us assume that you borrow $1,000 to buy a physical asset such as a machine or a house. To keep things simple, I will assume that the asset as a life of two years, after which it falls apart. I shall also assume that you contract to repay the loan in two years, and the rate of interest is 10 percent.

Next let us examine this situation from the point of view of the lender. He or she has a $1,000, which, if lent out at the rate of interest of 10 percent, will bring $1,000 $(1 + 0.10)^2 = $1,210$ at the end of two years. In a perfect market, which for our purposes is a market with no uncertainties involving either present or future transactions, the lender is only interested in getting $1,210 in return for giving up $1,000 two years earlier. Some skeptics claim that the world is not quite perfect, however, and these people have devised schemes whereby people who borrow money are compelled to pay it back periodically rather than in one sum at the time the loan becomes due. Later on we shall break these periodic payments down into *amortization* (or *depreciation*) payments and *interest* payments; but for now we simply lump them together. As for our period, this will be taken as one year.

At the end of the first and second year the borrower makes a payment to the lender. These payments are of such a magnitude that the lender is indifferent between these payments and the certainty of the sum of $1,210 received two years after lending $1,000. So far there should be no riddles, but the next step may require a detailed exposition, since it involves the stipulation that the payment made at the end of the first year in our two-year model is put into a savings institution (or invested in a financial asset) where it draws interest. To begin our explanation of this arrangement, let us employ a simple diagram showing what we have done so far (figure 5-2).

In the figure, P_w is the amount lent ($1,000), whereas P_s is the value of the yearly payment. In line with our earlier discussion, we must have:

$$P_s(1 + r) + P_s = P_w(1 + r)^2$$

or

$$P_s(1 + 0.10) + P_s = 1,000(1 + 0.1)^2 = 1,210.$$

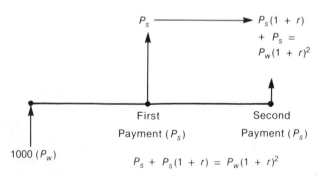

$$P_s + P_s(1 + r) = P_w(1 + r)^2$$

Source: Reprinted by permission of the publisher from Ferdinand E. Banks *Bauxite and Aluminum: An Introduction to the Economics of Nonfuel Minerals* (Lexington, Mass.: Lexington Books, D.C. Heath and Company, 1979), p. 54.

Figure 5-2. Payment Scheme for a Two-Period Model

These expressions simply indicate that the payment scheme, handled as in the preceding discussion, provides the same amount of money after two years as a perfectly safe bank deposit or bond offering a 10-percent interest rate. Solving the last equation, we get P_s = $576. In other words, if you borrow $1,000 that must be repaid in two years, you can repay the debt with two end-of-year payments of $576.

The following discussion will show that, given the conditions of the problem, the first payment must be put into an interest-bearing asset or instrument of one type or another. Let us suppose the first payment were not put into an interest-bearing asset. Then we would have to make the following payments P_s' so that our lender would end up with $1,210 two years after making the loan:

$$P_s' + P_s' = P_w(1 + r)^2 = \$1,210$$

or

$$P_s' = 605.$$

However, this cannot be satisfactory! If each payment were $605, the lender could take the first payment, put it in a bank where it would draw interest at 10 percent, and at the end of the second year the lender would have $605(1 + 0.10) + 605 = \$1,270$. But $1,270 at the end of two years corresponds to an interest rate of 12.7 percent, which we get by solving $1,000(1 + r)^2 = \$1,270$ for r. Since, on a single market with perfect information, we can have only one interest rate, the contradiction that we have arrived at indicates that the scheme involving two end-of-year payments that are equivalent to $1,210 requires that the first payment draw interest over one year.

In other words, two end-of-year payments of $576 in return for a loan of $1,000 implies an interest rate of 10 percent, which is the interest rate with which we began our labors. The reader should now, before continuing to the next paragraph, go through the preceding example using a three-period situation while keeping everything else the same. Draw the type of diagram shown here, putting in all the symbols; after calculating P_s, verify that a payment scheme of the sort we have just discussed actually provides the lender with $1,000(1 + 0.10)^3$ dollars three years after he makes a loan of $1,000.

Now let us approach this problem from another direction. Suppose you have a *stock* of money, and you want to turn it into a *flow* of income over a given number of years. With a given rate of interest, you can buy an annuity from, for example, an insurance company, which, in return for a given sum of money, will provide you with a stream of payments at specified intervals over a number of years. For instance, if the interest rate is 10 percent, a

$1,000, two-year annuity will yield payments of $576 at the end of each year for two years. You can check this result by examining an annuity table.

Now we take up the matter of interest charges along with depreciation and amortization. Referring back to the preceding example, if you borrow $1,000 and the rate of interest is 10 percent, then your interest charges are $100 per year. Thus in terms of the example, of the $576 in total yearly charges, $100 are interest charges, and the rest, $476, are amortization or depreciation charges. To clarify the last concept, let us note that if we have a physical asset, the usual practice is to put money aside every year to replace the asset at the end of a given period. For instance, if we have a machine costing $1,000 that lasts two years and the interest rate is 10 percent, if we make yearly depreciation payments (or specify depreciation charges) of P_d and if at the end of the first year the payment is put into a bank where it can draw interest, then if we are to replace the machine at the end of two years, we must have:

$$P_d(1 + r) + P_d = \$1,000$$

or

$$P_d(1 + 0.1) + P_d = \$1,000;$$

thus

$$P_d = \$476.$$

Note how this result agrees with the previous example in which, with $576 in total charges and $100 in interest charges, we arbitrarily called the difference between them ($476) the *depreciation* (or *amortization*) charge. Note that when we speak of a machine, we generally use the expression *depreciation charges;* whereas with a loan these charges are called *amortization charges* (we speak of *amortizing* a loan).

Finally, before we delve into a little algebra, the preceding discussion can be extended slightly. If we have $1,000 and the interest rate is 10 percent, and if with this money we buy a physical asset such as a machine that is used to produce some product, and if the life of the machine is two years—or, as we say, the machine is *fully depreciated* over two years—the *capital cost* or *price of capital services* of the machine is equal to $576, or the depreciation cost plus the interest charges. Also, in this example the *investment cost* is $1,000. It would prove useful to spend a few minutes reviewing this terminology because the investment cost and the capital cost are not the same thing, although some people make the mistake of using the terms interchangeably.

Generalizing the preceding discussion, if we take the matter of depreciation or amortization charges first, use the same notation as before, and assume that the asset is depreciated or amortized over T periods, then it follows that we must have:

$$P_d(1 + r)^{T-1} + P_d(1 + r)^{T-2} + \cdots + P_d = P_m. \qquad (5.1)$$

The reader should interpret this expression in the light of the previous numerical example; and then take $T = 3$ and $r = 0.1$, and write out equation 5.1, keeping in mind our previous discussion. We can now multiply both sides of equation 5.1 by $(1 + r)$, which yields:

$$P_d(1 + r)^T + P_d(1 + r)^{T-1} + \cdots + P_d(1 + r) = P_m(1 + r). \quad (5.2)$$

Subtracting 5.1 from 5.2 gives:

$$P_d(1 + r)^T - P_d = P_m(1 + r) - P_m = rP_m$$

or

$$P_d = \frac{rP_m}{(1 + r)^T - 1}. \qquad (5.3)$$

The reader should now observe what happens if we take the case of a physical asset with an infinite life: here $T \to \infty$, and on the basis of (3.7), $P_d = 0$. And this is as it should be: if a machine, house, or some other physical asset were never to wear out, then there would not be any depreciation charges. Similarly, if a borrower were to have forever to repay a loan, then his amortization charges would be zero. We can now take equation 5.3 and set $T = 2$, $r = 0.10$, and $P_m = 1,000$. As in our previous example, we get:

$$P_d = \frac{0.10 \times 1,000}{(1.1)^2 - 1} = \frac{100}{0.21} = \$476.$$

As explained earlier, our interest cost is rP_m. The total charge, which we have defined as the price of capital services, then becomes:

$$P_s = \frac{rP_m}{(1 + r)^T - 1} + \frac{rP_m}{1} = \frac{rP_m(1 + r)^T}{(1 + r)^T - 1} \qquad (5.4)$$

Again, if $T \to \infty$, then $P_s = rP_m$; for an asset with infinite life, the only charges are interest charges. Then, taking the figures from the preceding example, we get:

$$P_s = \frac{0.10 \times 1,000(1 + 0.1)^2}{(1 + 0.1)^2 - 1} = \$576.$$

The reader should also appreciate that these results hold regardless of whether we borrow or use our own money to buy a physical asset. In the latter case these charges are called *opportunity costs* since they signify what we could realize if we chose to lend out this money. Finally, let us derive the price of capital services using the convention that in a perfect market, all assets should have the same return. In other words, if we have P_m, which is loaned out for T periods at interest rate r, it should prove the same return as a stream of payment, each amounting to P_s. Thus:

$$P_s(1 + r)^{T-1} + P_s(1 + r)^{T-1} + \cdots + P_s(1 + r) + P_s = P_m(1 + r)^T. \quad (5.5)$$

Multiplying 5.5 by $(1 + r)$ gives:

$$P_s(1 + r)^T + P_s(1 + r)^{T-1} + \cdots + P_s(1 + r)$$
$$= P_m(1 + r)^{T+1}. \quad (5.6)$$

Subtracting 5.5 from 5.6 gives:

$$P_s(1 + r)^T - P_s = P_m(1 + r)^T(1 + r) - P_m(1 + r)^T$$

or

$$P_s = \frac{rP_m(1 + r)^T}{(1 + r)^T - 1}. \quad (5.7)$$

Now that we have our basic relationship, let us apply it to the capital cost of a nuclear installation. First, however, the reader should appreciate that capital costs are intended to reflect the cost of plant construction in the broadest sense. In the following example, which employs information from a report by Electricité de France, the time of construction of the installation discussed is five years; but construction periods today are usually in the vicinity of ten years; and—in the United States—interest charges have risen from 8 percent of capital costs for plants coming on stream in 1972, to 20 percent for those becoming operational in 1985.

It is also important for the reader to pay particular attention to the capacity at which the reactor operates. The capacity factor is the ratio of electricity actually produced during a time period to the amount that would be produced if the reactor operated at full capacity during the entire time period. As will be shown in the following example, the higher the capacity factor, the lower the capital charges. The finer points of the capacity factor cannot be gone into here, but it should be pointed out that, at least in the United States, the large nuclear units constructed recently show a tendency to have lower capacity factors than the medium-sized units that were installed during the 1960s and 1970s. Some observers take this to mean that we may have seen an end to the increasing returns to scale that nuclear facilities generally have shown since their introduction.

My example assumes a light-water reactor (specifically a 2 × 1,300 MW nuclear power plant) constructed in France, with an (installed) investment cost of $420/kilowatt (KW). The rate of interest will be taken as 9 percent, and the life of the equipment is twenty years. We can thus calculate the capital cost to be:

$$P_s = \frac{rP_m(1 + r)^T}{(1 + r)^T - 1} = \frac{0.09 \times 420(1 + 0.09)^{20}}{(1 + 0.09)^{20} - 1} = \$46.01/\text{year}.$$

As valuable as this figure is, it does not tell us enough: what we want is the cost per kilowatt hour. This is because although *power* is consumed in units called kilowatts, *energy* is consumed in units called kilowatt-hours. For example, a 500-watt bulb consumes power at the rate of 500 watts, or 0.5 KW. After five hours of use of energy that it has consumed is equal to (time) × (rate of power consumption) = 5 (hours) × 0.5 (KW) = 2.5 kilowatt-hours (KWh). Readers who have paid an electric bill know that they are charged for KWh.

As should be evident, in order to get KWh we have to have *time*. If the 1,000 watts (1 KW) in the preceding example operated twenty-four hours per day every day of the year, then we would be talking in terms of 365 (days) × 24 (hours/day) = 8,760 hours. The $46.01 capital cost obtained for 1 KW would be spread over 8,760 hours, and therefore would become $46.01/8,760 = $.00525/KWh. This figure is usually turned into *mills,* where a mill is one-tenth (.01) of one cent. Capital cost is therefore 5.25 mills/KWh.

If, however, we take a load factor of 0.75, then the 8,760 hours that a perfect piece of equipment could operate is reduced to 0.75 × 8,760 = 6,570 hours. The cost thus becomes $46.01/6,570 = $0.007/KWh, or 7 mills/KWh. Electricité de France has calculated the capital cost for this example at 7.76 mills/KWh; at least a part of the discrepancy between their results and mine, however, is due to the load factor they employed (which turned out to be 72.5 instead of 75), and the possible error in my estimate of the installed cost—or investment cost—of a kilowatt of generating equipment. I will now present Electricité de France's cost calculation for the type of equipment mentioned here, for 1976, and their calculation for generating equipment using fuel oil as an input.

Cost	Nuclear	Fuel Oil
Investment (i.e. capital)	7.76	4.69
Fuel	4.56	18.66
Operating	3.16	3.16
Total	15.48 mills/KWh	18.42 mills/KWh

The total number of hours of operation of the equipment (over the twenty-year life) was calculated by Electricité de France as 54,100. As far as I can tell, this figure can also be turned into an annuity, yielding 6,355 hours a year for twenty years; and from this a kind of average load factor can be calculated. This is $6,355/8,760 = 0.7254$.

Before finishing this discussion, a few comments should be made on the preceding entries covering fuel and operating costs. The key factor where fuel costs are concerned is the cost of uranium; as explained earlier in the chapter, this varies with the size of the uranium deposit, its location, and the mode of extraction. Operating costs include royalties, labor costs, the cost of materials, inventories, maintenance, overhead, and so on. Almost all these costs are increasing very rapidly, but not only for the nuclear industry. It seems extremely likely that the cost of nuclear energy is rising in real as well as monetary terms; the theory here is that if the extraordinarily high rates of interest that exist in the world at present do not recede drastically, the cost of nuclear energy will almost certainly rise more rapidly than the cost of energy that is generated from such things as oil.

One more important point should be made in this discussion. Because of the high capital cost, but low fuel cost, of nuclear (and coal) units, they are very useful as *base-load plants*. These are plants that generate electricity both during *peak* periods—such as the daytime, when plants and offices are in full operation—and during off-peak periods—such as the evenings. When demand exceeds the capacity of these base-load units, electricity suppliers switch on older, less efficient plants (*intermediate-load plants*) that have a higher operating cost than base-load plants, but are only used four or eight hours a day.

Last but not least, there are *peaking* plants. These operate only the few hours per day when the peak power demand occurs. The most important characteristic of these plants is that they can start up very rapidly and deliver power almost immediately. Many utilities employ gas-turbine plants for this purpose since they have a comparatively low capital cost, but a

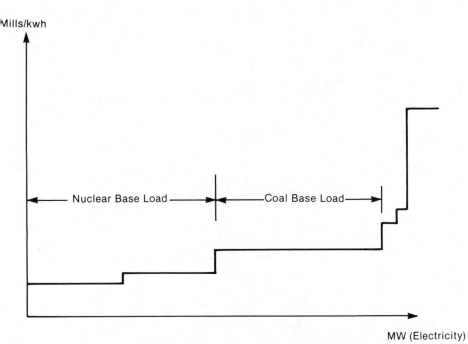

Figure 5-3. Simplified Cost Curve for the Supply of Electricity

high operating cost. Oil-fired installations also fall in this category, although obviously it was not very long ago that they handled a good share of the base load. We can now draw the important, but not altogether familiar, cost curve showing operating costs (mostly fuel and maintenance costs) as a function of electricity supply (see figure 5-3).

This cost curve shows the short-run picture, assuming as it does that the electricity-supplying authorities already possess nuclear, coal, gas, and other equipment. Thus it gives the incorrect impression that nuclear power must be less expensive than most other types of power. The reason this is not the case is that when equipment-purchasing decisions have to be made, capital costs are also important.

6

The Economics of Natural Resources: Nonfuel Minerals

Despite my inclinations, I will refrain from elaborating here on the subject of ultimate resource availability. Some so-called experts have promulgated the idea that not only are there sufficient mineral resources in or near the crust of the earth to permit all of humanity to live in some facsimile of bourgeois comfort, but this bounty is actually enriched as the number of people striving to enjoy it multiplies.

As I emphasized in *The Political Economy of Oil,* if the population-growth dilemma cannot be solved, then there will be no acceptable solutions to any economic problems—and certainly none that have to do with natural resources. I consider this a nonnegotiable position. As far as I could tell, all of the 1981 Nobel Prize winners in medicine, physics, and chemistry entertained similar beliefs in the matter. If present population growth trends continue and reach in the next century the 12–15-billion mark that the OECD Interfutures project and Professor Joseph Spengler see as a distinct possibility, then the sheer weight of human appetites is going to overwhelm any conceivable natural resource base, including the so-called crustal abundance base cited by Julian Simon and others. In addition, once conventional ores have been worked out and a new mineralogical threshold crossed, energy requirements per unit of mineral output may increase by a factor of at least 100, and perhaps as much as 1,000.

Although it may take some time for the overall picture to become clear, at present the world economy is on its way toward a new equilibrium, or a new international economic order—though not the one envisioned by many development economists and international civil servants. This new equilibrium involves limited material resources (definitely in a flow, but conceivably also in a stock sense) combined with a greater international competitiveness. The result is going to be very bad economic news for somebody. Eventually, of course, such an arrangement could lead to certain sociological and political discomforts that, if carried to an extreme, would make the mere change in consumption habits implied here seem trivial. The lessons of history are clear: of all the competitions that humans and animals engage in, the competition for space and natural resources is the most destructive.

Although it does not seem to be appreciated by the cognoscenti, the so-called energy crisis could be a preview of a coming age of scarcity. I say a

177

preview rather than the real thing because, in reality, there are enormous amounts of energy resources in the world, almost everywhere; it is only because of the sustained and apparently incurable ignorance of politicians in the industrial countries that the oil-price shocks resulted in such high levels of unemployment and inflation. On the other hand, adjusting to a sudden decrease in the availability of nonfuel minerals—when and if that takes place—would involve more than a display of political finesse. It would also require some miracles of abstention by so-called ordinary citizens—miracles that most people have no intention of performing. It is useful to note at this point that some economists claim that our small planet could eventually graduate to a sustainable paradise based on unlimited amounts of iron and aluminum in the crust of the earth. The truth is, however, that such a metamorphosis, if possible, could only be gradual, requiring scores or even hundreds of years. In the short run the adjustment would be painful to many, and more than a few of them would be prepared to use dramatic methods to transfer their misfortunes onto others.

What role, then, can technical change and the price system play in keeping the wolf away from the door? The answer is some, but not enough: as Professor Georgescu-Roegen has correctly observed, the solution to this quandary must be found within the ambit of ethics, and not the pseudoscientific manipulations of neoclassical economic theory. Although it is possible that technology can usher us into a Garden of Eden, what must be understood is that it does so at its own pace and under its own rules. Radar and the Supermarine Spitfire kept the German army away from the population centers of England, and as a result Hitler lost the Battle of Britain. But the French, Poles, Russians, and so on were not so lucky in their military research, although they were just as deserving and probably just as gifted. Moreover, had Hitler won the battle of Britain, such distractions as the North African campaigns would probably not have taken place, and the armies of fascism could have devoted 100 percent of their efforts to the USSR before the United States came into the war. Thus it is possible to argue that the total number of victims of World War II would have been considerably in excess of the 50 million who succumbed to bomb, bullet, and so on between 1939 and 1945. I can see no great undercurrent of technological progress or anything else acting during this period to save anything or anyone who, a priori, might be classified as absolutely worth saving—such as small children. Then as now, there was only the operation of some giant impersonal lottery that was completely beyond the control—and often the interest—of Mr. and Mrs. Consumer.

As for the price system, I have commented on this matter in some detail in my previous work, particularly in *The Political Economy of Oil,* and for that reason will keep my remarks brief here; but there is one thing that every reader needs to understand. The functioning of the price system in the manner prescribed in the elementary textbooks can only take place in the absence of uncertainty, and there are few varieties of uncertainty in all of economic theory that can compare with geological uncertainty. President Reagan was undoubtedly correct when, shortly before his election, he called the United States energy rich; but his confidence in the ability of the price system to bring about the location of enough oil to satisfy the appetites of the U.S. motorist at a reasonable cost was undoubtedly misplaced—as a number of oil-company executives were quick to inform him.

Even if we assume away the absence of geological uncertainty, however, there is still another type of uncertainty that cannot be ignored. This concerns the preferences of future generations and the constraints on their activities. It is generally assumed that the future will be better than the present or the past (which is provably false as a general proposition), and for anyone taking this attitude it hardly makes sense to be concerned with the comings and goings of unborn citizens. The same applies to anyone whose span of concern is a few weeks, a few years, or his or her own lifetime. If we assume, however, that people are genuinely concerned with the future of humanity beyond their own limited life spans, then it is true that prices determined with the aid of discount rates that effectively cut off all except ten or twenty years of the future could be the wrong prices for safeguarding the welfare of this generation, much less the next. (The concept of discount rates is discussed in some detail later in this chapter.)

An Elementary Supply-Demand Model

This section discusses the basic supply-demand model of the elementary textbooks in economic theory. This model takes the form shown in figure 6-1. This is what is known as a *flow model:* supply and demand are flow variables in that their dimensions are units per time period. In addition, we can speak of a pure flow model where d and s refer to current demand and production: d cannot involve goods intended for inventory, nor can any part of s originate in stocks. If we were dealing with copper, the dimension might be metric tons (tonnes)/year. Our equilibrium, where $s^* = d^*$, is the point at which the market is cleared in that, given the equilibrium price p^*, consumers can consume what they desire to consume, and producers can sell what they desire to put on the market. Thus there is no tendency for the price to change because consumers desire more than is available (excess demand), or producers find themselves with unsold goods (excess supply).

Anyone familiar with the copper—or aluminum or zinc—market, however, understands that a pure flow model cannot help us understand price formation on these markets; and we increase clarity only a little if we attempt to introduce other important elements, such as inventories, directly

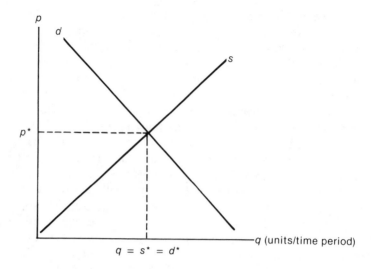

Figure 6–1. A Typical Flow Supply-Demand Scheme

into the flow model. Yet inventories must be present if the model is to correspond to the real world, because in the real world *short-run* prices move up and down in phase with inventory movements. For example, figure 6–2 indicates the movement of copper prices and inventories on the London Metal Exchange (LME). As discussed in Banks (1974, 1979b) there are many indications that inventory movements on the LME can be used as a *proxy* for worldwide inventory movements.

Our flow model will thus have to be augmented. A first step might involve no more than making changes in the price of the commodity a function of inventory changes; in the next chapter—where commodity and natural resource problems are discussed employing some slightly more advanced mathematical techniques—this type of extension will be carried out. At present, however, a simple graphic exposition will be presented that most readers of this book should attempt to follow because it is extremely important.

Diagramatically we have the situation shown in figure 6–3, where the demand for a commodity is divided up into a *stock* or inventory demand, as shown in the figure on the left; and a demand for the commodity as an input into current consumption, or a flow demand, as shown in figure 6–3b. Flow supply *s* is the current output of a commodity, but this output can be used either for satisfying current consumption requirements or for augmenting the stock of the commodity. On the other hand, the stock-supply curve *S* is a datum: it simply says that on such and such an occasion there is a given amount of the commodity in inventory. Though passive in the usual sense

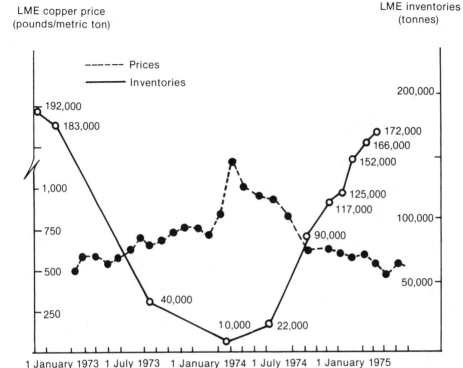

LME copper price
(pounds/metric ton)

LME inventories
(tonnes)

- - - - - Prices

———— Inventories

Source: Reprinted by permission of the publisher from Ferdinand E. Banks *Bauxite and Aluminum: An Introduction to the Economics of Nonfuel Minerals* D.C. Heath and Company, Copyright 1979, D.C. Heath and Company, p. 167.

Figure 6–2. Prices and Inventories on the London Metal Exchange

of the word, this curve is quite important; and movements in it are due to the following mechanism: when flow supply is larger than flow demand, stocks are increasing, and the stock supply curve is moving to the right. Obversely, when flow demand is larger than flow supply, stocks are falling because some of this flow demand is being satisfied by drawing down inventories. In this situation the stock-supply curve is moving to the left. These details should become clearer in the sequel.

Now let us start our investigation of this model with a full equilibrium, which here means both a stock equilibrium (at A, with $S = D$), and a flow equilibrium (at B, with $s = d$). The price to begin with is p_0. The initial level of inventories is S_0. Next we shall assume that for some reason there is an increase in the demand for inventories. Perhaps inventory holders feel there will be a large increase in the demand for this commodity at some point in the future and that as a result its price will escalate upwards. In this case it

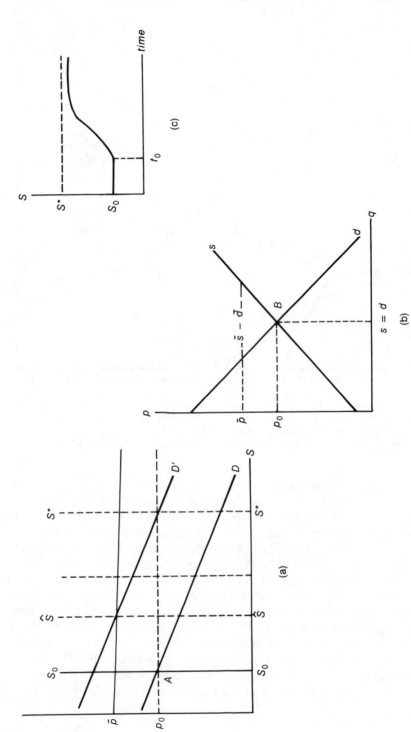

Figure 6-3. A Stock-Flow Model Showing Movement to a New Equilibrium after an Initial Disturbance Resulting in a Shift of the Stock-Demand Curve

would pay to assume the carrying charges associated with larger inventories. This behavior can be illustrated in the figure by moving the stock-demand curve to the right, from D to D'_1. Now, if the price remained at p_0, then the desired stock level would be S^*.

But there is no reason that the price should remain at p_0. We are dealing with an aggregate situation here, so when many buyers inform their suppliers that they require more of a commodity than they usually take during a given period, they are likely to be informed by these suppliers that if they want more of a commodity in the same amount of time (such as a month or a year), then they must pay more. This type of response corresponds to the rising flow-supply curve s shown in figure 6–3b. Naturally some buyers will greet this news by decreasing the rate at which they intend to increase their stocks. Had they, for instance, intended to raise their inventories from 100 to 150 units during a single time period, the news that prices are rising because of a general desire to raise inventories might make them decide to stretch this augmentation in stocks out over two or more periods.

At the same time, however, there are other buyers that would not raise their inventories by the intended amount in one period even if the price of the commodity did not increase as a result of the increased demand. They know that what appears to be one thing today might turn out to be something quite different tomorrow. Thus, they may decide to increase stocks, but they contemplate the entire stock increase taking place in the following period, or later.

These observations then bring us to the concept of an *investment function*. We have an *excess stock demand,* or $X = D - S$ ($S^* - S_0$ in figure 6–3), which refers to a desired increase in (aggregate) inventory holdings rather than to the *rate* at which additions would be made to existing inventories. The rate is provided for by an investment function that tells us by how much the excess stock demand is to be reduced in each period. For example, we might take as an investment function $x = kX$. Then if we have an excess stock demand of $X = 50$ (due to $S_0 = 100$ and $S^* = 150$), and a $k = \frac{1}{2}$, the investment function indicates that in the first period after the increase in the stock demand it is desired to increase inventories by $x = kX = (\frac{1}{2}) \times 50 = 25$ units. If inventories were in fact increased by 25, then in the following period the desired stock increase would be $x = kX = (\frac{1}{2})(S^* - S_1) = (\frac{1}{2})(150 - 125) = 12.5$, and so on. The approach to the new equilibrium, where $S = S^*$, is asymtotic and is shown in figure 6–3c.

Before continuing, the reader should make an effort to understand two significant features of the foregoing analysis. The first is that although investment functions always exist, they may not have the convenient form of the expression used here. Remember that in a market for a typical commodity there are usually a large number of buyers, each of which might have a different type of investment function. Mr. Y might decide to acquire so much of a commodity per period, regardless of its price; more realistic-ally, Ms. Z might be very sensitive to the price at which additional stocks can be obtained and thereby change her order at the slightest movement in the price. A function such as $x = kX$ in the model under discussion here thus represents the aggregate investment behavior of all these inventory holders, and clearly it is too unsophisticated to have any real-world signifi-cance. By the same token, even very straightforward investment behavior might be difficult to generalize. For example, investment behavior calling for an excess stock demand of 50 to be satisfied by acquiring 25 units in the present period and 25 units in the next period does not lend itself to a simple algebraic representation.

It is not important, however, that we have a simple algebraic depiction; nor is it important whether we ever see, or can formulate, actual investment functions. What *is* important is that we understand the concept of invest-ment demand—as explained here or elsewhere; that we can use the graphic apparatus shown in figure 6–3, and in particular can trace through a sequence of the type that will be discussed next; and finally that the concept (or concepts) of equilibrium, which will also be taken up in the sequel, are fully comprehended.

Returning now to figure 6–3, we should appreciate that if the increase in stock demand leads to a price increase of $\bar{p} - p_0$, to $\bar{p}$, then we have excess flow supply in the sense that flow supply becomes larger than flow demand, and the difference $(s - d)$ goes into stocks. As a result the stock-supply curve moves to the right during the first period by an amount equal to the first period excess flow supply, or $s - d$. This is shown in figure 6–3a, where the position of the stock-supply curve at the end of the first period is given by $\hat{S}-\hat{S}$.

What about the following period? Unless the increase in stocks was suf-ficient to bring total stocks up to the desired level, we must once again bring about an excess flow supply. As before, this will take place if the price rises above p_0, which decreases current consumption while increasing current production, thus leaving a margin of production that goes into stocks. This time, however, since many buyers increased their inventories by very large amounts during the previous period, they are less concerned with the speed at which they acquire additional stocks, and therefore would tend to restrain their demand if the price increased sharply. Thus it seems likely that in the second and succeeding periods the price would not reach $\bar{p}$. The reader should now continue the analysis, choosing a price slightly lower than $\bar{p}$; pointing out the excess flow supply; and pointing out how this excess flow supply influences the stock-supply curve. Obviously, employing the logic being developed here, it is possible to continue moving the stock-supply curve until it approaches, asymptotically, the desired inventory level S^*, and the price is back at p_0 and stationary.

Now for our equilibria. The first equilibrium that we must be cognizant of is a *total equilibrium* in which the desired inventory level is equal to the actual level. Referring to figure 6-3, this means that the stock-supply curve has moved to $S^*—S^*$, impelled by periodic surpluses of flow supply over flow demand of the type represented by $s - d$ at price $\bar{p}$. Now observe carefully that at a total equilibrium we have both stock and flow equilibria, because the price has reattained the level p_0 and remains at or very close to this level.

On the other hand, it is possible to specify a *market equilibrium* where, during any given period, supply and demand are equal in the sense that flow supply satisfies both flow demand and investment demand (demand for increased stocks) for that period. If we go to our previous discussion, and the situation existing right after inventory holders decide to increase their stocks, we have a market equilibrium at a price such as $\bar{p}$, where flow supply *exceeds* flow demand; but this excess corresponds to the amount by which buyers wish to increase their stocks during that period, given the price $\bar{p}$.

Some Practical Considerations

We now turn to some practical matters having to do with price formation in the markets for primary commodities. The first topic has to do with specifying exactly which price we are talking about. If, for example, the subject is copper, then the price under discussion is that of refined copper metal; if we are talking about lead, then the relevant price is that of refined lead metal. What about the ores of copper and lead, or for that matter the ores of copper and lead processed into something intermediate between ore and refined metal? These ores also have a price; but because of their lack of homogenity, it is extremely difficult to discuss this price systematically. Consider, for example, copper ore. There are dozens if not hundreds of copper ores, depending on where the ore comes from and its grade; fortunately, however, the refined copper metal that is created from this ore can be described in such a way that the characteristics of the ore are uninteresting. More important, at a given point in time copper metal on a given market carries only one price, which means that the people buying this metal do not have to be concerned with its genealogy.

Another crucial issue is the difference between a posted or a producer price, and a free-market price. A producer price is one that is set on the basis of what sellers (who are usually producers) believe to be the long-run equilibrium price—the price that will equate supply and demand as these develop over the foreseeable future. If we examine this price, what we see is that it often (but not always) moves in steps. On the other hand, a free-market price is usually (but not always) highly responsive to current supply and demand. An example of this latter phenomenon was shown in figure 6-2, where current supply and demand were reflected in inventory movements (with inventories falling as demand increased relative to supply, and vice versa). As the reader can easily verify, price movements are unequivocably synchronized with shifts in market conditions. In the commercial world a free-market price is sometimes called a dealer or merchant price; on metal exchanges such as the London Metal Exchange (LME) and the New York Commodity Exchange (Comex) it corresponds to the spot or cash price. Figure 6-4 shows the producer and dealer (free-market) prices for nickel over the period 1967-1971.

The materials presented here should be examined very carefully, because unfortunately few economists have even an elementary insight into pricing on actual metals and minerals markets. Perhaps this is as it should, since there are some phenomena treated here that are incomprehensible in terms of the analysis presented in elementary courses in economic theory.

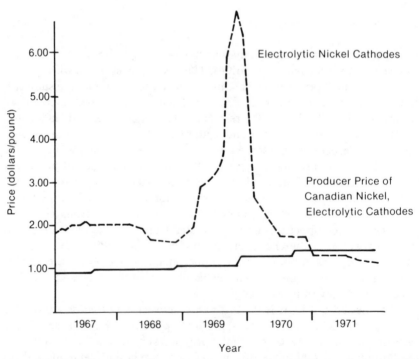

Source: Reprinted by permission of the publisher from Ferdinand E. Banks, *The International Economy: A Modern Approach* (Lexington, Mass.: Lexington Books, D.C. Heath and Company, 1979), p. 50.

Figure 6-4. Producer and Dealer Price for Nickel

How can one explain, for example, why the producer price for nickel remained immobile in the face of the extraordinarily high demand for nickel that prevailed in 1969?

A possible answer to this question is that the management of the most influential nickel firms felt that the free-market price reflected speculation, and that the producer prices they (and their colleagues) agreed on were *the* market clearing (equilibrium) prices, and would eventually be revealed as such. Then, too, during the period shown in figure 6–4 the management of these firms might have reckoned that if they increased the price of nickel to the vicinity of the free-market price, important clients might start looking for substitutes and finding them; in addition, more nickel mines and processing facilities would have appeared on the scene. This latter possibility is always a serious matter in primary commodity industries, although sellers of nickel have not generally revealed themselves to be particularly sensitive to the likelihood of losing their customers to substitutes. On the other hand, producers of copper must always keep this prospect in mind.

Reserves, Resources, and Exploration

The main topic that will be discussed in this section is exploration, but first a few things must be said about reserves and resources. If we look at bauxite, world reserves in 1977 came to almost 25,408 million metric tons (tonnes), which corresponds to 5,081 metric tons of aluminum. Accordingly, with a production rate of 75.12 million tonnes in 1977, these reserves would last for 297 years if this production rate was constant over time. Had bauxite consumption continued to grow at the trend rate of 9 percent/year, however, these reserves would only have lasted 37 years. This last figure is called the dynamic reserve/production (R/P) ratio.

World bauxite reserves are constantly being adjusted upward. These were estimated at one billion long tons (L-T) in 1945, 3 billion in 1955, 6 billion in 1965, and 25 billion in 1977. The jump between 1965 and 1977 can be accounted for by the increased amount of exploration taking place in Australia, Brazil, and the west coast of Africa. Everything considered, however, one must ask whether it is sensible to undergo large expenditures to search for more reserves just now. A dollar invested in a bond or bank deposit at an interest rate of 10 percent (which is probably a typical interest rate on a long-term deposit just now) would yield $(\$1 + \$0.1)^{35} = \$28$ after 35 years, assuming the dynamic R/P ratio to be 35 years. Consequently, in order to justify investing an extra dollar in creating another unit of bauxite reserves, the profit in money terms from selling these reserves would have to amount to $28. Given the recent movement of bauxite prices relative to the cost of extracting and selling bauxite, this may be asking too much. According to some economists, the ideal figure for the dynamic production/reserve ratio is 12–15 years: there is too much uncertainty associated with a time horizon longer than this, whereas if the R/P ratio is lower, there is a risk that the capacity of the bauxite extraction and processing equipment might be underutilized.

Some definitions follow. The expression *resources* includes all of a certain ore in or on the crust of the earth whose extraction is currently or potentially feasible. *Reserves,* on the other hand, are a subset of resources— those resources that are identified and whose extraction is economically justifiable. This last statement underlines the importance of the price at which an ore is sold, and the cost of obtaining it. For instance, in 1970 the U.S. Bureau of Mines listed U.S. reserves of copper as 164 million tonnes at a copper (metal) price of $2/pound—assuming a 12-percent required return on the capital invested in mining, concentrating, smelting, and refining this ore. On the other hand, with the same yield on invested capital but a copper price of only 50 cents/pound, reserves amounted to 76 million tonnes. This calculation takes into consideration the wage level existing in 1970 and also the main by-products found in copper ore, such as gold and molybdenum.

One of the most useful devices for discussing this subject is the classification system developed by the U.S. Geological Survey. This is shown in figure 6–5, and it functions as follows. Along the horizontal axis total supplies are listed, with the basic categories here being discovered and undiscovered supplies. These are described with respect to the information avail-

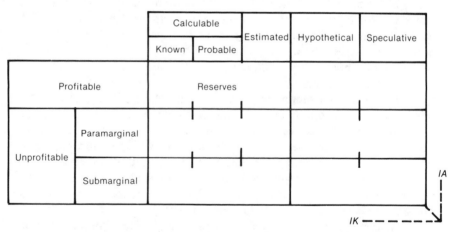

Source: U.S. Geological Service. Reprinted by permission of the publisher from Ferdinand E. Banks, *Bauxite and Aluminum: An Introduction to the Economics of Nonfuel Minerals* (Lexington, Mass.: Lexington Books, D.C. Heath and Company, 1979), p. 10.

Note: 1A: Increased availability. 1K: Increased knowledge.

Figure 6–5. The Classification of Mineral Supplies

able about them. On the vertical axis the variable of interest is economic value, and here the natural partition is between profitable and unprofitable. Thus in the upper left-hand corner of the system we can identify reserves as supplies that definitely or probably exist and are profitable to extract.

Now we can get down to the topic of exploration. The first thing to understand here is that there is no systematic theory of exploration to which we can refer. It is possible, however, to think in terms of a strategy of exploration that, though popular and widely used by mining companies, may be far from optimal from the social point of view in that there is a great deal of duplication of effort. This is perhaps best seen at the first stage of a typical multistage program, where the principal activities consist of a literature search, study of existing maps, preliminary field reconnaissance, selection of potential targets, and a preliminary evaluation of costs. The individual firm carrying out these routines might eventually realize some benefits, but since much of the information generated here can be found in the archives of other enterprises, a great deal of social waste is clearly involved.

The second stage involves detailed mapping, to include geologic, geo-chemical, and geophysical delineation of the targets. This is followed by a testing of the potential ore bodies with the aid of drilling and trenching operations. The third stage includes detailed drilling, underground explora-tion, the estimation of the most important parameters of the ore body—such as grade, tonnage, and configuration—and so on. This would be fol-lowed by a preliminary mine feasibility study designed to determine mining methods and costs. At this stage core samples would be given a metallurgi-cal examination.

The last stage involves a complete economic analysis of the operation of the mine, in the course of which it is determined whether the operation should actually be undertaken, at what level, and for what duration. Each of these stages involves certain costs, and since these costs increase progres-sively, a determination of the benefits that might be realized from a poten-tial investment is made at the end of each stage. If these benefits do not appear to be high enough, then the exploration program is terminated at that point, and the project is dropped. In general these programs do not go beyond the first stage, where costs can be held to the vicinity of a few hun-dred thousand dollars.

As would be expected, once an area has been designated a target area for geological purposes or has been deemed to possess some characteristic that goes along with minable deposits of an important natural resource, then it is inevitable that it is reexamined from time to time to ascertain whether, in the light of the most up-to-date technology and expected demand for the resource, its potential can be upgraded. Until recently the mineral nepheline—whose presence and location were well documented in connection with exploratory work involving other substances—was considered to be of little or no value. Now, however, a technique has been developed for using it as a source of aluminum, so it has been classified a valuable raw material. The same thing is true of taconites. Today they are an important source of iron ore, but before World War II there was no economical technique for extracting ore from these rocks. Reassessments of this nature will undoubtedly continue and, to a certain extent, counterbalance some of the more pessimistic forecasts of the depletion of mineral resources.

Now let us examine some information on aggregate exploratory costs. Canadian expenditure on mineral exploration increased by a factor of about seven from 1945 to 1970, and per-unit exploratory costs are climbing faster than ever—apparently doubling for each ton of metal produced. On the whole, however, the number of ore discoveries appears to be on a downward trend. The latest figures reveal that the average cost of a discovery rose from about $2 million in the period 1945–1955 to about $15 million by 1976. On the other hand, the value of an average discovery went from $245 million in 1954–1956 to $711 million in 1970, thus providing some compensation for the increase in discovery costs. A worldwide survey of mining companies shows that it is not unusual for it to take up to ten years to locate a commercially viable prospect; in Australia a study based on forty mines revealed that for the period 1958–1979, the average cost in 1975 dollars of making a discovery that resulted in a mine came to between $30 and $35 million. Bosson and Varon (1977) have cited the case of a major Canadian firm that spent $30 million on exploration. Over a thousand properties were examined, and of these only seven became the sites of profitable mines.

There is no evidence that diminishing returns to exploration have arrived in Canada; but as one can judge from the previous exposition, there may no longer be any basis for the kind of optimism that claims that the mining sector will be able to meet any and all demands placed on it over an indefinite future. In the United States exploration has seen an increase in the number of personnel involved, as well as a very large increase in the value of equipment put at the disposal of each of these individuals. The number of discoveries per year has not shown a long-run tendency to either increase or decrease, but the cost of finding a given amount of minerals is at least twice as much today as it was in 1955–1959, and is increasing rapidly. The problem here is that the continental United States has been thoroughly explored, and it is hardly likely that previous searches have missed many large deposits. This does not mean, however, that it will be impossible to increase annual production during the remainder of this century. As pointed out earlier, scientific progress should continue to function in such a manner as to increase the exploitability of many known deposits, and this includes an increasing number of those currently regarded as subeconomic.

Australia and southern Africa (to include Namibia) may deviate somewhat from this scenario. The Australian case is particularly interesting because important discoveries seem to take place in that country somewhat more often than elsewhere, and many people in the Australian mining industry feel that intensifying the efforts put into exploration would result in the discovery of a great deal of additional mineral wealth. Exploration costs in Australia are lower than in Canada and the United States, although exploration expenditures in Australia are accelerating. These expenditures have paralleled a rapid expansion in the supply of Australian minerals on world markets, which in turn resulted in a period of rapid economic growth and ascending levels of per-capita consumption for Australia. Recently, however, there has been a slowing down in these trends as the Australian economy reacted to the decline in momentum of the world economy, which among other things entailed decreased requirements of industrial raw materials.

The search for minerals is becoming increasingly sophisticated, and the computer is playing an increasingly important role in this activity. Whether it is playing a more important role than luck is difficult to say, however, particularly when we view the unevenness with which mineral resources are spread over the globe. Most of the world's metallic minerals are found on huge blocks or shields that have been stable components of the earth's geographical structure for almost 400 million years. In Paleozoic times the shields found in the highlands of Brazil, Africa, and Western Australia were conjoined, and the name that geographers gave this union was Gondwanaland. The breaking up of Gondwanaland is sometimes associated with the phenomena called *continental drift*. One of the reasons this process is so interesting is that, given the mineral riches of Brazil and Australia, it could be suspected that Antarctica is one of the most attractive regions in the world from a geological point of view, since it was sandwiched in between the other two during the Gondwanaland period. Other shields that deserve to be noticed here are the Precambrian Shield south of the Amazon River, and the Arabian Shield, which has many similarities to those of Canada and Australia. This shield is about 1,200 miles long and 1,000 miles wide, covers a large part of Saudi Arabia, and stretches across the Red Sea into Egypt. It contains at least twenty minerals, located in more than three-hundred places. These include iron, silver, gold, copper, and phosphates.

Mineral Processing

Industrial raw materials pass through a number of stages between their mining and use by final consumers. The mining itself takes place in underground or surface (open-pit or open-cut) installations, and in addition to ore removal we have such operations as separating, blasting, and loading.

Next comes the first stage of processing, which features such activities as crushing, grinding, and concentrating. With some commodities (such as uranium) this stage is called *milling*. The next stages are smelting and refining, each of which raises the purity of the particular commodity. These are two distinct engineering and metallurgical operations, as with copper; but in the case of aluminum there is only one operation, which some people call smelting and others call refining. Smelting is probably the correct description.

At this point we have a product that is almost a pure metal, and because of its homogenity is more important for economists than any preceding or succeeding stage of processing. As shown in figure 6-6, the next stage in processing is called *semifabrication*. Here the bars, ingots, and so on that come from the refinery (or smelter) are transformed into sheets, tubes, and various structural shapes. Some people would now list fabrication as the next step of the cycle, and for completeness I have done so in the figure; but it could be argued that the finished products resulting from this stage are so heterogeneous that logically they have no place in this part of our discussion.

One of the most interesting characteristics of this cycle is the increase in *value added* that takes place at each stage. This topic is extremely important, especially for countries whose overall economic development is dependent on minerals; it has already been discussed in chapter 2 in association with oil. Nonetheless, this concept will also be treated here, though in a more fundamental setting. To begin with, it is necessary to distinguish between two categories of inputs. First, there are primary inputs, comprising labor, capital, and land. These are primary in that they (along with minerals) form the very substructure of production. Until the reports of the Club of Rome and the first oil price shock, however, it was customary to ignore minerals, assuming instead that they would always be available when needed, and at reasonable prices.

The other category of input consists of natural resource and energy inputs, semifabricates, and other *intermediate inputs* or *intermediate goods*. To distinguish fully between these inputs, let us consider the following simple example. A superwoman, employing her bare hands, removes 100 bauxite rocks from a small mine that she inherited. She then sells these rocks to a man operating a processing plant directly adjacent to her property—so near, in fact, that there is no transportation cost between the two. She obtains $1 per rock. This processing plant, using only machines and labor, turns these rocks into 25 units of a valuable material called *aluminum*. Let us now assume that these 25 units of aluminum sell for $1,100.

With this information we can identify and discuss value added. For the superwoman the input is muscular exertion, and the physical output is 100 bauxite rocks/day. The output in value terms is $100/day, and since there

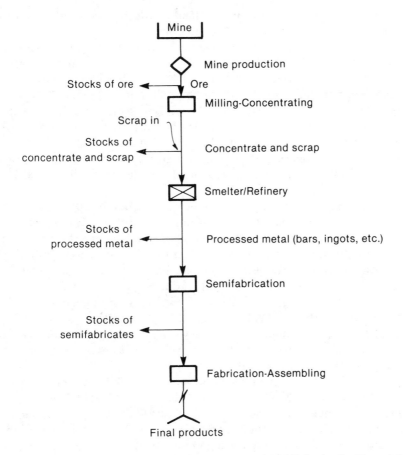

Source: Reprinted by permission of the publisher from Ferdinand E. Banks, *Bauxite and Aluminum* (Lexington, Mass.: Lexington Books, D.C. Heath and Company, 1979), p. 153.

Figure 6-6. Mining-Refining-Final-Use Cycle for a Typical Industrial Raw Material

are no intermediate inputs at this stage, the value added is \$100/day. An equivalent expression for value added is, then, return to the primary factors of production.

On the other hand, bauxite is an intermediate good for the aluminum processor. As given, the monetary return realized by the processor is \$1,100, but since \$100 was paid for the bauxite, the net output or value added by processing the bauxite into aluminum was \$1,000. As before, this \$1,000 can be identified as payments to the primary factors used to produce the aluminum, which in this case would be the wages and salaries of the owner of the processor and his employees, the rents paid the owners of the capital used in the installation and the land on which it was located, and any profits that might have been realized in selling the aluminum.

Actual aluminum-processing operations call for turning bauxite into *alumina,* and alumina into aluminum. In reaching aluminum, value added is multiplied up by almost a factor of 100: by 10 in the processing of bauxite to alumina, and once again by 10 in the processing of alumina into aluminum. This is why many less developed countries (as well as some developed ones) reject the role of merely supplying world markets with unprocessed raw materials. At the same time, however, it should be made clear that many mining companies are quite content to fill this latter function, since the *rate of profit* from mining has generally tended to be higher than that from processing minerals.

We can now pick up the principal theme being developed in this section. So far we have examined the disposition of so-called *primary* materials: inputs to processing activities that originated in the mining stage or, possibly, in inventories of ores. There is another source of industrial raw materials, however, in the form of shavings and similar unusables that are generated in processing and fabricating operations, which is called *new* scrap; and in the reprocessing of discarded or obsolete finished goods containing the particular materials. Scrap in general is called *secondary metal* in the case of metals, or secondary materials when we do not limit ourselves to metals (for example, paper can also be recycled). The input of scrap into the metal-producing cycle is also shown in figure 6-6.

In the United States about 140 million tons of solid waste is discarded every year, to include 3 billion pounds of aluminum (equal to the entire consumption of aluminum in the United States in 1958). When one remembers that recycling aluminum requires only about 5 percent of the energy input per unit used to produce primary aluminum, it becomes clear that recovering this metal saves 20 billion kilowatt-hours of electricity, which is 1 percent of the energy used in the United States in a year. At present the recycling of old and new aluminum scrap yields less than 20 percent of the yearly consumption of aluminum metal in the United States, although in theory recycling is capable of providing a much larger fraction of annual aluminum requirements. By way of contrast, 40 percent of the total copper consumption of the noncentrally planned economies has its origin in secondary copper.

The next topic is the disposition of an industrial raw material. Here we must not only employ our knowledge of the different phases that a material goes through in being processed from ore to metal, but must also be aware that the various activities making up the cycle can take place in different parts of the world. In the case of copper about 40 percent of ore production takes place in the four largest of the CIPEC countries (Chile, Zambia,

Zaire, and Peru), which at present smelt and refine about 50 percent of their output. Much of the remainder is smelted and refined in the United States, Japan, and the more highly industrialized countries of Europe. Semifabrication is hardly practiced at all in the less developed countries, apparently because most of the items produced at the semifabricating stage have to meet very precise specifications drawn up by their purchasers.

Before getting down to the details of this matter, let us peruse some aspects of the production cycle of copper (which is still regarded as the most important metal). First we have mining, which results in the production of ore with a copper content (or ore grade) of 0.5–2.5 percent. The next phase, milling or concentrating, results in a concentrate containing 30 percent copper. The concentrate is then roasted to remove sulphur, and smelted in a reverbatory furnace in order to obtain a *matte* containing about 32 percent copper. A converter then transforms the matte into blister copper with a purity of about 99 percent, at which point refining begins. This involves fire refining (in a refining furnace) and electrolytic refining. These are separate proceedings, however; unless some special properties are required for the copper, the refining process might be terminated after it is fire refined. Finally, the refined copper goes to brass mills, wire mills, and so on, where it is turned into such things as sheets, tubes, rods, and wire.

At this point an example will be constructed showing how the output of a copper-producing country might be disposed of. The figures used in table 6–1 resemble, but are not the same as, those for Peru in 1968. I have modified the original figures in order to eliminate some blank entries and to make it easier for readers to check the calculations.

A few clarifying remarks about this table are in order. First, we should understand that the figures given here signify copper content. If the ore grade in the country shown was 1 percent, then in order to get 200,000 tonnes of copper, 200,000/0.01 = 20,000,000 tonnes of material must be mined and treated. Observe also that with exports of ore equal to 4,000 tonnes, and stocks of ore increased by 1,000 tonnes, then 195,000 tonnes of the 200,000 tonnes of ore that were mined move to the first stage of processing; if there are no processing losses (which is unlikely in the real world), this 195,000 tonnes emerges as concentrates. The reader should note these entries in the table.

One of the more interesting aspects of this situation is the comparatively large exports of ore and concentrates. It is difficult to deduce the economic motivation for these transactions, since generally it is a very uneconomic matter to export ores—and the same can be said, though with less emphasis, about the exporting of concentrates. A possible explanation could be that there was a shortage of milling and smelting capacity in the exporting country, whereas there was an excess capacity in the country

Table 6–1
The Disposition of Copper Ore and Processed Products

Stage of Processing	Production	Exports	Export Destination
Ore	200,000	2,000	Japan
		1,000	Sweden
	1,000	1,000	Others
Stocks	1,000		
Exports	4,000		
Concentrates	195,000	17,000	Japan
		4,500	United States
		2,000	Sweden
		1,000	Others
Stocks	500		
Exports	24,500	80,000	United States
Blister	170,000	25,000	Belgium
		20,000	West Germany
		10,000	Others
Stocks	5,000		
Exports	135,000		
Refined copper	30,000	15,000	United States
		3,000	Japan
		2,000	Others
Stocks	2,000		
Exports	20,000		
Semimanufactures	8,000	100	Denmark
		100	Others

Note: Residual $= 8,000 - 200 = 7,800$ units to domestic use + stocks + unknown.

importing the ore and concentrates. Thus the price of ores and concentrates might have been very low, whereas excess capacity on the processing side would imply low processing costs.

Finally, a word of caution about inventory statistics: at all stages shown in the exercise, stocks have increased. The opposite could have been possible, however: copper could have been taken out of inventory and either exported or processed. If the amount of material coming out of stock exceeds that going in, then the stock change shown in the table would have carried a *plus* sign. For example, if 5,000 tonnes of blister copper had come out of stock instead of going in, then the input to the refining stage would have been 35,000 tonnes (assuming no processing losses).

Now we can look at the situation for a country that is a major consumer of industrial raw materials. The United States is of course one of the largest processors of ores and concentrates, and one of the largest consumers of industrial metals, so here we will examine supply-demand relationships for the U.S. aluminum industry. First, however, a few introductory remarks are in order.

The production of aluminum involves three major activities: first, the mining of bauxite, followed by some preliminary processing at or near the mine, whose purpose is to avoid the transport of large amounts of rock or earth containing only small quantities of the mineral. Next the bauxite is processed into alumina; finally, this alumina is smelted into aluminum. Usually it takes four to six tons of bauxite to produce two tons of alumina, and two tons of alumina to obtain one ton of aluminum. By taking into consideration the grade of bauxite purchased, we find that for the United States (on the average) x short tons of bauxite can be turned into its aluminum equivalent by dividing by 4.46. For example, in figure 6-7 the 3,330 short tons of bauxite imports are in terms of its aluminum equivalent. In terms of bauxite these imports amount to $3,330 \times 4.46 = 14,852$ short tons of bauxite. Now let us examine the aluminum supply-demand position of the United States in 1976.

The thing to be noted in figure 6-7 is the disposition of the supply of

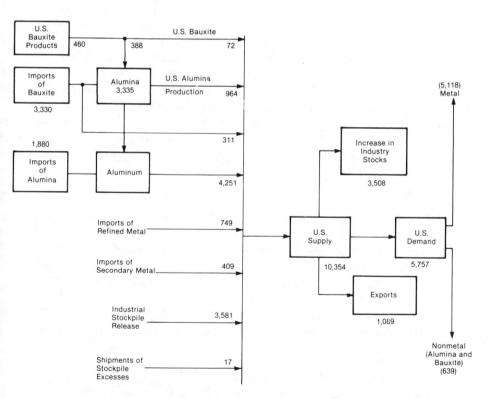

Figure 6-7. The Aluminum Supply-Demand Position of the United States, 1976

aluminum. A large part of it (3,508 short tons) goes into inventories, whereas 1,089 short tons are exported. The demand, or consumption, in the United States then comes to supply − change in stocks − exports = 5,757 short tons. This is an *aluminum-equivalent* figure, however, and includes both metal and nonmetals (alumina and bauxite).

The Trade in Primary Commodities

The large-scale consumption of natural resources seems to have begun somewhere in the middle of the industrial revolution and accelerated as technical change made it possible to produce ever-increasing amounts at decreasing *real* costs. During the last seven decades, the global demand for minerals of all types has increased by a factor of almost 13, as compared to an increase in population of 240 percent.

There have also been marked shifts in the world pattern of supply and demand. In the period from 1700 to 1900, England mined a large part of the world supply of lead, copper, and iron ore. Similarly, in the late 1920s, France was producing a quarter of the world's iron ore and 30 percent of its bauxite. Another 18 to 19 percent originated in Hungary. At present, however, these countries produce only minor quantities of various raw materials, although France still has some bauxite in metropolitan France and a great deal of nickel in New Caledonia. The chief suppliers of nonfuel minerals at present are the United States, Canada, Australia, the Republic of South Africa, and the USSR, but it seems likely that Brazil will soon join this list. On the demand side Japan has emerged as a huge market for raw materials of all kinds.

If we look at some trend growth rates, we see that the escalation in demand for various industrial raw materials has been interrupted by the world recession that began in 1973. Another factor of some interest just now is the tendency toward stagnation in the production of some nonfuel minerals by countries like the United States. In the year 1900, the United States produced about one-third of the world supply of the most important minerals, including petroleum, and by 1978 this share was cut by one-half. It also appears that LDCs are increasing their production of minerals at a faster rate than the developed market economies, though not so fast as the centrally planned economies, especially the USSR. In fact, the centrally planned economies accounted for 26 percent of total world mineral output in 1970, as compared to 14 percent in 1950.

Most of the growth in world mineral consumption has taken place in the developed market economies, although there has been a substantial growth in total and per capita consumption in the centrally planned countries in the past decade. In 1970, the LDCs were producing about 35 percent of the world mineral output but consuming no more than 6 percent. Many economists think that this arrangement will not change very much during the rest of this century, although individual countries like Brazil and perhaps some of the OPEC countries might raise their input of nonfuel minerals by a large amount.

In the matter of trade, the developed market economies have steadily increased the amount of minerals they are buying from other parts of the world. In 1974, they imported about 40 percent of their requirements, as compared to 33 percent in 1950. North America seems to have reduced its import requirements, but this can be explained by the large increase in Canadian exports to the United States. On the other hand, Western Germany and Japan have stepped up their imports of minerals; and if the present trend is not broken by a further deterioration in the international business cycle, Western Europe may soon be importing more than 80 percent of its nonfuel mineral supplies.

In the centrally planned economies, production is approximately equal to consumption, and there is no large amount of trade between this group of countries and the rest of the world. Whether this is an ideal arrangement remains to be seen, since the USSR and perhaps China possess considerable mineral resources and are potentially major exporters, although in the beginning China may import some raw material. The USSR may, in fact, occupy about the same place in the scheme of things as the United States in the recent past, in that it has a high degree of self-sufficiency that could increase as it perfects its extraction technology. Before continuing, we can look at some production and growth-rate statistics for the most important nonfuel minerals (see table 6-2).

Of the minerals listed in this table, only the reserves of nickel and tin are being depleted faster in the industrialized countries than in the LDCs. Copper and lead are being depleted at about the same rate in both groups of countries, while bauxite, zinc, and manganese are being depleted at a faster rate in the LDCs. Naturally, in the long run the issue is not reserves but resources—or hypothetical and speculative reserves—and technology. What these will mean for patterns of production and trade cannot be ascertained now, but it seems reasonable to assume that things should continue about as they are now during the rest of this century, with the possible exception of the emergence of Brazil as a mineral-exporting power of great significance.

The Production and Processing of Ores

Prior to continuing the above discussion, we must once more consider the production (that is, mining) of various ores and their processing. As it stands, the market economies and the centrally planned economies process

Table 6-2
Total Value of Nonfuel Mineral Production (1973)
(in millions of dollars)

	Total Value of Production (Millions of Dollars)	Average Yearly Growth of Production (%) 1950–1973	Most Important Mineral
1. Soviet Union	6800	7.0	—
2. United States	5100	4.0	—
3. Canada	3800	3.9	Copper, nickel, gold
4. South Africa	3800	6.5	Gold
5. Australia[a]	1700	8.8	Iron, bauxite, copper, nickel
6. Chile	1200	—	Copper
7. China	1200	—	Iron, copper, tin
8. Zambia	1000	—	Copper
9. Zaire	870	—	Copper
10. Peru	690	—	Copper
11. Brazil	640	—	Iron
12. Mexico	610	—	Copper, zinc, iron
13. France	490	—	—
14. India	420	—	Iron
15. The Phillippines	410	—	—
16. West Germany	400	—	—
17. Sweden	390	—	Iron
18. Poland	390	—	—
19. Japan	380	5.7	—
20. Yugoslavia	360	—	—
Others	6050	—	—

Source: USBM and World Bank documents. Various issues of the OECD biannual economic report. Also OECD country studies. Ferdinand E. Banks, *Scarcity, Energy, and Economic Progress* (Lexington, Mass.: Lexington Books, D.C. Heath and Company, 1977).
[a]Including Papua New Guinea.

most of their domestically extracted ores, while the LDCs ship a great deal of ores and concentrates abroad for transformation into metals. The big exceptions are the Republic of South Africa and Australia, which, while fully developed market economies, are huge suppliers of unprocessed or partially processed minerals to the industrial world, in particular Japan and Western Europe. Taken together, these countries process only 38 percent of their mined output. (By way of contrast, Japan processes 1046 percent; Western Europe, 295 percent; the United States and Canada, 179 percent;

and the centrally planned countries, 108 percent. The LDCs process 29 percent.)

The percentage of ore processed varies from country to country and from metal to metal. Between 75 and 80 percent of the tin mined in LDCs is processed locally. Next in line seems to be copper; but the problem here is that copper undergoes several stages of processing, and getting an exact figure for the processing quotient is a tricky job. Then we have lead, nickel, zinc, iron ore, phosphorus, and bauxite. Only 10 percent of the bauxite mined in LDCs is processed in the Third World, but here there is a great deal of talk about joint ventures involving several countries in processing operations. One possibility that has been named is Jamaican bauxite combined with Mexican or Venezuelan oil, since the processing of bauxite is extremely energy intensive.

The matter of who should and who should not process ore will be touched on below, but a number of factors must be considered before clearcut decisions can be made. Australia, which is a rich country, still finds it profitable to ship a great deal of ore to Japan for processing rather than building the facilities needed to perform these operations domestically. If one inquires into this, one learns that Australian mining firms still feel that it is more profitable to invest in extraction than in processing, even given the uncertain state of world mineral markets. There is no point in arguing against the right of companies to invest in projects offering high rather than low profit rates; however, Australia is a high-unemployment country and needs the employment opportunities that are associated with processing industries. It thus becomes obvious that some of the money the Australian government is using to finance the dole should be used to subsidize employment in processing (and other) industries. As should be evident, this is equivalent to raising the profit rate, since at a given level of revenue it reduces employer expenditure.

Where LDCs are concerned, it may be true that there are a number of countries in which there are many potential investments with a higher yield than alumina production, smelting, and refining—although, as pointed out earlier, this is probably not true in countries producing oil. On the other hand, once the social profit from on-the-job training and such intangibles as inculcating the correct work habits in people demoralized by long periods of unemployment are taken into account, a great deal more processing is probably justified in all mineral-producing LDCs.

The Trade in Primary Commodities

In 1970 the United States, Japan, and Western Europe imported about 60 percent of their requirements of the nine most important nonfuel minerals.

This figure was about 55 percent in 1960 and between 40 and 44 percent in 1950, according to U.S. Department of Interior statistics. Approximately 12 percent, by value, of these imports came from Australia, 16 percent from Canada, and 80 percent from the LDCs. Among the LDCs, Brazil seems to be on the verge of attaining a special position, and it is said in Australia that Brazilian exports of bauxite and iron ore may threaten the export revenue that Australia obtains from these commodities in the near future.

There is also the possibility that the USSR, and even China, will take a more active part in world nonfuel mineral trade. During recent years the USSR has supplied 32 percent of the U.S. imports of chromium, 32 percent of its imports of platinum, and 19 percent of its imports of titanium. Similarly, given the attitude of the present Chinese government toward importing foreign technology, China may eventually find itself exporting larger amounts of minerals in order to pay for this technology. It should be remembered in this respect that Japan gets more than 90 percent of its mineral supplies from overseas, and unless there is a radical setback in the world economy, it seems likely that Japanese requirements will grow. (Of the Japanese imports of nonfuel minerals, 40 percent come from North and South America, 15 to 20 percent from Europe and Africa, 20 to 25 percent from Asia, and 15 to 20 percent from Australasia.) If this is the case, it is possible that in the not too distant future Japan will attempt to enter into the kind of joint ventures with China that have been so successful elsewhere. Depending on the size of these arrangements, they could have a profound effect on the development of the Chinese economy. Table 6-3 gives an approximation of nonfuel mineral flows on the world markets in 1970.

Table 6-3
Nonfuel-Minerals Trade Flows for 1970
(percentages)

	Exports		Imports	
	Ores and Concentrates[a]	Metals[b]	Ores and Concentrates[a]	Metals[b]
Developed Market Economies	58	65	94	90
LDCs	41	31	2	8
Centrally Planned Economies	1	4	4	2
Total Value[c]	4.2	10.47	6.1	10.5

Source: Estimated from United Nations World Economic Survey.
[a]Iron, bauxite, copper nickel, manganese, zinc, chrome, and lead ores.
[b]Copper, tin, nickel, lead, zinc, aluminum.
[c]Billions of dollars.

By way of clarifying the world trade in nonfuel minerals, a closer look can be taken at the world market for iron ore. As indicated earlier, self-sufficiency in this commodity is diminishing rapidly for the major industrial countries. The United States produced 95 percent of its requirements in 1950 and was down to 69 percent in 1970—although it is not inconceivable that this could be raised in the future. Similarly, the European Economic Community (EEC) dropped from 79 to 49 percent during this period. Trade has progressively become more important, and it is interesting to see the character it has taken.

In 1960, 42 percent of all iron ore traded originated in captive mines, or mines that were controlled to one extent or another by steel companies. Some 19 percent was traded via the medium of long-term contracts, and the remainder was traded on what is sometimes called the free market. It should be noted, however, that the free market in this case is only a facsimile of a textbook free market, where a commodity is sold openly, with large numbers of buyers, and all concerned in full possession of every scrap of information pertaining to the future disposal of the commodity. A typical free market deal might involve Swedish sellers and Western European buyers, with the details of the transaction specified on a short-term contract running for one year.

In the case of long-term contracts, running times of twenty years have been known, although the usual period is ten to twelve years. Generally, these contracts give the quantities to be delivered annually, allowing a margin of flexibility of up to 10 percent in favor of the buyer. Prices are also specified long in advance of delivery, although in practice there is considerable room for adjusting these around the time of delivery and sometimes, on this type of contract, there is a fixed element in addition to a component that can be negotiated. For the Japanese these contracts provided considerable security until the oil price rises of 1973–1974, when some of the suppliers of iron ore found themselves in difficulty. Many of the prices appearing on these contracts were then revised upward; but recently this practice appears to have been discontinued.

Last, but not least, we come to captive mines, or mines that are linked to iron and steel makers by ownership. One of the traditional advantages of ownership is that it reduces uncertainty for the managers of processing facilities. This still holds true in the United States, where steel companies obtain about four-fifths of their ore through what are essentially intrafirm transactions. Much of the movement of ore is from Canada to the United States and involves Canadian mines that are owned by American companies.

As things have developed, the nationalization of mines in Venezuela, Peru, and Mauritania, as well as the advantageous aspects of noncaptive arrangements, has dulled the taste of many steel firms for ownership ties abroad. Actually, from most points of view, these ties are unnecessary. Risk can be spread much more efficiently through consortia or joint ventures act-ing through the medium of long-term contracts. In addition, there is so much iron ore in the world as to make it unlikely that the market power of suppliers will overwhelm that of buyers in the near future. The arrange-ments described above can be summarized in table 6–4.

We can conclude this section by looking at some projections of iron ore demand. Remembering that the demand for iron ore is a derived demand that follows from the demand for steel, we see that the estimates of the International Iron and Steel Institute in Brussels indicate that world steel production is headed for 1.1 billion tons in 1985 and a minimum of 1.7 bil-lion tons by the year 2000, with the estimate for the year 2000 based on an assumed growth rate of steel production of 3 to 4 percent in the period between 1985 and 2000.

UNIDO (United Nations Industrial Development Organization) esti-mates are in about the same range, predicting a world consumption of steel equal to 1.065 billion tons by 1985 and between 1.66 and 1.93 tons by the year 2000. In the UNIDO scenario the share of the LDCs is expected to rise from 12 percent in 1985 to 23 percent in the year 2000; and the developed market economies are, again according to the UNIDO picture, scheduled to take a big drop. It should be remembered, though, that UNIDO prognosti-cations are closely tied to the new international economic order mythology; and so, unless there is a gigantic upswing in the world economy, they are not worth the paper they are printed on. In fact, should the developed market economies find some way of getting their costs under control, there would no longer be a reason for Europe and North America to consider a further contraction of their steel industries, since in both theory and fact they are capable of constructing the most modern installations in the world. At the

Table 6–4
Purchasing Arrangements for Iron Ore, 1968

Area 1968	Captive Mines (%)	Long-Term Contracts (%)	Free Market (%)
United States	96	—	4
Japan	—	96	4
United Kingdom and EEC	31	—	69
Eastern Europe	—	87	13
World	30	36	34
World (1960)	42	19	39

same time, it seems clear that several countries in the Third World, such as Brazil and Mexico, are capable of greatly increasing their *output* of steel.

Approximately 1.2 tons of iron ore is required to produce 1 ton of steel. On the basis of UNIDO projections, this means that the demand for iron ore would rise from 780 million tons in 1975 to 1280 million tons by 1985 and 2150 million tons by the turn of the century. The question that must now be asked is: Just how are these increases in capacity to be financed, particularly since the present inflation rate indicates that the capital costs referred to above will double or triple in the next two decades? If we stick by the above estimates, either some potent price increases are necessary for iron ore (and this would be surprising considering the price trend over the past fifteen years) or the Euromarket must surpass itself in irresponsible lending to the producers of iron ore, which could easily happen. Another possibility, though, is that energy uncertainties will remain to the extent that the projections given above will be progressively scaled down.

Everything considered, the opinion here is that the supply of ore will continue to expand at such a rate as to obviate price rises. Moreover, at this time there is no evidence that, with the exception of Brazil, global trade flows will deviate greatly from those shown in table 6-5.

Table 6–5
Estimated Trade Matrix for Iron Ore, 1975
(millions of metric tons)

	Exports to				
Exports from	*United States*	*Japan*	*Western Europe*	*Others*	*Total Exports*
Australia	1	62	9	—	72
Brazil	2	25	24	6	57
Canada	22	7	13	—	42
Sweden	—	—	30	1	31
India	—	18	3	5	26
Liberia	3	—	19	2	24
Venezuela	16	—	4	—	20
Chile and Peru	3	12	1	—	16
Mauritania and Angola	—	4	9	—	13
Other Africa	—	3	7	1	11
Other Asia	—	3	—	2	5
Total Imports	47	135	122	17	321

Commodity Agreements

The talk about commodity policies and commodity agreements was exactly that, talk, until the OPEC breakthrough in the fall of 1973. A group of Third World countries, keeping their own counsel and pointedly without the approval of the developed market economies, initiated what has turned out to be a remarkable redistribution of world income, and power.

As made clear earlier, the rise in oil prices led to dramatic price increases for many of the nonfuel minerals, as the major consumers of these commodities expanded their inventories at record speed. The assumption here was that if the oil producers could do it, so could the producers of copper, bauxite, and iron ore, and perhaps some others. In the wake of OPEC a number of producers' associations appeared, or reappeared; and by 1977 there were producers' associations for no less than fourteen commodities. At the same time, only a small number of economists believed that the feats of OPEC could be duplicated by other primary-commodity producers on a large scale; and of these only a handful did not have something to gain personally in the way of consulting work with the United Nations or invitations to some long-winded palaver in exotic surroundings paid for by the organizations working for the "poor countries."

At present most of the talk about future OPEC-type coups has died down. Instead, both consumer and producer countries seem to be moving in the direction of commodity agreements whose ulterior purpose is to subsidize the export earning of LDCs. The assumption by the industrial countries is that eventually concessions in these matters will be essential if a series of confrontations with the Third World is to be avoided, and this is particularly true now that the most trivial incident could have serious consequences on the geopolitical front.

As pointed out in Banks (1977, 1979a), certain types of commodity agreements could be good business for the industrial world—the principal exception being the upper-class welfare scheme designed by UNCTAD (which will be taken up below). The key thing here is to begin scaling down so-called development aid and replacing it by forms of cooperation that are capable of making a real, rather than an illusory, contribution to the economic development of LDCs. Unfortunately, illusory contributions are quite acceptable to the employees of many U.N. and aid organizations, whose more-than-ample salaries are a function of their belief in aidsmanship and the rhetoric pouring out of the front offices of talk shops like UNCTAD.

The logic of substituting commodity agreements for so-called direct aid is as follows. There are certain countries that have received an enormous amount of aid over the past few decades and are no farther from the Stone Age today than they were several thousand years ago. By the same token, there are countries supplying the industrial world with invaluable primary commodities who are making a genuine effort to develop, but whose economies are periodically driven to the brink of ruin by unforeseen swings in the price of their exports. The question raised below is: Does it really make sense to continue showering money on this first group—as Sweden does, and as Holland wants the whole world to do—while denying relief to the other group on the grounds that commodity agreements might interfere with the international allocation of resources? Table 6-6 gives some idea of just what proportions of certain primary commodities are exported by LDCs.

The commodity agreements that have been proposed, discussed, and in some cases initiated comprise a wide variety of measures and schemes. Here we find such things as buffer stocks and funds, compensatory financing, indexation, and other price and income supporting and stabilization arrangements. Before we survey these devices, some background to commodity agreements is necessary, since the matter was first raised long before the expression *less-developed countries* had been coined.

The first serious attempts to regulate the price of primary commodities came in the interwar period. Tin and rubber were the objects of the first experiments, and the reader interested in the tin episode can consult Banks (1979a). Some people call these efforts a success, although it does not take a great deal of imagination to successfully operate one of these schemes in a period of high prosperity. In any event, the coming of the Great Depression reversed the trend price of most commodities, and no type of commodity

Table 6-6
The Leading Producers of Several Important Primary Commodities and the Percentage Produced by the Four Leading Producers among LDCs

Commodity	Leading Producing Countries	Percentage Output of Four Leading LDC Producers
Bauxite	Australia, Jamaica, Guinea, Surinam, U.S.S.R., Guyana	42
Copper	U.S., U.S.S.R., Chile, Canada, Zambia, Zaire, Peru	32
Iron Ore	U.S.S.R., Australia, U.S., Brazil, China, Canada	24
Lead	U.S., U.S.S.R., Australia, Canada, Mexico, Peru	18
Petroleum	U.S.S.R., U.S., Saudi Arabia, Iran, Venezuela, Kuwait	36
Tin	Malaysia, U.S.S.R., Bolivia, Indonesia, Thailand, Australia	62
Zinc	Canada, U.S.S.R., U.S., Australia, Peru, Mexico	18
Natural Rubber	Malaysia, Indonesia, Thailand, Sri Lanka, India, Liberia	85

Source: *Commodity Year Book,* various issues.

agreement sufficed to get them up again. Even supply restrictions on a fairly large scale were unable to keep prices from falling. Brazil, for example, resorted to burning large amounts of coffee in an attempt to support the price, but finally abandoned the program when it showed no success. During this period, both international cartels and the governments of countries producing primary commodities were involved in these activities.

After World War II the struggle to control commodity prices was resumed by the main producers of these commodities. The first postwar agreement was the International Wheat Agreement of 1949 and the International Tin Agreement of 1956. Negotiations for a coffee agreement were also commenced, but were finally given up. Even so, primary-product producers were beginning to pick up steam in their drive for higher prices. Among other things, they had a growing support at the United Nations, and various U.N. agencies provided them with theoretical arguments that were very useful when first propounded.

According to the first secretary general of UNCTAD, Dr. Raul Prebish, there is a long-run tendency for the price of raw materials to fall relative to those of manufactured goods. Prebish gave as the reason for this the low-income elasticity of demand of primary products: as the total world money demand increases, the demand for most primary products increases less than proportionally, which means that relatively less is spent on primary products than on industrial goods. [A short algebraic proof of this is given in Banks (1977).] The net result of all this is the turning of the terms of trade between primary commodities and manufactures against the former, which means that Third World countries find it increasingly difficult to finance their development.

Empirical work has not provided a clear confirmation of this argument, since the conclusions of econometric commodity models are generally tied to the prejudices of the people who finance them, but as far as I am concerned, it can definitely be proved for the most important nonfuel minerals over the postwar period up to 1977, and it is probably true for most primary commodities. But even without proof, there has been an outcry for more equitable treatment for primary producing countries—particularly when these countries begin to think in terms of cartels and production and/or export restriction. Prebish himself thought largely in terms of *indexation* as a tool to redress the balance: if the price of manufactures increased faster than the market price of primary commodities, then this latter price would be supplemented by a levy, or tax, which would effectively maintain the real purchasing power of Third World exports. As it turned out, the objection by the developed market economies to this type of arrangement bordered on the psychopathic. In addition it was suggested that the poorest of the LDCs would also be damaged by indexation, since they are heavily dependent on imported food and raw materials. As far as I am concerned, the worst indexation scheme ever devised is still better than "untied" transfers of cash to most LDCs. In addition, it may be possible to design indexation schemes that are satisfactory to the governments of many countries in the industrial world.

The concept of compensatory financing has been vigorously promoted by the U.S. government, particularly when Dr. Henry Kissinger was in charge of the U.S. State Department, and the opinion here is that the suggested program was well worth considering. It amounted to a guaranteed-income scheme under which LDCs would receive automatic compensation from an international loan agency whenever their revenues from the export of raw materials fell below a trend value. This compensation was divided into three categories: commercial loans, soft loans, and outright grants. (As yet, no comprehensive scheme has been designed which provides LDCs with the incentive to restructure their economies along more efficient lines; and, as far as many of us are concerned, this means that all proposed schemes are incomplete to some degree. But a few of them are a step in the right direction.) As for the LDCs, many of these reject the U.S. program as discretionary, which means that they want a blank check.

Buffer Stocks and the UNCTAD Integrated Program

The United States has offered its version of a compensatory financing scheme as an alternative to indexation and buffer stocks. This does not make compensatory financing superior to these other devices, although, as things now stand, it is definitely superior to all proposed buffer stock schemes. On the basis of the record, buffer stocks have not shown much (if any) success, while the UNCTAD buffer program should be regarded as a political rather than economic tool, and no less than an attempt to fabricate another organ of confrontation—presumably along OPEC lines. However, while the directors of OPEC have learned that in the long run all their goals can be achieved by using a little imagination and outwardly displaying what politicians in Western countries interpret as moderation, the bureaucrats sitting at or near the top of UNCTAD are looking for a stage on which to play out their frustrations and further their careers.

If we begin with a conventional buffer stock, such as the one operated under the tin agreement, the rationale is to support prices during periods of excess supply and limit price increases during periods of excess demand. The buffer-stock management is supplied with resources in the form of money and the commodity: when demand is low relative to supply, it buys the commodity; and in the opposite situation it sells. The way things have worked with the tin buffer stock, however, is that shortages of both tin and money have appeared at crucial times. Without sufficient tin, it has been impossible to maintain a ceiling price, although in truth there has not been a great deal of concern about this shortcoming. As for the floor price, when it appeared certain that this could not be held, export and/or production restrictions were imposed on the tin-producing countries. In fact, as explained in Banks (1977, 1979a) even a threat to resort to these measures was often sufficient to stop a falling price, as buyers began making speculative and precautionary purchases. For important insights into the effectiveness of the tin agreements, the reader should consult Gilbert (1977) and MacAvoy (1977). It is particularly interesting to observe that MacAvoy regards only *sisal* and *sugar* as suitable candidates for buffer stocks.

Now for the UNCTAD program. This involves a proposal to create buffer stocks for ten commodities, which would be financed through a common fund contributed to by producer and consumer countries. UNCTAD's goal is $3 billion for the first stage and between $10 and $13 billion by the time all ten *core* commodities are stocked in the prescribed amounts. It has been suggested that this is quite a bit of money to turn over to an organization that has never produced anything except conferences that even the once staid *London Economist* has referred to as "farces," stacks of useless memoranda, and astronomical travel bills for its directors and staff, some of whom are at this moment trying to sell the common fund to a growing number of disbelievers in the Third World.

What the reader should be aware of, though, is that the integrated program also wants to make some provision for production and/or export controls. Although it may not be realized, these controls are essential because even though $10 to $13 billion may sound like a lot of money, there is no guarantee that it is sufficient to do the job proposed. There have been suggestions that from $3 to $6 billion would be required just for a copper buffer stock if it were to operate without having to employ export and production controls.

According to existing prospectuses for the Common Fund, this program would operate a second window through which money would be made available to poor countries trying to diversify out of unprofitable activities, as an aid to marketing, and the promotion of what UNCTAD calls research. Obviously, however, with billions of dollars already being put into these endeavors by the United Nations, unilateral aid schemes, and OPEC, no more efforts of this type are really needed at present, and certainly none by an organization that has received the kind of criticism directed at UNCTAD.

Metal Exchanges and Exchange Pricing

The prices of some of the most important industrial raw materials are quoted on the two major commodity exchanges, the London Metal

Exchange (LME) and the New York Commodity Exchange (COMEX). Of these two, the LME is considered the most important in terms of turnover, physical deliveries, and its influence on the pricing of metal in general. On the other hand, until recently COMEX handled a wider variety of metals and in addition provides facilities for trading hides and rubber. It seems to be true that in the future, these exchanges will be even more important than they are at present since at least some American copper producers intend to price their copper on the basis of exchange prices, while aluminum and nickel contracts are being introduced on the LME.

Despite the inference found in the term *exchange,* the buyers and sellers of metals and other products do not usually come to the commodity exchanges to make their deals. Some traders agree on a formula for pricing the commodity in which they are interested, relating the price of the commodity to a price or prices on an exchange. One possibility is that the price at which the commodity will be traded will be taken as the average of the spot prices prevailing on the LME during the week before the scheduled date of delivery of the commodity. As for these spot (or cash) prices, they are determined by a small amount (by volume) of trading that takes place daily, during a limited time period, on the exchange.

The LME dates from 1882, although various metals had been quoted in London earlier on informal exchanges. During World War II and the immediate postwar period, the government controlled the price of raw materials, and so the exchange was closed; but it reopened in 1953, and since that time a steady increase in activity has been recorded. Once again it should be emphasized that this activity has only a small physical component. In 1968, with the world consumption of copper equal to approximately 7 million tons, approximately 2 million tons of copper futures were traded on the LME. These futures resulted in *physical deliveries* of only about 12,000 tons per month, to which we can add the minor deals of various agents to get a figure for total deliveries that is almost trivial. (Of course, it would hardly make sense for African or South American producers to deliver their products into an LME warehouse when the final consumer was in Japan or France. There are eight LME warehouses and delivery places in the United Kingdom, in addition to warehouses in Rotterdam, Hamburg, and Antwerp.) As for the categories of buyers and sellers on the exchange who are interested in physical transactions, the largest group of buyers usually consists of merchants acting on the part of customers desiring to make marginal adjustments in their stocks; while the principal sellers include large producers selling small quantities, small producers selling excess production, and fabricators selling excess stock.

Furthermore, the trading of futures on the LME has to do with hedging, or transferring price risk from buyers and sellers of physical commodities to speculators. This function is important since commodity prices are extremely unstable, and later in this chapter the mechanics of hedging will be explained in detail. As an introduction to this topic, we can now take up the difference between the futures, forward, and spot (or cash) market.

The first thing to remember is that with the possible exception of a futures market, none of these markets need exist in the form of an organized exchange where all buyers and sellers or their agents can congregate. As pointed out earlier, the spot market is not necessarily a market, but rather an arrangement that concerns delivery in the immediate future. The forward market, on the other hand, involves forward sales, or the sale of an item that will be delivered in the future at some mutually agreed on fixed price, or perhaps a price related to the price or prices on a metal exchange at or around the time of delivery. In addition, the forward market involves physical delivery, which is specified in a forward contract.

In contrast, a futures market features physical delivery in only a minority of cases. Strictly speaking, a futures contract *is* a forward contract, but this market is so organized that sales or purchases of these contracts can be offset, and future deliveries are unnecessary. The key element in this process is the presence of speculators who buy or sell futures contracts with the intention of making a profit on the difference between sales and purchase price. This matter will be clarified later with the help of some simple numerical examples, but in brief the arrangement functions as follows: A producer sells a physical commodity for forward delivery at a price related to the price of the commodity on a metal exchange at the time of delivery. At the same time he *sells* a futures contract. Then at the time of delivery, he *buys* a futures contract, offsetting his previous sale. If the price of the commodity has fallen, he loses on the physical transaction; but if the price of futures contracts fall, as they should, then he has a compensating gain on the futures (or paper) transaction. (The matter of how the price of futures contracts should move in relation to the price of the commodity will also be explained later.)

What has taken place in this example is that the seller of the physical commodity has turned over most of his price risk to the buyer of the futures contract, the speculator, who in this case might have thought that the price of the commodity was going to increase, in which case the price of the futures contract would also normally increase. Thus he would be able to sell this contract for a higher price than he paid for it. The difference between the producer and the speculator where this market is concerned is that the producer is primarily concerned with insurance and regards the buying and selling of futures contracts as being of secondary concern: in the long run, with extensive buying and/or selling of these contracts, he should break even. As for the speculator, his business is to make money on the difference between the price at which he buys the contract and that at which he sells it. Naturally, the speculator also sells contracts in the hope that he can later purchase similar contracts, and in the process make a profit (per contract) equal to the difference between the sales price of a contract and the purchasing price.

Speculation is also possible in physical items. If, for instance, the producer in the preceding example had produced a certain amount of metal and held it in inventory without hedging it, thinking that he could make a profit by selling it later at a higher price, then he could be called a speculator. The same applies to someone who buys a commodity on the spot market and holds it unhedged in hope of selling it later at a higher price. This position is called being *long* in the commodity. Another way of speculating is to be *short* in a commodity, which works as follows: A speculator agrees to sell a commodity in the future for a price above the price at which he thinks he can buy the commodity just before the date the purchaser is to receive the commodity. In other words, when the (forward) contract comes due, he buys on the spot market for a price that, if all goes as planned, is under the sales price on the contract. The term short simply means selling something that one does not own.

At various times exchange pricing has come under attack from producers, consumers, and interested outsiders. Some people say that the exchanges are responsible for the severe price swings that are typical for certain commodities. If we take the copper market, however, this contention overlooks the well-known fact that when the exchange was closed during the first part of the postwar period, all known quotations of copper prices showed a tendency to oscillate that was in no way different from what we have observed on the LME or COMEX over the past decade. Other complaints are that the exchanges are a tool of Big Capitalism; they can be infiltrated by the agents of international communism and prices rigged; they take the orders of high finance; they favor the Third World; and so on. Some of this, to a fairly small degree, may be true, although I feel that many of the charges against the exchanges originating from the academic world can be attributed to a profound unconversance with the way these institutions work. On the other hand, it seems to be true that many producers of industrial raw materials are peeved with the exchanges because they feel that they raise the price level of these materials which not only attracts outsiders to these industries but also leads to excessive investment in new capacity.

Exchange Contracts for Copper

The details of the proposed (and probably forthcoming) exchange contracts for aluminum are not available at the present time; however, since they will almost certainly function exactly like the existing copper contracts, we can still review briefly the various contracts for copper on the LME and COMEX. Here it should be remembered that copper is, for all its importance, a typical industrial raw material, and an exchange contract for copper is analogous to those of any commodity traded on these exchanges.

At present there are three standard forms for copper traded on the LME:

1. Electrolytic or fire-refined, high-conductivity wirebars in standard sizes and weights.
2. Electrolytic copper cathodes, with a copper content of not less than 99.90 percent, or first-quality, fire-refined ingot bars, with a copper content not below 99.7 percent.
3. Fire-refined ingot bars with a copper content not below 99.7 percent.

Anyone can buy or sell on the LME, and the only limitation is that the transaction must involve 25 L-T. The brand and place of delivery are chosen by the buyer, but it must be remembered that place of delivery means one of the LME warehouses. It is this provision that makes the LME of only limited interest as a physical market, although it does not reduce its attractiveness for hedging purposes, nor does it suffer a reduced efficiency as a market on which smaller amounts of metal are purchased for physical delivery. Each of the three types of refined copper traded has its own quotation for both cash and forward deals. The spot price is also referred to as a settlement price, while the forward quotation, which in practice is mostly a futures price, is a three-month price. In other words, if this contract is to be used as a futures contract, the offsetting arrangement must be made within three months of the time the contract is issued; or if it is to be used as a forward contract, then delivery must be made within the same period.

As for COMEX, only one standard contract form is available. The basic commodity is electrolytic copper in wirebars, slabs, billets, ingots, and ingot bars, of standard weights and sizes, with a copper content of not less than 99.90 percent. The standard unit for trading purposes is 50,000/lb. In addition to electrolytic copper, a number of other varieties of copper may be delivered at the option of the seller. These include fire-refined, high-conductivity copper, lake copper, and electrolytic copper cathodes. According to the regulations of the exchange, copper may be delivered from any warehouse in the United States that is licensed or designated by COMEX, but other warehouses are excluded. Deliveries must be to designated delivery points, and the period of forward trading must be within fourteen months. There are seven delivery months: January, March, May, July, September, October, and December. For more on this topic, and also for some of the effects of speculation in these markets, see Banks (1979b).

Hedging

This topic is important, and the reader who is intent upon mastering it in the shortest possible time needs first to grasp the following rule: Those wishing to insure against a fall in prices sell futures; those wishing to insure against a rise in prices buy futures. Individuals falling in the first category might be sellers of a commodity that will be delivered and paid for in the future while those in the latter category might be the purchasers of a commodity that will be received in the future and paid for at the same time. In both cases the relevant price is the spot price prevailing at the time of delivery. (Of course, price risk could also be eliminated here by using a forward contract with the price specified. But this is not always possible.)

Consider the following example. A producer sells a commodity for delivery two months in the future at the spot price prevailing on a certain commodity exchange on that day. Assume that 50 is the spot price of the commodity on the day of the sale, and 53 is the price of a five-month futures contract. The producer contacts his broker and orders him to *sell* a futures contract for 53. Two months later he delivers his commodity, with the spot price on the exchange on that day being 45 and the price of a futures contract at 46. He thus gets 45 for his commodity and pays 46 for a futures contract. The contract he sold earlier has now been offset, and his deals can be summed up as follows:

+ 53	Sales of futures contract
+ 45	Sale of commodity
− 46	Offsetting purchase of futures contract
+ 52	Realized on the sale of the commodity

The broker's commission should be subtracted from the + 52 to get the net value of the sale. There are some other technicalities associated with the example that the reader should be aware of. On the day the sale was arranged, the spot price was 50 and the price of a futures contract 53. The difference between these is called the *basis,* which in the example equals 3. Notice also that the futures price is higher than the spot price, which is the normal situation and is called a *contango.* When, on the other hand, the spot price is higher than the futures price, which happens from time to time, we have *backwardation.* The insurance aspect of the hedge can now be noted. Had the producer held this good unhedged, his revenue would have been 45. By hedging, he gets 52, from which the broker's fee is subtracted.

It should also be appreciated that for hedging to work satisfactorily, the basis must exhibit certain regular tendencies. There should be no frequent movements from contango to backwardation, nor should there be excessive movements in the value of the basis. This is what we meant earlier when we indicated that when the spot price increases, the futures price should also increase or, as in the case of the preceding example, when the spot price decreases, the futures price should follow it down. History seems to indicate that this is the usual arrangement, but it has happened that this pattern has been interrupted for short periods, causing hedgers some discomfort. Accordingly, the risk to hedgers is often called the *basis risk*.

By way of deepening our insight into this situation, let us change the preceding example so that the futures price on the date of arranging the sale was 51, and the price of a futures contract at the time of delivery is 52. The other prices—the spot price on the date of arranging the sale and the spot price at the time of delivery—will be left the same, 50 and 45, respectively. We then have a change in basis from 1 to 6, and the revenue from the sale, not including the broker's fee, is $51 + 45 - 52 = 44$ (sale of futures contract + sale of commodity − offsetting purchase of futures contract). The hedger has lost on this deal in the sense that, had he not hedged, he would have obtained 45 (or the spot price on the day of delivery). This loss may seem insignificant, but the reader should remember that it involves only one unit. Had the producer hedged 100,000 units, it would have been a serious matter. In practice, though, abnormal changes in the basis happen to be rare, and the regular hedger can be comforted by knowing that, statistically, he should break even in the long run.

As mentioned earlier, commodities are bought on the exchanges employing standard contracts. A standard contract on COMEX for copper is for 50,000 lb; thus if a deal involved 100,000 lb, two contracts would be bought or sold. There is also a time element associated with these contracts. In the first example, we mentioned that on the date of arranging the sale, the price of a 5-month contract was 53, and presumably this was the type of contract the producer sold. He delivered his commodity in two months at which time he bought an offsetting contract. Logically, the contract he bought was a three-month contract—logically because the rule is that when one sells a futures contract, it involves a certain month, and when the offsetting purchase is made it must be made for the same month. The same is true if we begin the process by buying a contract: We buy for a certain month, and the offsetting sale concerns contracts for the same month. It is also possible to think in terms of a certain date: the *maturity* date. If someone has sold or bought a contract with a certain date of maturity, then before the date he must make the offsetting purchase or sale of a contract referring to that date.

The next example is a situation in which a buyer of copper arranges for delivery of 50,000 lb of copper two months in the future at the price prevailing on the exchange at that time. Assume the spot price on the day the copper is bought is 30¢ per pound, and the futures price 32¢ per pound. The basis, which is a contango, then stands at two. The buyer then buys *one* contract, paying 32¢ per pound for 50,000 lb. If we assume that the spot price of copper increases to 40¢ on the day of delivery, with a futures price of 41, then the buyer makes the offsetting sale of one contract (or 50,000 lb) at 41¢ per pound and buys his copper at 40¢ per pound. His account now appear as follows:

− 32	Purchase of futures contract
+ 41	Sale of futures contract (offsetting)
− 40	Purchase of copper
− 31	Price paid for copper

He pays 31¢ per pound for 50,000 lb of copper (to which he must add his brokerage fee). Notice what would have happened had he not bought and sold this futures contract: He would have paid 40¢ per pound for his copper.

It might be instructive to alter this example slightly. Assume a situation in which copper is bought for a *fixed* price of 40¢ for delivery several months in the future. On the same day the futures price of copper is 42¢, and so the buyer *sells* a contract for 42¢. On the day the copper passes to his ownership, its spot price is 30¢, and the price of a futures contract 31¢. The buyer then makes his offsetting purchase of a futures contract at 31¢. These transactions can be summarized as follows:

− 40	Purchase of copper
+ 42	Sale of futures
− 31	Purchase of future (offsetting)
− 29	Price paid for copper (per pound)

Several points are important in the preceding example. Had the buyer not hedged, he would have bought copper at 40¢, while some of his fellow producers bought at 30¢ or 29¢, either by waiting to buy spot, or buying at 40¢ and hedging. There is also the matter of what the situation would have been had the buyer expected the price to rise dramatically. With this the case it could be argued that had the basis remained constant, it would have been better to buy at 40¢ and not hedge: Had the price risen to 65¢, and the basis stayed the same, then hedging would have meant selling a futures contract for 42¢ and making the offsetting purchase at 67¢. The cost of copper would then have been 40 + 67 − 42 = 65 cents. This situation could conceivably come about, but in general we expect that the large price rise would have been preceded by a substantial rise in the price of futures contracts. Instead of selling a futures contract for 42 cents, it might have been possible to sell one for 55 or 60.

Figure 6–8 shows the relationship between the free-market price and the producer price for four nonferrous metals over the first half of the 1970s. Although it cannot be discerned from these diagrams, the free-market arrangement gave on the average higher prices during the period shown for

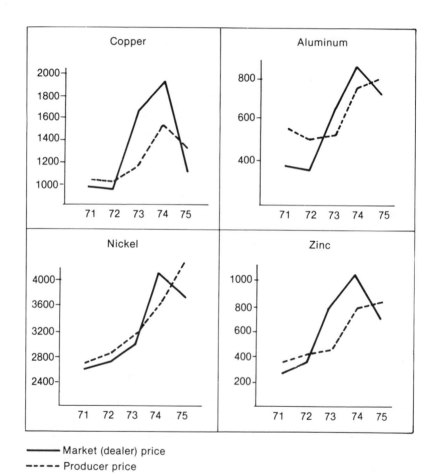

―――― Market (dealer) price
– – – – Producer price

Source: Reprinted by permission of the publisher from Ferdinand E. Banks, *Bauxite and Aluminum* (Lexington, Mass.: Lexington Books, D.C. Heath and Company, 1979), p. 118.

Figure 6–8. Producer and Market Prices for Copper, Aluminum, Nickel, and Zinc

these and other metals. Market prices were 8 percent higher for copper, 23 percent higher for zinc, and 7 percent higher for tin. Producer prices were 7 percent higher for aluminum and 2 percent for nickel. Statistically, there was no difference for lead. If, however, we go back to the period 1966–1970, we see that on the average the market price of copper was 45 percent higher than the producer price, and the market price of nickel 125 percent higher than the producer price.

Commodity Prices and Inventories

We turn now to the link between commodity prices and inventories, but first a few remarks are necessary about long- and short-run prices. Long-run prices are determined by trend movements in supply and demand. If, for example, over a long period the supply of a raw material expands at a more rapid rate than demand, then we should expect a downward pressure on price. This has been the case with a number of commodities over the past decade. In contrast, until recently it was felt that the demand for aluminum metal was increasing faster than the building of new facilities for producing this metal; and given that the gestation period of investments in aluminum capacity is a minimum of three or four years, it has been claimed that we are moving toward a sellers' market in aluminum that may feature strong price increases. As things stand now, only the prolonged business downturn that began in 1980–1981 postponed this scenario.

On the other hand, when we examine short-term prices, we see peaks and troughs that are separated by weeks or months instead of years, and that are to some extent independent of business activity. The explanation of these short-cycle price oscillations lies in speculative tides of bullishness and bearishness fueled by fantasy, naiveté, or just plain bad judgment. Thus a relatively minor surge in demand might cause metal prices to rise for a few weeks, leading an influential group of market analysts to glimpse what they think is the beginning of a new golden age. Something like this happened with copper in the United States in April 1978, when the price temporarily jumped from 56 cents/pound to 63 cents/pound. Its failure to remain at the higher level can be attributed to the well-documented fact that unfounded enthusiasm is still no match for the realities of supply and demand, which can be expected to surface as more attention is directed to deciphering various market signals and more information becomes available.

As soon as a substantial number of market participants and analysts conclude that the underlying economic situation points toward excess supply, a downward movement in prices is only a step away. Bullish (or good) news is now discounted or disregarded, and any bearish news that might have been given short shrift when the market was rising is reevaluated.

Speculators begin to sense a downswing and increase their sales of futures contracts, expecting to buy them back later at lower prices. This decreases the price of these instruments, and since some market participants take the price of futures contracts as a forecast of the spot price of various commodities *in the future,* the overall feeling of deterioration is reinforced.

As with the upswing, the downturn is brought to a halt by the filtering through of sufficient high-quality information to dispel the fog concealing what is actually happening on world markets. Much of this information comes in the form of interpretations of the significance for present and future prices of inventory sizes and movements. For example, during the past decade the zinc, copper, and nickel industries have been plagued by stock overhangs that tended to preclude any rapid upturn in price, even given temporary dislocations in production caused by strikes or by several producer countries experiencing difficulties in producing or shipping some commodity. This kind of price immobility can only be attributed to excessive world inventories.

For the most part the economic literature dealing with inventories has been confined to the learned journals, where it can be conveniently ignored. But for our purposes some of the more elementary concepts associated with this literature must be examined since, with a little luck, they give us the possibility of completing the readers' introduction into the logic of short- and intermediate-run price movements.

First, if a producer's or a consumer's inventories are low, then each extra unit held in stock reduces the possibility that production somewhere will have to be scaled down because of an unforeseen absence of some input. Remember that both producers and purchasers of industrial raw materials are bound by contractual obligations to their customers; as a result, inventories must be held even if there is an inverted relationship between the spot price of the commodities being stocked and all predictions of the future price. In other words, even if the expected money yield from holding and later selling a commodity does not cover such things as its storage cost, this negative aspect is counterbalanced by a convenience yield when the size of inventories is small relative to the amount of the commodity being used as current inputs in the production process. In this situation an effective price system must function in such a way as to ration existing stocks among existing and potential stockholders, which often calls for a departure from normality in the form of an inversion between present and expected future prices. This inversion is called *backwardation,* and on several occasions over the past decade it has existed on the world copper market in the wake of some misunderstandings about the future availability of copper metal.

The same type of reasoning makes it clear that if the inventories being held are large in relation to the amount of the commodity being used as a current input in the production process, then there is little incentive to hold more. In these circumstances convenience yields are small, and stockholders require that the expected future price of the commodities being held is such as to cover storage, handling, insurance, and other charges—unless it happens to be possible to sell futures contracts at a premium that, in the stockholder's opinion, would be sufficient to cover carrying costs. Otherwise these stocks are put on the market, driving down spot prices and widening the gap between present and expected future prices to an extent that holding the existing stock is justified. Figure 2A–2 in appendix 2A illustrates some of these notions for an unspecified commodity. (The reader who is particularly interested in futures markets would do well to examine the material on that subject in chapter 2.)

We can conclude this discussion with an example from the real world. In the *London Financial Times* of 6 February 1979, Mr. John Edwards, the commodity editor, pointed out that on the copper market at that time the gap between the cash price and the three-month quotation was much smaller than it should be to carry supplies at the then existing interest rate. According to Mr. Edwards, this situation implied that a shortage of supplies to the market could develop despite of the large inventories being held at that time.

Although many people do not realize it, the total amount of copper being stored in the world is determined not by the desires of inventory holders, but rather by the physics of production and the preferences of producers and consumers. The level of world inventories is what it is, and the purpose of the market mechanism is to ensure that if storers are dissatisfied with this level at one structure of prices, that price structure is changed to one where storers will be satisfied. With no change in expectations, in the case given by Mr. Edwards, this would mean the selling of enough stocks to make the spot price of the commodity fall to the extent that the gap between the present and expected future price justifies the holding of existing stocks. My conclusion, delivered in a lecture at the University of Geneva on 6 February, was that in the short run no shortage of supplies to the market should be expected. This is exactly as it turned out to be.

The World Tin Market

By way of discussing some of the topics taken up earlier in this chapter, I should like to present a short discussion of the world tin market. Tin is especially interesting because it is the only prominent mineral that is largely processed to the metal stage in the LDCs. Excluding China and the USSR, almost 90 percent of tin-mining and 75 percent of tin-smelting capacity is located in the Third World. The main producers of tin are Malaysia,

Bolivia, Thailand, and Indonesia; but important deposits are found in Nigeria and Australia. Malaysia is the world's largest producer, and it has more than 30 percent of both mining and smelting capacity. On the consumption side the major purchaser is the United States, taking on the average about one-third of the tin metal produced in the noncentrally planned countries. Tin is always found in alloy form, and its principal use is in the production of tinplate, which in turn is the basic raw material for tin cans. Some supply data is presented in table 6–7.

Although not indicated in table 6–7, there is now considerable excess processing capacity both in the less developed world and in the industrial countries. This is the result of the downturn in world economic activity following the first oil-price shock. This rise in oil (and energy) prices has also caused an upward pressure on tin-processing costs, despite the availability of excess capacity: on the average, processing charges doubled between 1974 and 1978. The contracts offered by tin processors are typical of those employed by the producers of most nonferrous metals. There are deductions for impurities in the concentrates; a definition of the basic price and how this price is modified to take into consideration special circumstances; specification of the tin content, where the best known tin is Straits or Grade A, which means that it is commercially pure with a minimum tin content of 99.8 percent; and so on.

The ownership of tin-mining and -processing facilities is mainly in the hands of the governments of the countries in which tin is produced. State-owned enterprises also tend to handle the marketing of tin to consumers and to the London Metal Exchange; in fact, a Malaysian-government holding company acquired control of the most important British operations in Malaysia by way of a takeover on the London Stock Exchange. Joint ventures and similar operations featuring government control are also quite common (for example, in Thailand). There is, however, still a prominent role for transnational companies to play on the world tin scene—particularly where processing and marketing are concerned. Even so, it should be appreciated that although these firms still own a significant amount of processing capacity, most of these facilities are in the tin-producing countries; thus value added in the form of wages and taxes remains in these countries. Table 6–8 shows the capacity at the disposal of the main tin-producing firms.

We can now take a look at the complicated matter of tin pricing. Most tin is purchased in metal form directly from producers, but some is obtained from merchants who buy metal directly from smelters in the producing areas and usually handle all the transactions and operations necessary to deposit this metal in the storage facilities of tin consumers (for example, fabricators). Many of these merchants also have a direct access to the LME; as a result, they can, if necessary, obtain some metal from this source.

Table 6-7
The Supply of Tin by the Major Tin-Producing Countries

	1977 Production	1983 Capacity[a]	1977 Production	1983 Capacity[a]	1977 Reserves	Production Cost[b]
Australia	10.7	12.0	5.8	8.0	370	363
Bolivia	32.6	38.0	13.3	30.0	1000	403
Brazil	6.4	8.0	7.0	12.0	610	—
Indonesia	25.0	35.0	24.0	35.0	2400	360
Malaysia	58.7	95.0	66.4	95.0	830	364
Nigeria	3.3	10.0	3.5	4.0	280	—
Thailand	24.2	30.0	23.0	30.0	1200	312
Zaire	3.6	6.0	0.6	1.0	200	—
Others	25.5	23.0	40.4	63.0	1190	—

Note: Production, capacity, and reserves figure in thousands of tons.
[a]Estimated.
[b]Cents/pound.

Table 6–8

The Major Tin Producers and Their Production Capacities
(in thousands of tons)

Patino NV (Australia, Brazil, Malaysia, Nigeria)	69,500
USSR state-owned plants	60,000
Overseas Chinese Banking Group (Malaysia)	40,000
Chinese state-owned plants	35,000
Shell-Billiton (Thailand)	25,000
PT Timah (Government of Indonesia)	26,000
Rio-Tinto Zinc-Copper Pass (U.K.)	20,000
Comibol (Bolivia)	14,000
Gulf Chemicals (U.S.)	9,000
Metallurgie Hoboken-Overpelt (Belgium)	5,000

Source: UNIDO documents.

The pivot of the world tin market is located in Penang (Malaysia). Here almost 1,000 mines sell to the Straits Trading Company and a syndicate of smelters. The concentrates supplied by these mines are subsequently converted to metal and sold on the basis of contracts that stipulate delivery within sixty days of the finalization of a contract. Although many observers do not consider Penang a true free market (because of the heavy concentration on the seller side), it definitely sets the world price for tin because of the quantities involved; and there seems to be statistical evidence that the LME price for large transactions is basically the Penang price, adjusted for shipping, insurance, and interest costs.

If we look back in time, we cannot avoid seeing that the struggle to keep the tin price above a certain level has been one of the main preoccupations of tin producers and, lately, of a certain type of international bureaucrat. As with most metals, tin has occasionally been in excess supply, but rarely to the extent exhibited immediately after World War II, when the price fell to £170 (sterling) per ton. The main producing companies at the time were located in Malaya and Indonesia (then the Dutch East Indies); being relatively few, they found it easy to come to the conclusion that the best way out of their difficulties was to restrict the supply of tin. They then proceeded to buy and store 20,000 tons of tin, and only in 1923 did they began returning this hoard to the market, at the rate of 1,000 tons/month. This strategy succeeded in getting the price of tin up to £240/ton and maintaining it at that level.

The post–World War I depression ended in 1924, and during the 1924–1925 upswing in the international business cycle a great many new tin-producing facilities were established, with most of them incorporating the latest production techniques. As it turned out, too many mines and smelters came into existence within a relatively short period. When the Great Depression arrived at the end of the 1920s, falling demand in conjunction with a huge surplus capacity led to an accelerated descent in the price of tin. As a result, by 1930 the Tin Producers' Association once again introduced an output-restricting scheme for the purpose of stabilizing prices. On this occasion the tin-control scheme, as it was called, had to be introduced on a worldwide basis because of the growth of the market. Even so, the cartel was able not only to stop the fall in the tin price, but to keep it at the desired level during the entire depression—which would have to be rated something of a miracle, considering the opinion held by most renowned economists on the inherent fragility of any type of cartel. It must be appreciated, however, that during this period there was no expansion in world tin-producing capacity as a result of two important factors: first, that in the main tin-producing countries almost all the major producers were in the cartel, and potential resources outside these countries were not very impressive; and second, that there was no money available to finance new capacity. The tin-control scheme was phased out in the late 1930s, right after the armament industries began tuning up for World War II; and in 1938 the price of tin rose well beyond the £200–230/ton range, where ideally it was supposed to be stabilized.

The days of low-priced tin were now over, although, as was the case with most metals, the price of tin was fixed during the war. There were, however, no shortage of conversations during this period about what the price should be after the war. In the course of these discussions, some of which involved representatives of the Allied governments, the concept of a *buffer stock* was treated seriously as a potentially usable instrument of economic policy—at least for modern times. The setup established was an inventory of tin and cash, conceptually distinct from the main channels of commerce, whose management would buy tin when the price was unacceptably low, and sell when the price was too high. There was naturally some difficulty in establishing the minimum or *floor price* for this buffer stock. Anticipating a full-fledged postwar depression, the tin producers wanted a floor price of £880/ton, whereas the U.S. government considered £520/ton (approximately $2,300/ton) a reasonable figure. As things turned out, this controversy was quite irrelevant. The postwar depression was surprisingly mild, and since it arrived at the same time as the Cold War (in 1948), the U.S. Strategic Stockpile—which had been created to ensure that in national emergencies the United States would not be cut off from essential raw materials—easily absorbed excess production, thereby holding up the price. Finally, to the great joy of tin producers everywhere (and many other producers of strategic metals also), just as the Strategic Stockpile was reaching its designated capacity, the Korean War began. The subsequent powerful increase in the demand for tin drove its price to a record high.

It was during this period that a floor price of £730/ton was established, and a buffer stock of 15,000 tons was agreed on. Note, however, that neither the floor nor the *ceiling* price (the price that the buffer-stock management is supposed to regard as the absolute maximum allowable) nor the size of the buffer stock were intended to be permanent: they could be adjusted at virtually any time by the buffer-stock management. Where prices were concerned, the natural direction of adjustment was upward. The rules of operation of the buffer stock are also of interest: there is a high range of prices in which the buffer-stock management must sell tin, a middle range where management remains impassive, and a lower range where purchases of tin are mandatory. Obviously, *mandatory* is a flexible word in this context, because the buffer-stock management is not supplied with an unlimited amount of money; thus it was decided that if the price fell too low (well past the floor price) quantitative export restrictions would be practiced. Naturally, there was no policy of this type that could be resorted to if the ceiling price was exceeded by a large amount, but no one worried too much about this particular eventuality: producers were not hostile to high prices, and in times of prosperity consumers did not notice them.

A great deal of space will not be used here to survey the actual operations of the tin buffer stock. As I pointed out some years ago in a note in *Econometrica* (1972) it was the buying and selling of the U.S. Strategic Stockpile (operating under the General Services Administration) in phase with the movements of the world business cycle that caused the Tin Agreement—as it came to be designated—to function satisfactorily. Other researchers have since confirmed this observation (see Gilbert (1977) and Smith and Shink 1976). I have also pointed out (Banks 1979a) that on at least one occasion, knowledge of the impending exhaustion of the buffer stock during a period of rising tin prices caused some individuals close to the Tin Agreement's management to do some very successful speculating in tin futures.

Tin prices fell by so much during the recession of 1958 that the buffer-stock management ran out of money trying to stem the tide. Physical restrictions were therefore invoked and maintained intermittently through 1960. A problem of the opposite character came into existence during the first part of the 1960s, when the tin market was treated to a long run, steady price upswing of such underlying strength as to bring about extensive investment in new capacity, along with a great deal of modernization of existing capacity. When prices began to slide in the mid-1960s, helped along by a slight recession in the United States and Europe, the presence of this new, modernized capacity eventually caused a price decline of such magnitude that export controls were unavoidable. Things would have been much worse had not the General Services Administration agreed to enter into consultation with the Tin Council before making large sales of tin. This type of arrangement should be kept in mind because there is a widespread belief in some quarters that a buffer stock is an ideal instrument for the regulation of commodity prices, when in fact the tin buffer stock—which is often taken as the model for some proposed buffer stocks—has only been able to function with a modicum of efficiency because of export controls, side agreements with other authorities, and the presence of the U.S. Strategic Stockpile—which was initially fifteen times larger than the Tin Agreement's inventory, and whose operations almost always reinforced those of the Tin Agreement.

Beginning in 1969 the demand for tin moved up rapidly, but tin production increased by enough to keep prices down. As the Vietnam War moved toward its conclusion, and world markets anticipated depression, tin prices began to decline, and once more export controls were resorted to. Now, however, with world production capacity very much in excess, and large inventories of tin available everywhere, prices could not make an immediate comeback. The recovery, which was only temporary, came in 1973–1974 in the wake of the first oil-price shock. On that occasion it was assumed by speculators—incorrectly, it seems—that *all* industrial raw materials were on the way to short supply. Table 6–9 shows the tin price from 1950 to 1978; since these prices are in *current dollars,* the movements referred to earlier can be easily traced.

In recent years the tin price has started to move up again, an interesting phenomenon in the light of the movement of other primary commodity prices. The explanation here, quite simply, is that supply is now under a much stricter control than earlier: the main producers of tin, particularly in the Far East, have begun to think a little more closely about their long-term interests; as a result, they have taken steps to ensure that supply does not outrun demand, and prices at all times reflect production costs. There might also be another reason—one due to a fundamental long-run shortage of tin ore. That takes us to the next topic.

At present (Spring 1982), a battle is being waged for control of the world tin market. During a period of deep recession, between three- and four-hundred million pounds have been used to purchase (and stockpile) tin, thus pushing the tin price to the highest level ever. The exact identity of the group doing this buying is ostensibly unknown, and the same is true of the source of the money they are using for this adventure; but the stakes are extremely high.

The principal fear of the tin establishment is that the structure of the world tin market will be irrevocably undermined, and the prestige of the London Metal Exchange (LME), which provides the only futures market for tin, irreparably damaged. (Side effects may include the demise of the Tin Agreement, the formation of a tin producers' cartel that would serve as an example to the producers of other commodities, and so on.)

Table 6-9
Tin Prices, 1950-1978

	N.Y. Market LME (Dollars/Ton)	(Dollars/Ton)
1950	2,055	2,106
1951	2,976	2,802
1952	2,658	2,656
1953	2,014	2,113
1954	1,981	2,025
1955	2,040	2,089
1956	2,171	2,236
1957	2,080	2,122
1958	2,026	2,097
1959	2,164	2,250
1960	2,195	2,236
1961	2,447	2,498
1962	2,471	2,528
1963	2,507	2,572
1964	3,409	3,474
1965	3,893	3,929
1966	3,574	3,617
1967	3,330	3,383
1968	3,127	3,266
1969	3,428	3,627
1970	3,674	3,841
1971	3,500	3,689
1972	3,765	3,913
1973	4,826	5,016
1974	8,199	8,736
1975	6,869	7,492
1976	7,582	8,373
1977	10,761	11,786
1978	12,000	12,990

Note: In *current* dollars/ton.

Taking the first of these, it is clear that for all the oligopoly and oligopsony on world commodity markets, it is probably best for all concerned if the fiction of an ideally functioning market—guided by scarcity prices—is maintained. To a certain extent, of course, this is the way things actually are, since the marginal amounts bought and sold on the LME do, as a rule, reflect supply-demand conditions in the physical markets. But the LME cannot function if, as is now the case, people who have sold short (that is, sold a commodity that they do not have, but intend to buy prior to the date on which they are scheduled to deliver the commodity to its buyer) have to deal with people who are in posession of this commodity and who have cornered the market. In these circumstances, before releasing the commodity to the short sellers, the holders of the commodity might elect to charge the short sellers a very high premium. Assuming that some of these premiums could be so high as to force the suspension of trading, the LME has ruled that a premium of no more than 120 pounds/tonne can be charged for cash (spot) tin due to be delivered the following day. This move has met with sympathy only from those short sellers now facing the possibility of bankruptcy. Free-market purists are uninterested in the odd bankruptcy—unless they happen to find themselves in that role; while the general position of the major tin-producing countries is that the changing of the rules by the LME is a typical piece of hypocrisy, since that organization does not make a practice of helping tin producers when the tin price collapses.

Readers who have examined the material in this book dealing with futures markets may be curious about the present relation between the spot price of tin, and the price of three-month tin contracts. (That is, contracts for tin that is to be delivered in three months). The situation is that the spot price is much larger: *backwardation* now prevails! This, as explained earlier, is fully acceptable if there is a shortage of the commodity; but just now the world production of tin of approximately 200,000 tonnes/year exceeds annual demand by at least 20,000 tonnes. World stocks, in relation to consumption, are well over the normal level; and the amount of tin stored in LME warehouses is close to a record.

To make the situation more grotesque, the United States Strategic Stockpile is attempting to get rid of 30,000 tonnes of tin over a three-year period. For these reasons the denouement of this unusual drama is not yet in sight. The main question here, though, is just who would buy tin at the present time to keep in inventory, given that the world economy is in full decline, storage charges are probably larger than ever, and the cash tied up in these stocks could be so much more profitably employed if it were routed into the financial markets. One answer is that whoever *is* doing it, is doing it for long-range (that is, political) purposes. It has been said that the syndicate doing this purchasing is well known to the governments of Indonesia, Thailand, Bolivia, and Malaysia—in fact, rumor has it that Malaysia might be the sponsor of this scheme, since the amount of money involved and the tenacity of the purchasing syndicate make it clear that this is more than just a speculators scam.

There are two conventional ripostes to this kind of gambit. The first is simply to cut consumption of the commodity, and let burgeoning inventories bring the purchasing syndicate to their senses. Tin users conventionally do a great deal of thinking about substitution, and obviously a great deal of tin can be replaced by aluminum, stainless steel, plastics, and glass. But it is dubious if the pace of substitution can be stepped up in the short run. The other response is to get more tin into the market. The most obvious source of this alternative tin is the United States Strategic Stockpile, but many observers consider it unlikely that the U.S. government would go out of its way to crush a cartel that may have been formed under the auspices of some of the most anti-Communist governments in the Third World.

My own belief is that if the United States Strategic Stockpile can be persuaded to cancel their plans to release more tin (they have already sold about 8,500 tonnes of the proposed 30,000 tonnes scheduled to be sold), then it will be possible to reduce the huge existing stockpiles in such a way as not to spoil the market. But this will take time, and unless the world economy returns to normality—or at least a semblance of normality—some losses are inevitable.

Elementary Discounting and the Elementary Economics of the Mining Firm

This section will provide an elementary discussion of the economics of exhaustible resources, even though I attempt to take up some matters associated with this topic that are usually overlooked in the more advanced literature. The reader desiring a more mathematical treatment will find it in the next chapter.

Before going to the theory of exhaustible resources, a brief introduction will be provided to the theory of discounting. The approach here, which will involve only the most elementary secondary-school mathematics, turns on the well-known preference for current pleasures as opposed to the same amount of pleasure in the future, all else remaining equal. For instance, if we have an asset now, then we also have the option of enjoying its services if we so desire; otherwise, we face uncertainties concerning our desires and appetites up to and including the date on which the asset is received. (Consider, for example, an automobile or a television set.) Usually, if people have a choice, they demand some sort of compensation for postponing present satisfaction. Thus a sum of money (which represents generalized purchasing power) today is related to a sum in the future through a discount or interest rate that says something about the actual consumption available for someone prepared to defer a unit of consumption for a given period. This period is generally taken to be a year. For instance, if I have $100 today and the interest rate is 10 percent, then it is customary to say that this $100 is equivalent to $100(1 + r) = $100(1 + 0.10) = 110 in a year's time. If I give up $100 on 8 January 1980, I can obtain a bond or bank deposit that will provide me with $110 on 8 January 1981: the $10 is my reward for waiting.

It is enlightening to turn this formulation around. $110 received after one year is worth $110/(1 + r) = $110/(1 + 0.10) = $100 today. The general expression that we are moving toward is:

$$A_{t+1} = A_t(1 + r) \quad \text{or} \quad A_t = \frac{A_{t+1}}{(1 + r)}.$$

In this expression t signifies the time period. Similarly, if we are interested in the relationship between A_t and A_{t+2}, we have:

$$A_{t+2} = A_{t+1}(1 + r) = A_t(1 + r)(1 + r) = A_t(1 + r)^2.$$

If, for instance, the rate of interest is 10 percent, $100 today becomes $100(1 + 0.1)^2 = $121 in two years: $21 is the premium for giving up control over $100 of present purchasing power for a period of two years.

We began this exposition with a brief reference to the agony of having to wait for our pleasure, and now we are talking about how money grows when used to buy a bond or is put in a savings account. The connection is roughly as follows: the interest rate is an objective criterion. If you visit your local bank or brokerage office, you can be quoted the interest rate on such and such a type of deposit, or a security possessing a certain maturity. Let us assume that this interest rate is 10 percent. At the same time the typical individual will, in the light of his or her subjective preferences, possess a subjective discount rate that concerns his or her willingness to surrender money today in return for money tomorrow. Some people might regard 10 percent as a nice reward for waiting and thus be inclined to put a sizable portion of their monthly paycheck in a savings account offering this rate of interest, but the subjective discount rate of others might be so high that they would never consider postponing any consumption unless they were rewarded with a 100-percent interest rate.

Unfortunately, neither economists or psychologists have had much luck measuring individual discount rates, and no luck at all with aggregating them. Instead the practice has been to use interest rates as a proxy for aggregate subjective discount rates, and so these terms are often used interchangeably in the literature of economics. Thus a society with a low average rate of interest, such as Switzerland, might be regarded one in which the average person has only a moderate preference for today's goods compared with tomorrow's.

Finally—and this is a very important point—note that if we receive various amounts of money at different times in the future, these future income flows can be discounted and summed to obtain a *present value*. For example, we can inquire into the present value of $110 a year from now *and* $121 two years from now. Assuming an interest rate of 10 percent, we obtain:

$$PV = \frac{110}{(1 + r)} + \frac{121}{(1 + r)^2} = \frac{110}{(1 + 0.1)} + \frac{121}{(1 + 0.1)^2} = 200.$$

This concept can be generalized. Letting A_i represent a money flow received at i periods in the future and r the discount rate (which for simplicity is in many cases taken as the interest rate), we have as the present value of a stream of n payments:

$$PV = \frac{A_1}{(1 + r)} + \frac{A_2}{(1 + r)^2} + \cdots + \frac{A_j}{(1 + r)^j} + \cdots + \frac{A_n}{(1 + r)^n}.$$

Income streams are compared, and generally ranked, on the basis of their present value. At this point the reader should experiment with discounting and comparing various income streams with the same and different lengths, employing different rates of interest. He or she will notice that as interest rates increase, present values decrease, assuming no change in the payment streams. This merely signifies a rise in impatience: the greater importance attached to near as opposed to distant payments, or the downgrading of more remote satisfactions.

The Theory of the Mining Firm

Now we can go to the elementary theory of the mining firm, which is a subdivision of the theory of exhaustible resources. In fact, the emphasis here will be on concepts involving exhaustible resources; certain important topics involving a mining firm (or industry) will be ignored for the simple reason that they have no place in an elementary exposition. Still, I will discuss several matters that are extremely important to the mining firm and its customers (such as the inevitability of long-term contracts and the importance of large capital costs), and act as a significant constraint on the firm's ability to vary output in the short or medium run. These topics are also almost entirely overlooked in both the pedagogical and the theoretical literature. I also work in terms of a fixed time horizon; although this may appear to be a rather extreme simplification, mining-company officials often think in terms of a specific length of life for a mine and its basic processing facilities.

We can begin by observing that if a particular extraction firm anticipates that the cost of extracting a given amount of output from a mineral deposit will remain constant over an indefinite future, and it is believed that the price of this output cannot be influenced either now or later by altering the production of the firm, then there is nothing to be gained—at least conceptually—by treating this mining firm as a different type of phenomenon from other firms. Even if we are talking about resources that are obviously finite and exhaustible, they can be regarded as inexhaustible by the management of this particular firm. In this case profit maximizing behavior is characterized by producing at the rate at which the marginal cost (MC) of output equals marginal revenue (MR) from sales, which is the usual prescription from elementary economic theory. Once management has taken care of this matter, the owners of such a firm have only to concern themselves with the transformation of their existing deposit of minerals into other types of industrial assets; other minerals; more minerals (through, for example, exploration); financial assets; or, for that matter, champagne suppers in the shadow of Notre Dame.

For example, let us consider a very simple example in which marginal cost (MC) = 8 up to a production of 100, when it becomes infinite, and where this is true both for the present and for all future periods. Let us also say that the expected price over a three-period time horizon is 10, and the firm in question has 250 units of reserves. As is easily deduced, production should be 100 units in each of the first two periods, and 50 units in the third period. Revenue (pq) is $100 \times 10 = 1{,}000$ in the first two periods; and since marginal cost is constant (and therefore equal to the average cost), total cost in each of these two periods is $100 \times 8 = 800$. The profit per period is therefore 200. Similarly, in the third period revenue is $50 \times 10 = 500$, and cost is $50 \times 8 = 400$. Now, with a discount rate of 10 percent, the present value of profit is:

$$PV = \frac{200}{1 + 0.1} + \frac{200}{(1 + 0.1)^2} + \frac{100}{(1 + 0.1)^3} = 421.$$

At this juncture the reader must do two things. He must first attempt to verify that there is no way to shift production between these three periods and increase PV—for instance by increasing production in the third period; and he must take particular note of the fact that nothing has been said about what will happen to the price of the mineral being produced if changes take place in the intertemporal pattern of production. The implicit assumption is that changes of this type do not influence either the present price or expectations about the structure of future prices.

This last point is crucial because a discussion of exhaustibility that does not introduce a rising pattern of intertemporal prices is comparable to

Hamlet with the Prince of Wales instead of the Prince of Denmark. The reader should therefore dwell on the following example. If there are 3,000 firms in a certain mining industry—all the same size as the firm in the example just given, with the same time horizon, the same price expectations, and so on—then the attempt by all of them to move production as far forward as possible would probably cause the present price of mining output to decrease relative to *expected* future price, since with unchanged present demand (or with the demand curve in the present period immobile) an increase in the present supply would result in a fall in the present price. As the reader probably recognizes, this situation has many of the characteristics of the single-period model of perfect competition, in that every market participant has the same information and reacts to this information in the same way.

What about the previous example? What does it represent? It represents real life in an industry such as the world copper industry, where there are several thousand firms but where most of the world's reserves are held by twenty or thirty majors, which also mine most of the world's copper. The firm in the example was one of the fourth- or fifth-raters, whose mineral resources are indeed exhaustible but whose actions have little or nothing to do with the theory of exhaustible resources, or with prices and overall supply in mineral markets, except when by chance their actions reflect those of the largest firms in the industry. Thus, when we discuss exhaustibility in terms of real-world situations, we are talking about a relatively few firms whose behavior can bring about the well-known results associated with the theory of exhaustible resources. (If these results are not well known to the reader, they should be very shortly.) This also implies that most of these large firms have similar behavior patterns, particularly in their response to expected prices, and that they have access to the same market information.

What has just been said is that although most firms in an extraction industry may, as in a textbook market of perfect competition, function as price takers—and in addition many of these smaller firms are indifferent to the exhaustibility of the mineral deposit they are exploiting—a correct description of optimal firm behavior in a specific mining industry entails an explicit recognition of the possibility that in terms of conventional time horizons, the resource may be exhaustible. In other words, even if most firms are price takers, the prices they take should result from the optimal behavior of the larger firms in the industry, which—in light of the information at their disposal—distribute their intertemporal production in such a way as to maximize the present value of discounted profit over finite time horizons, and in doing so influence both present and future prices—and perhaps also expected future prices.

Even if the time horizons of the management of the major firms are so short that they studiously ignore any possibility of resource exhaustion in the foreseeable future (and so end up producing where $MC = MR$, an outcome whose significance will be discussed later), society might still be inclined to think of the particular resource as exhaustible, particularly if it is considered essential. This could be the case if society believes there will be a sizable demand for the resource long after the most optimistic forecast of its exhaustion date—a demand that would exist if expensive durable goods had been constructed whose operation was contingent on large inputs of this resource. Under these circumstances attempts may be made to get the industry to lengthen its time horizon, which is another way of saying that it should attach more significance to temporally remote profit streams. In the more abstract literature this often involves the possibility of imposing a different (lower) discount rate on industry decision makers than the one they may consider appropriate. As a result, demand in the present or near future is reduced, which means not only that more of the resource is preserved for later use, but also that the price rise that accompanies the displacement of production toward the horizon could serve as an incentive for further exploration and for the development of substitutes. Another possibility, of course, is simply to pass laws that amount to a direct lengthening of time horizons. A case in point here are the restrictions on the flaring of associated natural gas in the United States.

Now we look at the technical aspects of this matter with the intention of getting a simple rule for the optimum behavior of a mining firm. *Optimum behavior* in this context means maximizing the present value (*PV*) of profits over a finite time horizon. This maximization takes place when, with a given pattern of intertemporal production, the transfer of a unit of the mineral from one period to another (regardless of which periods we are talking about) leads to a fall in the present value of profits. Thus conceptually we can start out with any pattern of production and reach our optimum by transferring units of output across periods in such a way that each transfer results in an increased *PV*—stopping only when transfers of this nature become impossible. The basic assumption here is that the mining firm is in possession of the resource deposit, and already has the physical capital and structures required to extract and sell any portion of the remaining deposit in any period. Our expression for present value is then:

$$PV = \frac{p_1 q_1 - c_1 q_1}{(1 + r)} + \frac{p_2 q_2 - c_2 q_2}{(1 + r)^2} + \cdots$$

$$+ \cdots + \frac{p_i q_i - c_i q_i}{(1 + r)^i} + \cdots + \frac{p_T q_T - c_T q_T}{(1 + r)^T}.$$

Here p_i is the expected price in period i, and c_i the expected unit variable cost of producing an amount q_i. The expression $p_i q_i - c_i q_i$ will be called the total profit for period i, although stricly speaking it is a *quasi-rent* because c_i is a variable cost. Both p_i and c_i are usually functions of q_i, in which case $p_i = p_i(q_i)$ represents the expected demand curve for the mineral in period i. $(p_iq_i - c_iq_i)/(1 + r)^i$ is thus the discounted profit (the discounted quasi-rent). Now we can introduce the important expression *marginal profit* (*MP*), where $MP = MR - MC$, and discounted marginal profit (M'), which is $MR_i - MC_i/(1 + r)^i$ for period i. MR, of course, signifies marginal revenue. Note that we are introducing the possibility of having a nonzero MP at the point where profits are maximized.

We now go to the rule mentioned earlier. Suppose that PV is a maximum, but

$$\frac{MP_i}{(1 + r)^i} > \frac{MP_j}{(1 + r)^j}.$$

The reader experiencing difficulty in working in terms of periods i and j should simply work in terms of two adjacent periods—for example, 0 and 1 or 1 and 2. Now, with the aforementioned relationship holding, if we taken one unit of the minerals and tranfer it from period j to period i we get:

$$\Delta PV = \frac{(MP_i)(+1)}{(1 + r)^i} - \frac{(MP_j)(-1)}{(1 + r)^j} > 0.$$

Thus PV was *not* a maximum. This immediately implies that the profit-maximizing condition is:

$$\frac{MP_1}{(1 + r)} = \frac{MP_2}{(1 + r)^2} = \ldots = \frac{MP_i}{(1 + r)^i} = \ldots$$

$$= \frac{MP_T}{(1 + r)^T} = \lambda.$$

This arrangement does not preclude the possibility that $MP = 0$ for all periods; but it must be made clear that we cannot have $MP = 0$ for some periods, and $MP > 0$ for the remainder. What we can have, however, is $q = 0$ for some, but not all, periods. This is so since it is fully conceivable that, in line with the foregoing demonstration, all the output has been transferred out of one or more periods. Moreover, as output is transferred out, it can happen that MP will increase, since MR can increase and MC decrease. Continuing, we see that if $MP_m \neq 0$ for some period m, then $\lambda \neq 0$. Accordingly:

$$\lambda = \frac{MP_m}{(1 + r)^m},$$

or:

$$MP_m = \lambda(1 + r)^m \qquad \text{or} \qquad MR_m - MC_m = \lambda(1 + r)^m.$$

That is,

$$MR_m = MC_m + \lambda(1 + r)^m.$$

WE see here that we have departed from the familiar profit-maximizing condition $MR = MC$ if $\lambda(1 + r)^m \neq 0$. The reason is that in the world of exhaustible resources, a unit extracted and used today is unavailable tomorrow. This unfortunate situation is taken into consideration by assigning a *scarcity rent* or *opportunity cost* to the resource. In the foregoing $\lambda(1 + r)^m$ is the current value of the scarcity rent for period m, and λ is the present value. If, within a given time horizon, the commodity is not subject to exhaustion, then it is not scarce, and the scarcity rent is zero. Accordingly, $MR = MC$.

Some Practical Matters

At this point it can be emphasized once more that in many natural-resource industries there are hundreds of small firms in control of diminutive amounts of a mineral, which certainly function as price takers. The firms that must necessarily interest us, however, are the large, oligopolistic organizations (including government-owned companies) whose production decisions can influence the market price. Equally important, these firms own not only mines but also the facilities for processing mine output and distributing it to markets all over the world. In the context of our previous discussion, this means that most owners of major processing and distribution facilities are in no position to permit these facilities to close down during certain periods—that is, to set q in the previous analysis equal to zero during one or more periods—just because of a forecast of sky-high mineral prices at some remote point in the future. The fixed or capital costs associated with these installations must be met regardless of output, and in the imperfect capital markets of the real world it is often impossible to borrow in order to cover these costs.

Uncertainty also plays a major part in determining just how much is going to be produced in the near (as opposed to the distant) future, as well as the price for which mining output is sold. Consider a mineral firm that is a one-man operation in which ore is removed with a bulldozer and shipped to a processor by truck. Assume, moreover, that both the bulldozer and the truck are owned by the proprietor of this firm. If he decided not to produce during a given period, he could hire both himself and his equipment out to another firm.

Despite his apparent flexibility, the owner of this firm might prefer to downgrade forecasts of high prices in some future year or years, and enter into commitments to deliver known quantities of ore at known or unknown prices over a given time horizon as a way of spreading the risk associated with uncertain future outcomes. Going to larger firms in the real world, we can also add that the ability to raise or lower production in the immediate future is generally quite restricted; and startup costs are often so high that many firms are prepared to accept substantial losses over a fairly long time in order to avoid closing down entirely. By the same token, since the prices used in the previous exercise were *expected* prices, around which there is a great deal of uncertainty, even in the short run, the individual firm would usually be uninterested in initiating large changes in production each time its price expectations changed.

A factor overlooked by most of the theoretical literature dealing with the mining firm is the selling of a great deal of mining output via long-term contracts: contracts that in some cases have a running time of twenty years or longer and that contain provisions for only minor adjustments in price and/or quantity. Just as mining firms have fixed costs that must be covered over a long period—and thus are in no position to seek customers and negotiate prices and quantities on a year-to-year basis—long-term commitments from buyers are necessary in order to cover the huge costs that are usually involved in obtaining and developing a deposit.

This means that the foregoing algebraic exercises, despite their elegance, are capable of explaining only a very small part of the behavior of real-world mining firms. In fact, taken at face value this type of analysis is palpably misleading. In the real world what might be called the average output per period is decided on for a mining firm, and often this is supposed to cover the entire life of the mine. Only minor provisions are made for sharp increases in production in response to sudden increases in actual or expected prices—although, conceivably, very poor market conditions could lead to a drastic scaling down of production. As mentioned, a large part of the output of a mine is sold on the basis of nearly fixed conditions in long-term contracts; although management tends to be extremely interested in long-term price forecasting, forecasts of this nature have little influence on production programs.

The World Iron and Steel Industry

From an historical point of view the steel industry can be placed at the heart of the industrial revolution. Almost from the inception of this industry (in England during the eighteenth century), it provided the main impetus for the accelerated industrialization of western Europe and North America. By 1913 slightly more than 50 percent of global steel-making capacity was located in western Europe, 43 percent in North America, and most of the rest in Russia. At that time Japan produced less than 1 percent of world output, although things have now changed considerably. Between 1950 and 1974 Japanese steel production increased at an average annual rate of 13 percent. As a result, Japan now has the second-largest—and the most efficient—steel industry in the world. Today western Europe and North America together possess 40 percent of the world steel industry, Japan has 20 percent, and other industrial countries and the Third World have the rest. As will be made clear later, Brazil has become the Saudi Arabia of iron ore, which means that at some point in the future its steel-making potentialities could rival those of Japan. As Getulio Vargas made clear in 1931: "The most basic problem of our economy is that of steel. For Brazil, the age of iron will signify its economic opulence." The largest producer of steel in the world is the USSR.

Bad news descended on the world steel industry along with the energy crisis. With aggregate rates of economic growth halved throughout the industrial world, excess steel-making capacity appeared almost everywhere. Initially many firms reacted to this phenomenon by increasing their capacity—that is, modernizing existing capacity and constructing new plant in an attempt to gain a competitive advantage over the rest of the industry. This kind of response, which perhaps made sense during previous postwar business-cycle downturns, turned out to be disastrous since the expected reversal in world economic fortunes has not taken place and does not appear to be just over the horizon.

The United States, Japan, and Brazil

The United States, Japan, and Brazil are the past, present, and perhaps the future leaders of the world steel industry, respectively. The U.S. steel industry is in deep trouble. Some say that it is a dying industry in the process of partial liquidation. From the point of view of many steel-industry employees, this may as well be true, since if the steel industry survives in the United States it will be only because many operations now carried out by human beings will be handled by machinery or electronic apparatus. This is so because increasing amounts of low-price steel are being imported into the United States. Much of this steel originates in the Third (or Second) World, which has both inexpensive labor and modern technology, and more steel imports are on the way.

Actually, the problems of the U.S. steel industry began in the 1960s, when enough capital suddenly became available to finance state-of-the-art installations outside North America. First, money moved to countries with weak unions and low wages—and it moved in huge amounts. Citibank (or Citicorp), Chase Manhattan, Manufacturers Hanover, Bankers Trust, and other financial giants in the United States gave firms like Nippon Steel, Nippon Kokan, Kawasaki Steel, Sumitomo Metal, Kobe Steel, and Nisshin Steel all the credit they needed to establish more than 100 million tons of supermodern steel-making capacity. Similar financial services are now available for countries like South Korea and Taiwan. Together these low-wage, high-efficiency countries, with few, if any, distractions such as labor unions, may have the same steel output as Japan by 1985 if the international consumption of steel regains even a fraction of the momentum it had before 1973.

There have been other difficulties for this industry, however. Steel-industry executives sum these up as inflation (which has had an unfavorable influence on wages and salaries), inadequate depreciation allowances, and low capital investment. The last two of these ostensibly go together, since had the U.S. government permitted a faster writing off of capital equipment, more money would have been available for investment in new facilities or for modernization. Some observers have claimed, however, that there has not been an impressive relationship between steel-industry cash flow and/or profits, and investment in new equipment. Besides, the multinational banks that provided the capital to reconstruct the German and Japanese steel industries could (or should) have been willing to underwrite the resuscitation of such industrial giants as Bethlehem or U.S. Steel. It appears that the basic shortcoming of the U.S. steel industry had a great deal to do with certain defects of steel-company management, which made the mistake of not concerning itself too deeply with what was taking place on foreign shores—and, when it did concern itself, was overcome by a crisis of comprehension.

We now come to Japan, which has a totally different approach to the making of steel or, for that matter, to any major industrial or commercial undertaking. The first thing to make clear is that only a portion of the amazing economic performance of that country can be attributed to the unhampered functioning of the free market, or the so-called invisible hand. Hardly had Japan moved out from under the tutelege of General Douglas MacArthur and his military government when the old cartel masters emerged from their semiretirement. Though espousing conventional capitalism with its emphasis on private initiative, the survival of the fittest, and the glories of the consumer society, they obviously felt a few refinements were in order if the Japanese economy was to be kept in high gear.

Huge mergers became the order of the day, sometimes with the good wishes of the concerned managements and stockholders, but just as often without, since the majority of these mergers were blessed in advance by the Ministry of Trade and Industry, and/or by one of Japan's major financial institutions. A typical bank-inspired fusion resulted in three of Japan's major producers of special steels forming a new company, Daido Special Steel, which became one of the largest firms in the world in this branch and one of the most efficient. The logic behind this fusion was self-evident: unchecked competition would only lead to Japanese companies destroying one another and opening Japanese markets to foreigners.

Although the original Adam Smith made it quite plain that businessmen tended to use this kind of logic, he personally thought it deplorable, and vehemently assured his readers that no good could ever come out of the restraint or regulation of trade. Still, there can be little doubt of the therapeutic effect of this kind of reasoning on the Japanese economy, and in particular on the incentive to invest. In the period 1962–1973 Japan invested, on the average, about 33 percent of its gross national product. Corresponding figures for other countries were: France, 27 percent; West Germany, 26 percent; Canada, 22 percent; Italy, 21 percent; the United Kingdom, 20.5 percent; and the United States, 15 percent. As a reward for their diligence, the Japanese obtained in this period a total increase in GNP that was more than twice that of France and West Germany, and about four times that of the United States. Similarly, productivity increased in Japan at an average annual rate of more than 10 percent, as opposed to 5.5 percent in Germany and only 3.25 percent in the United States. Basically, what was happening in Japan was that economic policy functioned to slow down the rate of growth of consumption and ensure that the right firms were well supplied with financial resources. Firms carrying the designation *right* were principally those that showed a solid record of success in production and marketing and were quick to import and exploit foreign technology. Of particular importance was the willingness and ability to open up new sources of raw materials.

Furthermore, according to Professors Dale Jorgenson and Mieko Nishimizu, the level of technology as a whole in Japan overtook that of the United States in the beginning of the 1970s. Even earlier, however, the Japanese steel industry was far superior to that of North America because of a faster rate of investment that led to a much higher output per employee, which in turn led to higher profits, which led to still more investment in the latest and most efficient technology. According to a 1977 CIA report, Japanese blast-furnace design is the best in the world, and Japanese operations are the envy of the world's steel makers. The correctness of this judgment has been certified by the presence of Japanese experts in both the United States and Western Europe (for example, in Sweden) to advise on the design of new installations.

One characteristic of the steel industry that is not sufficiently appreciated is the enormous investment required just to maintain production at existing levels. In the United States this came to about $5 billion per year in the late 1970s, and can be explained by the enormous (and growing) capital intensity of the steel industry (which, in the United States, is about 2.5 times the manufacturing average). In a declining economy, however, there is no possibility of steel companies generating this kind of cash from their own operations, and little or no incentive to borrow it in capital markets where interest rates are now considerably higher than the yield on invested capital. On the other hand, the Japanese economy is still expanding at the highest rate in the industrial world, and therefore is still in a position to create the resources needed to keep the Japanese steel industry well ahead of its competitors in the other industrial countries.

What about competitors in the nonindustrial or Second World countries? A report by a spokesman of the U.S. Steel Corporation claimed that another Japanese miracle cannot take place in the near future, and that eventually foreign intrusions into the domestic U.S. steel market can be stabilized at a level that leaves room for a very large domestic steel industry. Nevertheless, it seems likely that the only thing that can shield the U.S. steel market from the overzealous advances of the Brazilian, South Korean, and Taiwanese steel industries are tariffs and other import restrictions of the type with which the Japanese have been threatened, at least at the present time. It is true that these countries use a great deal of steel locally, but at the same time large exports are necessary in order to repay foreign debts and to purchase such things as coking coal and iron ore. Brazil is also concerned with achieving complete self-sufficiency in steel; but if the world economy begins to function normally again, it seems likely that the Brazilians will begin a systematic drive to expand their share of international markets.

The principal advantages possessed by the Brazilians are the following: first, a large slice of their industry features the latest technology, much of which was obtained from the Japanese—who, after the United States and West Germany, are the third-largest investors in Brazil. In addition, labor in Brazil is both inexpensive and technically competent. Here we face the other side of the Brazilian miracle: a situation where real wages often decline over long periods, and job-related rights that are taken for granted in the United States and Western Europe are distinguished by their absence. Brazil also has an enormous amount of high-quality iron ore. Brazilian reserves of this indispensable commodity are the largest in the world outside the USSR, but they are probably of better quality than Soviet reserves and cheaper to exploit. Later on it may be possible for Brazil to tie the export of iron ore to the export of steel, in the same manner that some Middle Eastern countries intend to associate the sale of oil with that of refined products and petrochemicals.

The only flaw in the pattern has to do with energy. For all its wealth of natural resources, Brazil is poor in energy materials. The most important deficiency is the almost total absence of high-quality coking coal; but the steel industry is a big consumer of energy in all forms, both directly and indirectly. Foreign arrangements such as long-term contracts with Australia and South Africa may solve the problem in the short run; technological advances that involve direct reduction processes and/or use some nuclear heating in the steel-making cycle might eventually provide a long-term solution.

To get a very distinct picture of just what can be done in the Second World, however, we have to look at the Pohang Iron and Steel Works in South Korea. The experts said that this installation (the eleventh largest in the world, and larger than any steel plant in the United States) could not be profitable, but it has in fact shown a profit almost from the beginning. Technical expertise for designing and building the plant came, naturally, from Japan; and at present the Pohang plant offers steel to international buyers at a price that is appreciably lower than that listed by the main Japanese steel exporters. The principal shortcoming in Korea, as in Brazil, is the absence of local energy inputs; in addition, South Korea must import more than 90 percent of its iron ore. Just now this is no problem, since the surplus of iron ore on world markets is even larger than that of oil; but it could cause some difficulties later on.

Steel and Iron

So far we have discussed the steel industry without saying exactly what steel is. Essentially, steel is iron alloyed with carbon or some other minerals, and it is produced by two basic techniques. Figure 6–9 shows the *hot-metal* method, which involves the production of high-grade molten iron (pig iron) by firing iron ore, sinter, pellets, or briquettes with coke (metallurgical coal raised to a high temperature) in a blast furnace. Other inputs at this stage are limestone, fuel oil, pulverized coal, and natural gas (these materials are sometimes called *fluxes*). The molten iron is then combined with carbon, other alloys, and scrap steel at temperatures up to 3,000° F. This activity takes place in Bessemer furnaces, open-hearth furnaces, or basic iron furnaces (BOF). The BOF (and of late the so-called super BOF) are considerably more efficient than the others; in particular, their operation and service requires much less labor per unit of output than the others. One popular

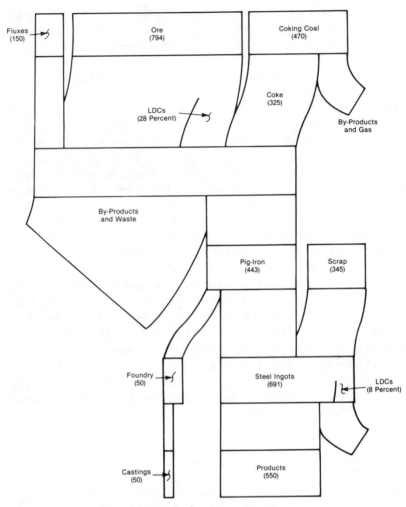

Source: United Nations Industrial Organization.
Note: In millions of tons.
Figure 6-9. Simple Flow Diagram of the World Iron and Steel Industry, 1974

measure of the modernity of a particular country's steel industry is the percentage of BOF in its stock of furnaces. In the mid-1970s it required 1.7 tons of iron ore to produce 1 ton of pig iron, and 0.7 tons of pig iron to produce 1 ton of steel.

As shown in figure 6-9, scrap is important in the steel-making process. Of the 345 million tons of scrap shown, 200 million were purchased from external sources and 150 million tons were generated in the industry. (This latter component is called home scrap.) The technology of steel making generally permits a broad tolerance in the mix of scrap and pig iron, but steel mills have been developed whose principal input is scrap. These use a *cold-metal* technique for producing steel, in which the blast furnace and coking ovens are eliminated (through eliminating the pig-iron stage), and scrap steel or some form of solid-state iron is directly refined into molten steel in an electric-arc furnace. This sort of technology is of particular interest to LDCs because of the comparatively low cost of shipping scrap, and for that matter is of interest to any country that might have some problem in finding enough space for a fully integrated steel mill. Electric-arc furnaces are the principal component of so called mini mills, which are smaller and less costly than full-fledged steel mills.

We can now examine some terminology. The form in which steel leaves the furnace is that of an *ingot,* although it is sometimes cast directly into molds for shapes that are very large or complex. *Finishing* consists of working steel ingots into their final shape via *hammering,* pressing, and rolling. All these operations require very high temperatures. In the process of rolling, the ingot is first passed through a cogging mill and then sheared into blooms, or slabs. These are then rolled into *billets, small sections, bars, plates, rods, wide strips, and sheets.*

Billets, rails, bars, and some rods are usually cut into standard lengths, whereas plates are cut to specification. Billets can be used for the manufacture of tubes or pipes by being rerolled as bars and pierced. They can also be made into such things as motor axles by forging or drop forging. Nuts and bolts, parts for automobiles, bicycles, machinery, shafts and gears, and seamless tubes are made from bars. Rods are turned into wire, rivets, needles, springs, and so on. Plates are employed as such in many heavy industries and in the making of tubes. Wide strip is sheared and after further processing is turned into tinplate. It is also used for car bodies. Pipes are manufactured from such items as strip, sheets, and plates. They can also be rolled from pierced or solid billets and bars.

The possibility exists for making steel using a nuclear reactor instead of a blast furnace. Japan, in particular, is interested in this technique, and may already have one or more of these installations in operation. The Japanese are very concerned with finding a cheaper source of energy for the steel industry, and it seems that electricity generated by nuclear reactors is more economical than the fossil-fuel inputs of blast furnaces. The Japanese also want to sidestep a shortage of high-grade coking coal, should that eventually arise; and there is a great deal of concern with the rising level of sulphur dioxide in the atmosphere that is caused by the technology now used to make steel.

We can now go to the consumption of iron ore. The nine member states of the European Economic Community (EEC), Japan, the USSR, and the

United States accounted for 90 percent of the iron ore consumed by developed countries in the 1955–1975 period. A similar situation exists with LDCs, with India, Brazil, and Mexico accounting for a large share of the Third World's ore consumption. In the period 1955–1965 the consumption of iron ore grew at an average annual rate of 6.2 percent; this figure slumped to 4 percent during 1965–1975.

Technological change in this industry has worked largely to economize on and improve the efficiency of blast-furnace inputs. A case in point is coke, whose cost is rising rapidly. The ratio of oven coke production to raw-steel output declined from 0,81 in 1960 to 0.59 in 1970, and is expected to reach 0.48 in 1980. Similarly, appreciable gains have been made in raw-material blending and beneficiation. Beneficiation principally covers enriching or concentrating the ore, which in turn means increasing its iron content while eliminating many impurities. It also has to do with agglomerating the ore through *sintering* and *pelletizing*.

Ore enrichment through crushing, grinding, and screening has been standard practice in the iron-ore mining and using industries for years, but of late the economies of blast-furnace operations have tended in the direction of requiring an even higher iron content of ore, uniform particle size, and so on. This in turn has necessitated more sophisticated processing operations. Sintering consists of agglomerating, or lumping together, the fines of an ore, whereas pelletizing involves reducing the ore to small, ball-shaped particles or pellets of uniform size and composition. These processes have facilitated the exploitation of poorer-quality ore and have made extensive economies possible in the transportation of blast-furnace inputs. Ten years ago very fine ores, those with a size of about 10 millimeters, were generally considered unsalable. At present there is a very high demand for this type of ore. Correspondingly, the percentage intake of the world's blast furnaces of sinterized and pelletized ore has increased from one-half to three-quarters in the last dozen years, and the use of iron ore at a lower stage of processing has declined accordingly.

On the supply side it has now been established that there are huge reserves and resources scattered about the world. The USSR possesses the largest amount of reserves and resources. North America is thought to possess one-third of world resources, whereas Latin America and Asia (mainly India and China) have 10 percent. Australia also has vast resources, whereas Africa is the only large region in which substantial resources have not been charted. This may be due to the fact that geological investigations have not been as extensive in Africa as elsewhere. Table 6–10 provides some information about iron ore production and reserves.

It is interesting to observe that Australia banned the export of iron ore between 1938 and 1960 because the governments of that country thought that domestic reserves were insufficient. As a result of more detailed explo-

Table 6-10
Reserves, Production, and Exports of Iron Ore, 1977

Country	Iron-Ore Reserves			Production			Exports[a] (Million Tons)
	Billion Tons (Ore)	Share (%)	Billion Tons (Iron Content)	Billion Tons (Ore)	Share (%)		
USSR	110.0	43	28.1	244	29		43.6
Canada	36.6	14	10.9	54	6		36.7
Brazil	27.2	10	16.3	87	10		71.8
Australia	17.8	7	10.7	93	11		80.3
United States	17.3	7	3.7	58[b]	7		—
India	9.1	4	5.1	43	5		—
China	6.1	2	2.7	56	6		—
France	4.0	2	1.6	42	5		—
Sweden	3.4	1	2.0	26	3		20.4
Venezuela	3.3	1	1.5	15	2		—
Others	23.4	9	10.0	135	16		122.5
Total	258.7		93.1	853			

Source: U.S. Bureau of Mines.

[a]For the period 1975–1976.

[b]Production was abnormally low in 1977 because of a strike. Production in 1977 should have been 80–85 million tons.

ration, however, as well as the unveiling of a huge potential market in Japan for iron ore, the embargo was lifted. By 1970 Australia was the largest iron-ore-exporting country in the world.

Price, Costs, and Trade

Iron ore is not traded on any of the world's commodity exchanges. The price is determined on the basis of negotiations between sellers and buyers, taking into consideration global supply and demand. Contracts are drawn up specifying dates of delivery, quality of ore, and so on, these are of both the short- and long-run variety. *Short* means one year, whereas *long run* can mean up to twenty years. In general prices are fixed only for shorter periods; long-term contracts usually stipulate periodic renegotiation of the agreement.

There is no such thing as a representative price for iron ore. The problem here is that iron ore is not a homogeneous commodity; thus each contract requires a reference to specific grades of ore, each of which has a distinct price that is determined by its quality or qualities. Loading conditions and transport costs are also important in determining the price. The price of iron ore has fallen in real terms since the early 1970s, and it seems unlikely that this trend can be reversed in the near future—particularly when we consider the tremendous oversupply of ore now existing in the world, and the stagnation in demand that has been caused by the slowdown in aggregate rates of economic growth in the major industrial countries.

Steel prices have evolved differently, for the most part increasing steadily during the past thirty years. In 1977 the price of steel in the United States was 260 percent higher than it was in 1950, and the corresponding figure for West Germany was 491 percent. Naturally, LDCs producing iron ore are not happy about the discrepancy between iron and steel price trends, but there does not seem to be much they can do about it at present, given the large amount of ore located outside the LDCs. For a country like Brazil, however, it is clear that the export of iron ore is a poor substitute for the export of steel and steel products. This, as a matter of fact, is one of the reasons behind Brazil's intention to become a major exporter of steel before the close of this century.

As far as many steel companies are concerned, even the aforementioned price developments are unsatisfactory. The shrinkage of the world steel industry is, according to the directors of these firms, the logical result of steel prices not increasing fast enough relative to costs. The U.S. steel industry, for example, reveals the following cost structures: ore, 20 percent; energy (coal, fuel, and so on) 20–25 percent; wages and salaries, 35–40 percent; and capital charges, 6–10 percent. Given the rise over the past decade of such things as energy prices and the remuneration of steel-company employees, it would be surprising if profit margins were not placed under considerable stress. In truth, the financial statistics of such organizations as Bethlehem Steel and U.S. Steel have made rather sober reading over the past five years or so—though not quite as discouraging as the balance sheets of some of the firms in the U.S. automobile industry.

This brings us to an interesting accounting problem. Many steel firms in the United States claim that one of the most depressing burdens they have had to carry is an outmoded depreciation schedule. In 1979 the U.S. Treasury reduced the depreciation life of steel plants from 18 years to 14.5 years. The last section in chapter 5 shows that with a given investment, a reduced depreciation period increases the capital cost. Capital cost (or an analogy) can be subtracted from gross income to obtain net (or taxable) income; and so, as a result, the cash flow of the U.S. steel industry was increased by $60 million in 1979. Still, this was judged to be inadequate, given that the depreciation schedules in some of the other leading steel-making countries are even more favorable from the point of view of producers. For instance, a mill was projected at Conneaut (on the shore of Lake Erie) that provided for a labor saving of one-third when compared with other facilities in the United States; even so, the huge cost of the installation (about $4 billion) and such things as the fact that a new steel mill being built in Ontario, Canada (about fifty miles north of the proposed Conneaut mill), could write off its investment in one-fifth the time possible for the U.S. firm, led to the postponement of the project—perhaps indefinitely.

Several other observations seem appropriate before we say a few words about trade. First, with the increased unpredictability of inflation and exchange rates, some steel firms (as well as other types of firms) seem to have gone over to maximizing profit per unit of output, rather than total profits. In practice this involves for the most part high margin–low volume combinations rather than the conventional low margin–high volume pattern. This is a fundamental departure from the arrangement most textbooks deem typical of modern industrial society, although the reasoning of producers under this regime is simple and would probably fit into some kind of theory of profit maximization under uncertainty: if the expected profit margins per unit sold are very large, then management considers itself adequately protected against adverse movements in prices or exchange rates.

Then too, steel scrap—and to a certain extent some other steel products—is a good example of what are known as *flexprice* products. In flexprice markets prices react to changes in supply and demand without appreciable lags; in recession situations, as at present, they can decline very sharply. For example, the price of steel scrap fell by more than 25 percent in 1981; expectations are that when the recession begins to lift, the price of this—and other—flexprice commodities will recover very rapidly. This proved to be true during the 1974–1975 recession.

By way of contrast, in *fixprice* markets—where many final products are sold, as well as a wide range of intermediate inputs—prices change very slowly. Moreover, prices in this market react sluggishly to changes in the price of their flexprice inputs; in particular, they are very reluctant to fall. As a result, the curing of inflation with recessions generally tends to be costly, although unfortunately the record indicates that it is relatively effective, and governments often make a practice of resorting to this medicine when some other might be just as appropriate. In the United States, Britain, West Germany, and Japan, there have been twenty-seven slowdowns in economic growth since 1950, and in twenty-six of these the rate of inflation slowed significantly.

Since I have already discussed the trade in iron ore earlier in this chapter, I will confine myself to steel here. The leading industrial countries, taken together, are self-sufficient in steel; within this group, however, there have been some notable changes in status over the last decade. North America has changed from a net exporter to a net importer, and the opposite has happened to Japan. After 1968 many industrial countries became very concerned about lessening the impact of foreign competition on their steel industries, and apparently this attitude is more pronounced than ever at the present time. The United States concluded some voluntary restraint agreements with Japanese and European producers that were replaced, in 1977, by a system of trigger prices based on Japanese production costs. Nonetheless, these trigger-price arrangements have not sufficed to depress U.S. imports noticeably, and steel makers in that country insist that more restrictive measures are in order.

Because of the increase in intensity of this behavior, the Japanese have taken steps to slow down the growth of their steel industry. Of course, had they not accepted export restraint on the terms at which it was offered, another less favorable variety might have been forced on them. The loss of a steel industry is a serious matter for an industrial country with ambitions to retain that distinction. Of late, more formal agreements are being concluded under the aegis of the so-called Davignon plan. According to Davignon, the only thing that makes sense just now is to shave the world steel industry down to viable proportions. Just what this will mean in the long run cannot be gone into here, although the world steel industry lost 80,000 jobs in 1981.

Western Europe

The situation in Western Europe is similar to that in the United States in outline, though perhaps not in detail; it would be a mistake, however, to claim that the overall outlook is brighter. Capacity utilization was only about 60 percent in 1977, and no one really knows just how far output will fall eventually if present trends are not reversed soon.

The issue is not just international competitiveness, or its lack, but the simple fall in demand for steel and steel products that is being displayed all over the world. Moreover, cognizance must be paid to the well-established historical tendency for steel output and demand to grow much more rapidly in the initial states of industrialization than in later stages. This means that it is natural for Second World countries such as South Korea and Taiwan to increase their share of world steel production and to face the possibility of generating an exportable surplus. Obviously, this is a cumulative process in which fast growth is associated with increasing modernization, which in turn brings about still faster growth. The loser in all this is the western European steel industry, which is not growing, but shrinking; whose tempo of modernization is not particularly impressive, at least in a relative sense; and which must purchase the large amounts of energy required in this industry at ever increasing prices.

A highly visible indicator of the hard times that have descended on the European steel industry is its faltering profitability. According to the U.S. Federal Trade Commission, data collected by the International Iron and Steel Institute shows that for the world steel industry "the United States has the highest profit rate, and the European Community the lowest, when profit is measured by net income divided by sales. However when profit is measured by net income divided by stockholders' equity, the profit rates of the United States and Japan are approximately equal, and that of the European Community is again the lowest."

An interesting observation here is that if we concern ourselves with the movement of profitability over time, we find a steady and very similar rate of decline for the United States, Japan, and western Europe. Even if the *relative* profitability of the western European steel industry did not deteriorate with respect to Japan and the United States, however, western European market shares declined spectacularly during the 1960s, with exports to third countries as a percentage of world exports (net of interregional trade) falling from 45 percent in 1960 to 22 percent in 1970. During the same period imports to western Europe from third countries increased from 3 percent of total steel consumption to 8 percent.

Leaving these aggregate indicators, and going over to some microeconomic phenomena, it has been well established that the steel industry displays significant economies of scale. As a result, it is often claimed that something can be said about the modernity of a country's steel industry by observing the size structure of its firms—particularly the capacity of individual plants. On a worldwide basis the data show that in 1975 the average integrated steel firm in Japan produced 12.8 million tons of crude steel, compared with 3.4 million tons for a firm in the EEC. The corresponding figure for the United States was 5.3 million tons, and 5.2 million tons for the USSR.

If we take a closer look at steel production, we observe that another important indicator of plant modernity is the size of the blast furnace, since iron making involves the most capital-intensive process in steel making, and, in addition, this is the activity where the most important economies of scale can be realized. In 1960 blast furnaces in the USSR were the largest in the world, but by 1975 they were on the average less than half the size of Japanese blast furnaces. Most investigations show that large-scale economies are extremely important for the overall cost picture, and in pig-iron production very significant reductions are being obtained at outputs well in excess of one million tonnes per year. In fact, some investigations suggest that scale economies are not exhausted at annual outputs of almost 10 million tonnes.

Two other criteria for judging modernity are the share of open-hearth steel production in total production, and the share of crude steel output that is *continuously cast*. More than half the world's steel production takes place in basic oxygen furnaces (BOF), and the increase in the use of the BOF has generally been at the expense of open-hearth systems. The complete absence of open-hearth capacity thus seems to indicate that the Japanese steel industry, along with that of Belgium, is the most modern in the world with respect to this particular criterion. (If the recent riots by Belgium steelworkers have any significance, however, the steel industry of Belgium has other serious shortcomings.) Here it should be mentioned that this index of modernity is far from unambiguous because of the large electric-furnace capacity of many steel-making countries; these furnaces have certain advantages that are not possessed by other systems. In particular, they permit a more intensive use of scrap; and they are particularly useful to countries like Sweden that produce a large assortment of special steels.

As for continuous casting, this technique reached the breakthrough stage in Europe in the early 1950s. What we have here is steel being cast directly and continuously into strands, thus eliminating the need for mold casting of individual ingots. Yields are higher and costs lower when this process is used, but at present no more than 25 percent of the noncentrally planned world's steel production is continuously cast. As would be expected, Japan has the highest percentage of continuously cast steel in total output, whereas the United States has generally been lagging in the introduction of this technology.

The final indicator considered here is energy consumption per unit of output. In a world in which the cost of energy inputs appears to be growing faster than the cost of other inputs, and in which energy costs account for 20–25 percent of the total production cost of finished steel products, some equipment is abandoned much earlier than is usually the case in order to introduce facilities having a greater energy economy. Germany and Sweden have made considerable progress in reducing the input of energy in the steel-making process, but not as much progress as the Japanese. The Japanese may also have been more successful in such things as workers' protection and pollution control, which in the main industrial countries now make up 5–10 percent of total production costs. In the future these particular expenses could add 10–20 percent to the cost of new investment.

Conclusion

The reader can write his or her own conclusion to the saga of iron and steel. In my opinion, the deindustrialization of a large part of the industrial world is now underway. Some—perhaps most—of the industrial jobs and activities that are being lost will be lost forever. Thus some argue that it is only fair that steel, for many years a symbol of the industrial revolution and of the prosperity of the inhabitants of the industrial world, should lead the retreat.

Appendix 6A:
A Note on Economic Profit and Depletion Allowances and Taxes

The first problem that will be addressed in this appendix concerns *economic profit. To begin, let us take a situation in which money is borrowed to purchase an asset such as a machine whose cost is B.* Take *r* as the rate of interest, and assume that there are variable costs *E* connected with keeping the machine in operation. These variable costs might be such things as labor costs, raw-material inputs, electrical power, and so on. To keep this example simple, let us also assume that the asset depreciates (that is, falls apart) completely at the end of the first period (for example, first year) of operation. Then assuming that all payments associated with operating this asset are made at the end of the period, with the asset earning *R* in revenue during the period, the discounted economic profit on this asset during its 1-period lifetime is

$$V = \frac{R - B - rB - E}{1 + r}$$

As observed above, *B* is the cost of the asset; but this sum of money also represents the amount of debt that will have to be amortized if the asset was purchased with borrowed money. Similarly, *rB* is the interest charge on *B*. Rearranging the expression, we get:

$$V = \frac{R - E}{1 + r} - B$$

The term $R - E$ is called the quasi-rent. Then if $(R - E)/(1 + r)$ is greater than *B*, it must be true that *V* is greater than zero. This in turn indicates a positive profit on the asset. In other words, the yield of the asset is greater than the yield on a financial asset (such as a bond or bank account) in a perfect market, and thus the asset should be purchased. If, for instance, we took $r = 10$ percent, then if $(R - E)/(1 + r) > B$, the yield on this physical asset is greater than 10 percent.

In looking at, for example, costs and benefits in the preceding expression, we see that the interest charges (rB) have disappeared. In this particular form of the equation, it is unnecessary to designate B as the amortization charge, and sufficient to identify it as the cost of the asset. Of course, an accountant interested in business expenses—but not enjoying a grounding in economic theory—might find all this a bit peculiar; but remember that what we want to do is compare the yield of a capital asset with that of a safe financial asset, and assuming that the first equation in this appendix, which contains amortization and interest charges, is logically sound, there is no point in finding fault with the second equation, since this is derived from the first by simple manipulation.

Since the quasi-rent, which is obviously some kind of net return, does not involve either the interest or amortization charges, it does not matter whether the money to buy the asset was borrowed or, for example, originated internally in a firm as retained profits. Thus, in the example above, the source of the funds for purchasing the asset is irrelevant. It is sufficient to know that in a perfect market, which in particular is distinguished by the absence of uncertainty, economic profit is positive or zero for those assets actually acquired.

Where depletion allowances and taxes are concerned, I am basically concerned with some simple rules that are employed in the United States. Readers desiring a more theoretical exposition of the topic are referred to my forthcoming book, *The Political Economy of Coal*.

Depletion provides for the recovery of money invested in a wasting asset in the same manner that depreciation does for a capital asset. (Although, obviously, both oil deposits and machines share many similarities and in a comprehensive theory of capital could probably be treated in the same way). The way this is done is to subtract an annual depletion allowance from the gross income realized by a mineral deposit when computing taxable income. In the United States a formula known as *depletion by cost* has been developed for determining depletion in any given year. Explicitly, we have:

Depletion for a given year = depletion rate $\cdot$ units sold

$$= \frac{\text{cost of property}}{\text{total units in property}} \cdot \text{units sold}$$

Another method for getting the annual depletion for tax purposes for oil, gas, and mineral properties is to take a percentage of gross income, providing that the amount allowed for depletion is less than 50 percent of the taxable income of the property before depietion. For obvious reasons, percentage depletion tends to be used if it results in less tax than depletion based on cost. In the United States some typical depletion allowances are

Oil, gas wells, sulphur, uranium, bauxite, cobalt, lead, manganese, nickel, tin, tungsten, vanadium, zinc	22%
Gold, silver, copper, iron ore	15%
Coal, lignite	10%

I will now construct an example in which it is possible to see the working of these two rules. Let us take a mining property that was purchased for $2,500. The property contains 50,000 units of ore that will be extracted over a period of two years at a rate of 25,000 units per year. The revenue from the sale of this ore is $12,000. Total operating expenses are divided into salaries and wages, and other inputs. The first of these is taken as $5,000 and the second $1,000. There is also some equipment used to mine the ore, and annual depreciation on it has been calculated as $250. Now we can calculate depletion.

On the basis of the preceding formula, the depletion rate is $2,500/50,000 = 0.05$. The depletion allowance for the first year is thus $0.05 \cdot 25,000 = \$1,250$, and we have an identical calculation for the second year. Thus, having computed depletion by cost, we can use it to obtain one value of the taxable income:

Gross Income		12,000
Wages, salaries, other inputs	$-6,000$	
Depreciation	-250	
Depletion	$-1,250$	
	$-7,500$	$-7,500$
Taxable income	$+4,500$	

With a tax rate of 20 percent, taxes are $900. However, if we use a percentage of gross income to obtain our annual depletion and, for the purpose of this exercise, assume that this percentage is 22 percent, we get for the first year:

Gross Income		12,000
Wages, salaries, other inputs	− 6,000	
Depreciation	− 250	
	− 6,250	− 6,250
Taxable Income before depletion		+ 5,750
Depletion allowance = 0.22 · 12,000 (This is 22 percent of gross income, and since it is less than 50 percent of taxable income before depletion, this method of depletion may be used.)		− 2,640
Taxable income		+ 3,110

If the tax rate is 20 percent, taxes are $622. In this example, the 22-percent depletion rule is more favorable for tax purposes.

7

The Econometrics of Primary Commodities

This chapter will consider the econometrics of primary commodities. Primarily, my intention is to provide the reader with a survey that is easily read but covers the most important aspects of the topic. I will also make some theoretical comments on flow and stock models and on the theory of exhaustible resources.

The assumption is that the reader is familiar with elementary econometrics. In securing this background virtually any beginning textbook will suffice, for it should be clear by now that econometrics is the most limited tool in the economist's armory. Although econometrics is not explicitly "cynical and destructive," to use David Stockman's description, it is very often cynical and counterproductive; and I personally am aware of occasions on which it has been downright fraudulent. This sort of phenomenon will be avoided here by limiting the level of sophistication of the analysis, because it also happens to be generally true in econometrics that *more means less*.

Still, some econometric models have taught us a great deal about commodity markets. The great breakthrough here was perhaps the copper model of Fisher, Cootner, and Bailey (1972), although the model of the wool market by Witherell (1967) was very instructive. Copper models were, in fact, the best-thought-out models in the commodity field for a long time, but of late some impressive work has been done on aluminum by Fisher and Owen (1981) and by Hojman (1980, 1981). As an aside I can mention that many of us who had spent a great deal of time working with metals were convinced that aluminum did not offer good prospects to potential modelers—but obviously we were mistaken.

Dynamic Analysis

The type of analysis that will interest us in this chapter falls under the heading of dynamic analysis, because the observation of such things as demand and price take place over time rather than at a specific point. As conventionally defined a system becomes dynamic as soon as dated variables relate different time periods, or link the past to the present and future

259

through trends, time lags, rates-of-change variables, and expectations hypotheses or adjustment processes. Most of this should become clear later in the chapter.

One of the principal purposes of dynamic analysis is to study the time path of variables that are passing from one equilibrium to another, and that were displaced from the original equilibrium because of the shift in a parameter or exogenous variable. Consider, for example, a simple demand curve for coffee. Demand is clearly a function of price, but it is also dependent on income: if income increases, so does the demand for coffee at every price. This means that an increase in income shifts the demand curve for coffee to the right. Here income is an *exogenous* variable where coffee consumption is concerned because—except in the coffee-producing countries—it influences the consumption (and price) of coffee without being influenced itself. On the other hand, price and quantity are endogenous variables because they are determined *in* the model on the basis of behavioral relasionships and assumptions about market clearing. A model of this type might take on the following appearance:

$$s_t = \beta_0 + \beta_1 p_{t-1} \qquad \beta_1 > 0. \tag{7.1}$$

$$d_t = \alpha_0 + \alpha_1 p_t + \alpha_2 y_t \qquad \alpha_2 > 0, \alpha < 0 \tag{7.2}$$

$$s_t = d_t. \tag{7.3}$$

Through simple manipulation we get:

$$\alpha_1 p_t - \beta_1 p_{t-1} = \beta_0 - \alpha_0 - \alpha_2 y_t. \tag{7.4}$$

This is a first-order difference equation; it took this form because we assumed that supply is a lagged function of price—that is, suppliers react to the market price attained in the previous period. To examine equation 7.4, we can begin by inquiring into the nature of the equilibrium in the event that one exists, which is not always certain. We must recognize here that in our most common form of equilibrium, as in physics, we have stationarity. Here this means that we have $p_{t-1} = p_t = p_{t+1} = p_{t+2} = \ldots$, which means that we can call the equilibrium price $\bar{p}(= p_{t-1} = p_t)$. Substituting this into equation 7.4 gives:

$$\bar{p} = \frac{\beta_0 - (\alpha_0 + \alpha_2 y_t)}{\alpha_1 - \beta_1}. \tag{7.5}$$

In dynamic economics we usually concern ourselves with the train of events from one equilibrium to another. If we assume that the αs and βs in these equations are constant, then disturbances that cause a movement from one equilibrium to another originate in changes in y. For the moment consider a situation where we have a value of $y = y'$, which at time t_0 goes to y''. Taking $\bar{p}'$ as the value of p at the initial equilibrium, we can immediately establish the price at the new equilibrium as $\bar{p}''$ from equation 7.5, but until we have a complete solution for equation 7.4, which is a first-order difference equation, we do not know if or how we attain the new equilibrium. Figure 7-1b, 7-1c, and 7-1d demonstrate three possible outcomes of the change in y shown in figure 7-1a. Note that in one of the last two cases we have an oscillatory movement that converges to a new equilibrium, whereas in the other we have an oscillatory movement that converges away from the new equilibrium.

It is also interesing to note that from equation 7.5 a kind of long-run multiplier can be obtained. This is:

$$\frac{d\bar{p}}{dy} = \frac{-\alpha_2}{\alpha_1 - \beta_1}. \tag{7.6}$$

Since α_1 is normally negative, an increase in y leads to an increase in p. Again, remember that the equilibrium implicit in this multiplier does not have to be attained. It is a possible equilibrium, dependent on the dynamic properties of equation 7.4.

Difference Equation

Solving the first-order difference equation 7.4 is simple. We already have our steady-state solution, as given in equation 7.5, and what we need now is a complementary solution. This solution has to do with the movement of the variable around the equilibrium, or its damping properties. To obtain this solution we deal only with the homogeneous part of equation 7.4 or:

$$\alpha_1 p'_t - \beta_1 p'_{t-1} = 0 \tag{7.7}$$

This could be solved through simple iteration, taking $p'_{t-1} = p_0$, and thus $p'_t = (\beta_1/\alpha_1)p_0$, $p'_{t+1} = (\beta_1/\alpha_1)\,p'_t = (\beta_1/\alpha_1)^2 p_0$, $p'_{t+2} = (\beta_1/\alpha_1)^3 p_0$ and so on. We thus see that if we take as a solution $p'_t = (\beta_1/\alpha_1)^t p_0$ and substitute this into equation 7.7 we get $\alpha_1(\beta_1/\alpha_1)^t p_0 - \beta_1(\beta_1/\alpha_1)^{t-1}p_0$, which equals zero. Our problem is, however, that we are solving equation 7.4, not just equation 7.7 and so must have a solution that will also satisfy 7.4.

This solution has two parts: the steady-state solution plus the complementary solution, or $p_t = \bar{p} + p'_t$. Unfortunately the solution to equation

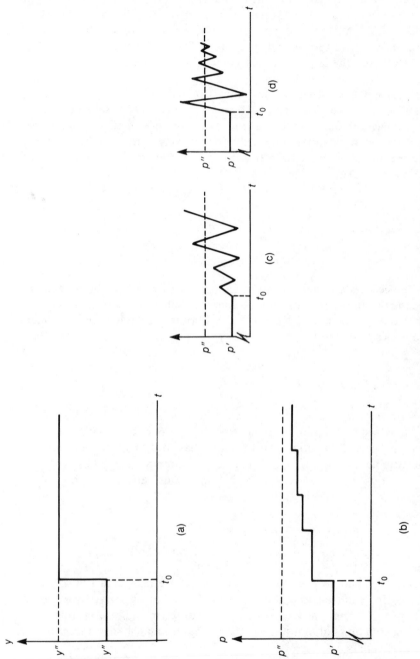

Figure 7-1. Price Movements Following a Change in the Variable y

7.7 will not work as a solution to 7.4 as it now stands, but it can be easily modified. Take as a solution to equation 7.7, $p_t' = A(\beta_1/\alpha_1)^t$, which the reader can easily show to satisfy 7.7 by direct substitution. Now put this in $p_t = \bar{p} + p_t'$, which yields:

$$p_t = \bar{p} + p_t' = \bar{p} + A\left(\frac{\beta_1}{\alpha_1}\right)^t \qquad (7.8)$$

We now need a value for the unspecified constant A. If we take $p_t = p_0$ when $t = 0$ (where $t = 0$ is the beginning time) we get from equation 7.8, $p_0 = \bar{p} + A$, or $A = p_0 - \bar{p}$. Our solution is then:

$$p_t = \bar{p} + (p_0 - \bar{p})\left(\frac{\beta_1}{\alpha_1}\right)^t \qquad (7.9)$$

Using equation 7.9 and the value of $\bar{p}$ from equation 7.5, we can show by direct substitution that this is a solution for equation 7.4. We should also notice that if $0 < (\beta_1/\alpha_1) < 1$ we have a movement from one equilibrium to another like that shown in figure 7-1d. In this case the initial equilibrium is at p_0, and the new equilibrium at $\bar{p}$. If we have $(\beta_1/\alpha_1) < -1$, then we get the situation shown in figure 7-1c, where the movement is away from equilibrium.

We have dealt here with a first-order difference equation with constant coefficients. Second-order (linear) difference equations are only slightly more complicated, and readers who want a thorough insight into their intricacies are referred to Baumol (1970). However, difference equations of higher than second order and nonlinear difference equations present a number of difficulties, which is why they are studiously avoided. Thus within certain branches of economics—such as dynamic inflation theory—we seldom find a higher-order difference equation, although we have no a priori reason to believe that first- and second-order linear difference equations were specifically ordained to describe the real world.

A Stock-Flow Model

Some elementary aspects of stock, flow, and stock-flow models have already been discussed in chapter 6, and the reader is strongly advised to review that basic material—although the following exposition is only slightly more complicated.

To begin, remember that a *flow* designates a quantity per unit of time at a point in time, while a *stock* is a quantity at a point in time. In the markets for many primary commodities, there is a demand for a stock to be held for speculative and precautionary purposes, and there is also a demand for these commodities for flow purposes—that is, as current inputs in the production process. We also have found it to be true that we can infer directly from actual markets that as the ratio of the inventory of an item to its current consumption increases, its price decreases. In chapter 6 I say that the basic flow supply-demand model of the elementary textbooks did not make these things clear, and as a result suggested an extension of this model. What we want to do here is consider a further extension—specifically, one that will include all those factors necessary for understanding price formulation on the markets for nonfuel minerals.

We can begin by looking at a stock demand and stock supply curve. Stock supply is a datum. It is the amount of an item that exists at a given time and therefore is not a function of price or anything else. In contrast, stock demand is normally a function of present price, *ceteris paribus*. In other words, at any given time market participants have certain expectations about the future price of a commodity; and so if the present (or spot) price decreases, the spread between present and future price increases, thus making it more profitable to acquire a larger inventory. Eliminating the *ceteris paribus* clause, we should also expect a relationship between stock demand and, for example, the interest rate, expected future prices, and even flow demand.

Normally if the interest rate were to fall, the cost of financing a given amount of inventory would decrease, and thus the stock demand (which is the demand for inventory) would tend to increase. Similarly, if the present price were constant, an increase in the expected future price would cause the expected profit from each unit held in inventory to increase, and stock demand would therefore increase. Finally, it seems that the larger the amount of a commodity used for current production, the larger the amount held in inventory. Typical stock supply and demand curves are shows as $\bar{S}'$ and D in figure 7-2a.

The stock supply curve is simply designated $S = \bar{S}'$, while the equation for the stock demand curve is written $D = D(p_t; p_t^e, r_t)$. Here p_t is the price in period t, p_t^e is the price expected in some future period, where the time at which we fashion the expectation is t; and r_t is the interest rate prevaling at time t. Obviously, many other variables could fit into the argument of this function. Flow supply and demand are shown in figure 7-2b and are algebraically designated $s_t = s_t(p_t)$ and $d_t = d_t(p_t)$.

Notice the situation at A and A'. Here we have a full stationary equilibrium, with flow supply equal to flow demand and stock supply equal to stock demand. In each period q_0 is produced, which is consumed in the current production process. The stock of the item is constant at $\bar{S}'$, and the price is also constant and equal to $\bar{p}'$. Next assume that the expected future price increases. We show this by shifting the stock demand curve to the

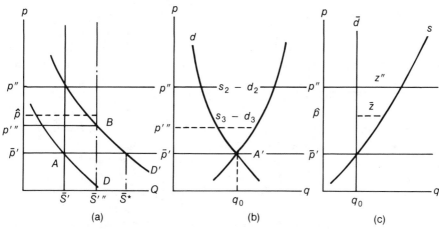

Source: Reprinted by permission of the publisher from Ferdinand E. Banks, *Scarcity, Energy, and Economic Progress* (Lexington, Mass.: Lexington Books, D.C. Heath and Company, 1977), p. 161.

Figure 7-2. Stock and Flow Demand and Supply Curves

right: At every value of the present price, more of the item is demanded by profit-conscious stockholders. This increase in stock demand causes an increase in price, which leads to an augmentation in the stock of the commodity. This works in the following way.

With an increase in price, flow supply becomes larger than flow demand. As shown in the diagram, the price increase caused the present consumption of the commodity to fall, while profit-maximizing producers increased production. We now face a situation in which every period the stock is increasing, since as long as price is greater than $\bar{p}'$ we have an excess of production over present consumption. In the diagram price initially goes up *toward p''*, where the excess of production over consumption is shown by $s_2 - d_2$. Given the model, this amount would be added to stocks during the first period *if* the price reached *and* stayed at p''; but remember that as various consumers obtain their stocks, the pressure on the market decreases and the price falls. For instance, by the end of the period, the price *might* have fallen to p''', in which case the addition to stocks would be some amount between $s_2 - d_2$ and $s_3 - d_3$. In figure 7-2a the exact amount is shown as being equal to $\bar{S}''' - \hat{S}'$. What the reader should pay attention to here is the vagueness with which the value of price is delineated. The reason for the vagueness is that an extensive study of the price dynamics of this model, using linear supply and demand relations, has convinced me that it is impossible, a priori, to say exactly where the price will be after the initial displacement from equilibrium. Ideally, however, things will function more or less as they are being described here.

At the beginning of the next period we are at point B on the stock demand curve, and we go through the same sequence as before: excess flow supply, increase in stocks, and a decreasing price. Eventually, *if all goes well*—which means no oscillatory price movements – stocks will reach the desired level. For figure 7-2a this means that the stock supply curve asymtotically, and smoothly, approaches the new equilibrium value $\bar{S}^*$. The price has again returned to the *full equilibrium value $\bar{p}'$* where flow demand is once again equal to flow supply, and all production goes to current use. In other words, we have stationarity on all markets with flow demand equal to flow supply and stock demand equal to stock supply. Even though we do not have a full equilibrium at prices above $\bar{p}'$, we may have some *market equilibriums* because in some periods producers are producing exactly the amounts that buyers desire for current consumption and for additions to inventories in those periods.

Because of the time element, a mathematical exposition of the preceding situation can become quite messy. But for simplicity, suppose that instead of the falling flow demand curve shown in figure 7-2b, we have a verticle flow demand curve as shown in figure 7-2c. This type of flow demand curve is certainly appropriate for such items as aluminum or zinc because of their low price elasticity of demand: in the short run we would not expect changes in price to influence the use of the item. We can then postulate two equilibrium situations. The first is a market equilibrium, and the relevant equations for this arrangement are given under column A in table 7-1. The second is also a market equilibrium, but at the same time it is a full equilibrium, with flow demand equal to flow supply, and stock demand equal to stock supply. This system is designated B in table 7-1.

Note that in this table the last equation under column A is not given, it is simply designated. As it happens, this is probably the most important equation of the system, and later I will present what I consider to be a feasible candidate for this equation and solve it explicitly for price in terms of some of the system parameters. Apart from this exercise, mathematical manipulations with this model will be avoided since a verbal discussion of certain key issues is much more rewarding. To begin, the reader should look at equation (c) in table 7-1, which is an equation for the investment demand i_t. To grasp what this is and how it is used, we can begin once more with a full equilibrium in figure 7-2, with $S_t = D_t$. Now assume that an increase in the expected price causes the stock demand curve to shift to the right.

This means that initially, before the price increases, we have an excess stock demand that, in figure 7-2a, is equal to $\bar{S}^* - \bar{S} = \bar{Z}$. The question that must be asked is if the price were to remain $\bar{p}'$, would we want all of $\bar{Z}$ now (or some now and some later)? Is *investment demand* equal to our *excess stock demand* in the initial period? If it is, then k in equation (c) is equal to unity. On the other hand it could be that we want only a portion of Z in the initial period, in which case k would be somewhere between unity and zero, and the investment demand would be less than the excess stock demand.

Table 7-1
Simple Stock-Flow Model

A	B
(a) $s_t = s_t(p_t)$	(a') $s_t = s_t(p_t)$
(b)* $d_t = d + i_t$	(b') $d_t = d = s_t$
(c) $i_t = k(D_t - S_t) = kZ_t$	(c') $D_t = D_t(p_t, p_t^e)$
(d) $D_t = D_t(p_t, p_t^e)$	(d') $S_t = D_t$
(e) $S_{t+1} = S_t + (s_t - d_t)$	
(f) $S_t = D_t$	
(g) Price equation	

Source: Reprinted by permission of the publisher from Ferdinand E. Banks, *Bauxite and Aluminum* (Lexington, Mass.: Lexington Books, D.C. Heath and Company, 1979), p. 128.
Note: The flow demand curve here is vertical, as in figure 7-2c. Total flow demand is shown in (b) (b').

Thus, regardless of price, whenever we have excess stock demand we must determine investment demand. It is investment demand rather than excess stock demand that is relevant in a given period. But once investment demand takes over, it becomes very difficult, if not impossible, to discuss this model with the help of simple diagrammatics. It is investment demand that determines how fast price rises and how far; and although algebraically this situation is not intractable, we may not be able to use the apparatus developed in figure 7-2. The following example gives some idea of why this is so.

Assume an excess stock demand at the beginning of period t. Also assume that investment demand functions so that buyers do not want any of the commodity in period t. This is the same as saying that there will not be any investment demand in period t: The excess stock demand will begin to make itself felt only as an investment demand in later periods. Thus although we would show a shift in the stock demand curve in figure 7-2a, we could not justify an increase in price in period t. To repeat: We have an increase in the excess stock demand in period t, but no increase in price in that period so the arrangement shown in figure 7-2a is inapplicable. What we are taking up now is a special situation, it is true, but it is certainly conceivable. Note, however, that this situation is not covered by equation (c) in table 7-1 because with k constant and not equal to zero, there is a positive investment demand in every period that we have an excess stock demand.

We can continue in this vein with a few more examples since it is important that the reader get some insight into just how complicated the dynamics of these equations can be. Assume that $k = 1$ in equation (c), which means that investment demand is equal to excess stock demand. Now assume an increase in the expected price of a commodity at time t, leading to an instantaneous increase in the stock demand and excess stock demand. In the very small interval immediately after this change in the expected price, the present price is still $\bar{p}'$, and in figure 7-2 we have an excess stock demand (and investment demand) of $\bar{S}^* - \bar{S}'$. Then the effect of this investment demand hits the market, and the price goes up very rapidly toward p''.

Note that this investment (or excess stock) demand impinges first on existing stocks. In the very short run there is no possibility of increasing production, so those individuals who are certain that prices will increase buy from those who are less certain; those who think that the price will go very high buy from those who think that the price will increase by only a small amount, and so on and so forth. Observe also that in figure 7-2a the price *could* eventually increase by enough to choke off excess stock demand. As shown, this price is equal to p''.

Assume now that producers go into action, taking the rapid increase in price as a signal that considerable production is warranted. But if they increase production by the amount z'' and attempt to sell it at price p'', they may not find any buyers. But if the price is slightly less than p'', say, $p'' - e$, then an amount can be sold equal to the difference between the stock demand and supply curve at price $p'' - e$. Continuing in this fashion, we can imagine an excess flow supply $\bar{z}$ which, at price $\bar{p}$, is equal to the excess stock demand at that price—and they are fortuitously brought together in the same period. This price, excess stock demand (which equals investment demand since $k = 1$), and excess flow demand constitute a market equilibrium, but not stationarity or a full equilibrium. Instead, the stock supply curve displaces to the right by the amount $\bar{z}$, and we begin our sequence all over again. If things continue in this fashion, the price eventually approaches $\bar{p}'$, and stocks approach $\bar{S}^*$. In other words, we approach a full equilibrium, with all production being used as a current input. The reader should take note of the following, however. In this example we have assumed, implicitly, some very simple inventory behavior on the part of producers. For example, if at the end of a period, they saw that they had overproduced, they simply reduce the price until they sold all of that period's production. But since, as we have made clear earlier in this book producers *are* in the habit of holding stocks, we might assume a type of behavior that had producers' unwilling to reduce prices below a certain level in the short run and thus putting their excess production into inventory. Making an assumption of this nature would mean, once again, that figure 7-1—or indeed any diagram—would be of only limited value in tracing the course of events.

Before going to the algebra, we can consider a simple variation on the preceding theme. Remember that the initial price signal to producers, following the increase in stock demand, was p''; but at this price it was postulated that excess stock demand had been eliminated. But consider a situation in which it is possible to increase production very rapidly, and so when the price reaches the vicinity of p'', production is increased by z''. A mistake has thus been made: Based on an assumption that price will be p'', production has been raised by an amount that, had the price actually turned our to be p'' (and stayed at this level during the period in question), could have been sold without producers taking a loss. But now, to sell all or part of z'' in the present period, the price will have to be reduced below p'', and with the given flow supply curve, producers will be taking a loss. Thus part of this production might be sold and the remainder put into inventory. But what part will receive what treatment? Even a simple measure such as this involves inventory costs as well as costs attributable to the unforeseen alterations in production that might have to take place as a result of unplanned inventory changes. As the reader might guess by now, this simple variation is actually quite intricate. It has led us into a situation of frustrated expectations, undesired producer stocks, and various disequilibria that probably can be treated only verbally case by case. Facile generalizations carried out with the aid of simple algebra and elegant diagrams are out of the question.

Now let us go to the price equation that was listed, but not spelled out, under (g) of table 7-1. Essentially what we need here is a simple behavioral assumption, and the one I will use that the change in price is directly proportional to the investment demand, increasing when this increases and decreasing when investment demand decreases. Ideally we would like to have a nonlinear relation here that could take into consideration such factors as excess capacity and faster rates of price increases as capacity becomes tight; but unfortunately I find the gap between wanting to derive such a relation and my ability to do so too much to bridge at the present time. The proposed equation is:

$$p_t = p_{t-1} + \lambda(D_t - S_t) \qquad (7.10)$$

Note especially that λ can be a more complicated variable than it appears in this equation. (For instance, since price movements should be a function of the investment demand, we might have $\lambda = \theta k$, in which as shown in table 7-1, $k(D_t - S_t)$ is the investment demand in period t.) We can now lag the previous expression to get:

$$p_{t-1} = p_{t-2} + \lambda(D_{t-1} - S_{t-1}) \qquad (7.11)$$

Subtracting, we get

$$p_t - p_{t-1} = p_{t-1} - p_{t-2} + \lambda[(D_t - S_t) - (D_{t-1} - S_{t-1})]$$

or

$$p_t - 2p_{t-1} + p_{t-2} = \lambda[(D_t - S_t) - (D_{t-1} - S_{t-1})] \qquad (7.12)$$

and

$$S_t = S_{t-1} + X_t \qquad (\text{where } X_t = s_t - d_t) \qquad (7.13)$$

Thus substituting this identity into the previous expression:

$$p_t - 2p_{t-1} + p_{t-2} = \lambda[(D_t - D_{t-1}) - X_t] \qquad (7.14)$$

Now we need an expression for $X_t = s_t - d_t$. As mentioned earlier, all supply and demand curves in this analysis will be taken as linear; thus we have $s_t = e + fp_t$ and $d_t = g + hp_t$. In addition, let us take $D_t = \alpha + \beta p_t$, which implies that $D_{t-1} = \alpha + \beta p_{t-1}$. Thus:

$$p_t - 2p_{t-1} + p_{t-2} = \lambda[\beta(p_t - p_{t-1}) - (e - g) - p_t(f - h)]$$

or

$$p_t[1 - \lambda(h - f + \beta)] + p_{t-1}(\lambda\beta - 2) + p_{t-2} = \lambda(g - e) \qquad (7.15)$$

This is a simple second-order difference equation of the type $p_t + a_1p_{t-1} + a_2p_{t-2} = a$, and it can be easily solved. Without going into this very elementary matter, it is apparent that depending on the values of e, f, λ, and so on, we can get such information as oscillatory price movements since the roots of this equation could be complex. In other words, we do not have the smooth progression from disequilibrium to equilibrium discussed in connection with the previous diagram. Finally, our equilibrium (when it exists) is the situation in which we have $p_t = p_{t-1} = \cdots = \bar{p}$. Thus, from the previous equation:

$$\bar{p} = \frac{g - e}{f - h} \qquad (7.16)$$

On the basis of the preceding equations, we see that this describes a *flow* equilibrium. But from the first equation in our exposition, we have $0 = \lambda(D_t - S_t)$, and since $\lambda \neq 0$, $D_t = S_t$. Thus we also have a stock equilibrium.

I would like to emphasize that the lag scheme in the above analysis is very simple, and the same is true for investment behavior. But at the same time I am not sure that a great many insights will be gained if we make these elements more complicated. The important thing here, in my opinion, is to master the preceding graphical analysis.

Simple Distributed Lags

A basic issue in this section is the reaction *over time* of one variable to a change in the value of another—in other words, the dynamic response of a variable given changes in the value of another in some period. For example, demand in period t might be a function of income in that period as well as of the incomes existing in a number of previous periods. Thus we might use the relationship $d_t = f(y_t, y_{t-1}, y_{t-2}, \ldots)$; and if this relationship was linear, we would have:

$$d_t = \alpha' + \alpha_0 y_t + \alpha_1 y_{t-1} + \alpha_2 y_{t-2} + \cdots$$

$$= \alpha' + \sum_{i=0}^{\infty} \alpha_i y_{t-i}. \qquad (7.17)$$

A static model implies $\alpha_0 \neq 0$, but $\alpha_i = 0$ for all $i = 1, 2, 3, \ldots$. In a model of the type given in equation 7.17, we obviously have $\partial d_t / \partial y_{t-1} = \alpha_i$, and the set of coefficients $\alpha_0, \alpha_1, \alpha_2, \ldots$ show how the reaction of d to a change in y is distributed over time. To get a better idea of what is happening here, let us assume that at time t_0 there is a *unit* impulse in y (which simply means that y increases by unity and almost immediately declines to its previous value). We see this arrangement in figure 7–3a, and it results in the change in the value of d shown in that figure: d has a constant value

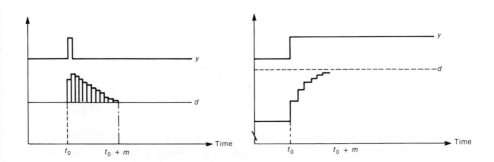

Figure 7–3. Reaction of a Variable d to an Increase in y in the Form of a Unit Impulse and a Unit Step

until time t_0, at which instant it responds to the unit impulse in y and continues to be displaced from its long-run level for a further m periods. (In reality, it approaches its long-run level only asymptotically.)

Next, instead of using a unit impulse, we employ a unit *step* with an increase in y at time t_0 of one unit, which is maintained at that level. The immediate (*impact*) reaction of d is α_0, whereas in the following period α_1 is added to this to give a reaction after two periods of $\alpha_0 + \alpha_1$. Similarly, after the third period the total reaction is $\alpha_0 + \alpha_1 + \alpha_2$; and in general after n periods the change in d is simply the partial sum that is given by $\alpha_0 + \alpha_1 + \alpha_2 + \ldots + \alpha_{n-1}$. If the α coefficients are all positive, then d increases until it reaches its new long-run value at time $t + m$, where once again we have an asymptotic convergence. This situation is shown in figure 7-3b. With a unit step initiating the sequence, the initial value of d is multiplied up by the complete sum $\alpha_0 + \alpha_1 + \ldots + \alpha_j + \ldots \alpha_z$, which can be called the *long-run multiplier* (with α_0 the impact multiplier). In some models the maximum lag z is set equal to infinity, but in order for the long-run multiplier to be finite, it is necessary that:

$$\alpha_i \to 0 \,(\text{as})\, i \to \infty \qquad \text{and} \qquad \sum_{i=0}^{\infty} \alpha_i < \infty.$$

Although in general it is unnecessay to specify that all the αs are positive, this is usually the case in economics. The obvious analogy between some of the foregoing and probability distributions also suggests that some kind of mean lag might be computed. One possibility is to make the mean lag the weighted average of the individual lags, with a lag of i periods entering with a weight of α_i, or with relative weight $\alpha_i / \Sigma \alpha_i$. Thus we get:

$$\text{Mean lag} = \frac{\displaystyle\sum_{i=0}^{\infty} i\alpha_i}{\displaystyle\sum_{i=0}^{\infty} \alpha_i}. \qquad (7.18)$$

In examining equation 7.17, it is clear that because of the infinite past of the y series on the right-hand side of the equation, it would be very difficult to postulate a regression equation that could be estimated from a finite sample of data. Fortunately, however, Koyck was able to prescribe a logical scenario for the behavior of the successive values of α: he assumed that they declined geometrically. The logic here seems reasonable: with variables such as demand and income, a consumer planning his consumption would almost certainly be more influenced by his recent income than by the income received fifiteen or twenty years earlier. This means that as i becomes very large, we would expect α_i to get very close to zero. Analytically, we have:

$$\alpha_i = \lambda^i \alpha \quad i = 0, 1, 2, \ldots \quad \text{and} \quad 0 < \lambda < 1.$$

We can now take these expressions and substitute them for the αs in equation 7.17. This gives us:

$$d_t = \alpha' + \alpha \sum_{i=0}^{\infty} \lambda^i y_{t-i}. \tag{7.19}$$

If we lag this by one period and multiply by λ, we get:

$$\lambda d_{t-1} = \lambda \alpha' + \lambda \alpha \sum_{i=0}^{\infty} \lambda^i y_{t-i-1}. \tag{7.20}$$

Subtracting equation 7.20 from equation 7.19 gives us:

$$d_t - \lambda d_{t-1} = \alpha' - \lambda \alpha' + \alpha y_t,$$

or:

$$d_t = (1 - \lambda)\alpha' + \lambda d_{t-1} + \alpha y_t. \tag{7.21}$$

We shall be making the acquaintance of this type of equation in the sequel, but its dynamic properties can be mentioned here. This is a first-order difference quation with an equilibrium value of $\alpha' + \alpha y_t/(1 - \lambda)$. Changes in y unleash the dynamic process; and, as the reader can easily verify, with values of λ between unity and zero we have an asymptotic movement from one equilibrium to another. It should be apparent that we can get our long-run multiplier from this equilibrium value, and that this is:

$$\frac{d\bar{d}}{dy_t} = \frac{\alpha}{1 - \lambda}. \tag{7.22}$$

A few other things are relevant here. The αs in the foregoing equations are weights that we attach to past values of the independent variable.

It could be argued that the time form of the lag should be constrained so that we have $\alpha_0 + \alpha_1 + \alpha_2 + \ldots = 1$, since technically this would make d (in an expression such as equation 7.17) a proper average of the y terms if there is no systematic trend. This arrangement obviously makes sense when we have output related to demand via a distributed lag, or $q_t = \alpha_0 d_t + \alpha_1 d_{t-1} + \alpha_2 d_{t-2} + \ldots$, and all past demands are equal to the current level, or $q_t = (\alpha_0 + \alpha_1 + \alpha_2 + \ldots)d_t$. In this case we would expect output to equal demand in the limit, or $q_t = d_t$, and thus $\alpha_0 + \alpha_1 + \alpha_2 + \ldots = 1$. If we use the Koyck simplification here, we get $\alpha = 1 - \lambda$.

Another Stock-Flow Model

We can conclude this section by examining figure 7–4. The thick line shows the movement of product from supply s to flow demand d and inventories. These inventories are labeled AI to signify actual physical inventories as opposed to desired inventories. These desired inventories DI are a function of such things as the current price, the expected price in one or more future periods, the interest rate, and so on. This is indicated in the figure. In line with the discussion in the previous section we expect that if $AI = DI$, then $p_{t+1} = p_t$; and, for example, if $DI > AI$, then $p_{t+1} > p_t$. In general we can postulate a relationship between price and excess stock demand of the type $p_{t+1} = p_t + \theta'(DI_t - AI_t)$, where $\theta' > 0$.

Similarly, if we are interested in the movement of inventories then we might postulate $AI_{t+1} = AI_t +$ (the change in inventories) $= AI_t + \theta(DI_t - AI_t)$, with $0 < \theta \leq 1$. The reader should note that θ is not generally equal to θ'. θ' translates an excess stock demand into an increase in price, while θ is an adjustment coefficient related to the behavior of inventory holders and the way they regulate stocks in response to a gap between desired and actual inventories. We can now rewrite this last expression to give $AI_{t+1} = (1 - \theta)AI_t + \theta DI_t$. This relationship is analogous to $d_t = \lambda d_{t-1} + (1 - \lambda)y_t$, and suggests that the discussion preceding the latter equation is applicable to an inventory model of the type introduced here.

One important reminder. Distributed lags can cause considerable problems in analytical work. For a thorough and not very difficult discussion of these matters see Gupta (1982) and Pindyck (1978).

Partial and Stock Adjustment Hypotheses

As indicated above, the demand for a given commodity turns largely on the demand for a stock. The problem is to work this concept into some kind of usable mechanism. We end up with the type of demand curve found in basic economics textbooks, but with a more complicated movement from one

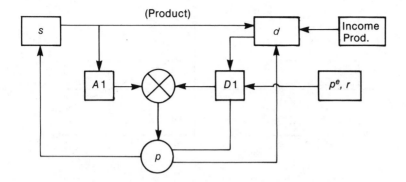

Source: Reprinted by permission of the publisher from Ferdinand E. Banks, *Scarcity, Energy, and Economic Progress* (Lexington, Mass.: Lexington Books, D.C. Heath and Company, 1977), p. 168.

Figure 7-4. Supply-Demand System Showing the Interaction of Desired and Actual Inventories

equilibrium to another. This can be explained by the fact that the consumption or demand variable d_t involves both current input and stocks; for example, a decrease in the price of a typical primary commodity may not increase demand for the commodity as a current input, but may call for an increase in inventories. On the basis of the previous discussion, we know that this could be expected to take place gradually rather than instantaneously. This type of reasoning applies almost without modification to durable goods: if, for example, there was a very sharp fall in the price of automobiles, many potential buyers might delay purchasing a vehicle until they were satisfied that there would not be a drastic rise in the price of fuel. Similarly, even if we are considering only flow demand, it might be that a change in the price would induce only a very small change in consumption in the short run because of such things as the force of habit—although there might be a substantial change in the long run.

A thorough diagrammatic exposition of the adjustment process will be presented in this section. This material is so important, however, that I have decided to begin from the beginning in the discussion of elasticities. The first step is to review the concept of elasticity in the context of the theory of demand. Thus a demand curve shows the demand for a commodity, given its price. The thing to realize is that *market demand curves* are an aggregate of individual demand curves, and basically these individual curves need not resemble one another. Take, for instance, the situation shown in figure 7-5, where we have portions of two individual demand curves that we aggregate into a market-demand schedule.

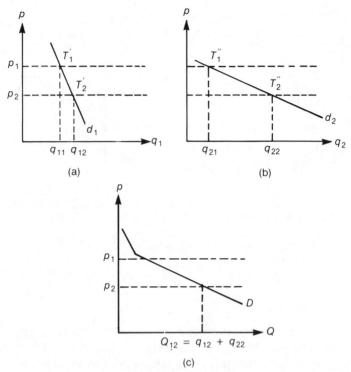

Source: Reprinted by permission of the publisher from Ferdinand E. Banks, *The Political Economy of Oil* (Lexington, Mass.: Lexington Books, D.C. Heath and Company, 1982), p. 91.

Figure 7-5. Typical Demand Curves and Their Aggregation

The aggregation here is straightforward, and it takes place horizontally. The value of Q_{12} in figure 7-5c is simply the sum of q_{12} and q_{22} in figure 7-5a and 7-5b, respectively, given a value of the price. What we should take particular care to notice are the elasticities of the individual demand curves. In figure 7-5a a decrease in price from p_1 to p_2 causes a relative small increase in demand q_{11} to q_{12}; in figure 7-5b the increase in demand, for the same price decrease, is much larger: $q_{22} - q_{21}$. For illustrative purposes, take $p_1 = 100$, $p_2 = 99$, $q_{11} = 1,900$, $q_{12} = 1,910$, $q_{21} = 1,900$, and $q_{22} = 2,000$. In this case we have a 1 percent *decrease* in price [(100 − 99)/100 × 100 percent] leading to a 0.526 percent [(1,910 − 1,900)/1,900 × 100 percent] increase in quantity in figure 7-5a.

We can now calculate one version of elasticity, the *arc* elasticity. In all cases elasticity is defined as the percentage change in quantity divided by the percentage change in price (or, the percentage change in quantity resulting from a 1 percent change in price). Since in the normal case these changes take place in opposite directions, the *price elasticity of demand* is defined as being negative. Thus in figure 7-5a the elasticity is $0.526/(-1) = -0.526$. This is greater than -1, indicating that a 1 percent decrease in price is not compensated by an increase in demand, and thus total revenue pq falls. Specifically total revenue goes from $p_1q_{11} = 190,000$ to $p_2q_{12} = 189,000$. Over the stretch, or arc, on which the elasticity has been calculated, demand is *inelastic*. By way of contrast, in figure 7-5b elasticity is $5.26/(-1) = -5.26$, and the demand is *elastic*. The reader can check what this means for the change in revenue.

We now face a slight problem as a result of calculating the elasticity over an arc rather than at a point. If, instead of a price fall, we had a price increase, say from 99 to 100, then we would get a different value of the elasticity. The issue here can be clarified by writing the formula for the arc elasticity:

$$e = \frac{\Delta q/q}{\Delta p/p} = \frac{(q_{11} - q_{12})/q}{(p_1 - p_2)/p} = \frac{q_{11} - q_{12}}{p_1 - p_2}\left(\frac{p}{q}\right) \qquad (7.23)$$

Notice that regardless of whether we have an increase (or a decrease) in the price equal to $p_1 - p_2$ (or $p_2 - p_1$), there is no change in the *absolute value* of $q_{11} - q_{12}$. It is p and q (or p/q) that change, depending on whether we have an increase or a decrease in price. For example, if the price decreases, we have $p/q = p_1/q_{11} = 100/1,900 = 0.0526$ in figure 7-5a. In the same figure, with an increase in price we would have $p/q = p_2/q_{12} = 99/1,910 = 0.0518$. This leads to different value of the elasticity over the same arc. At this point the reader should go through the same type of calculation employing the data given for figure 7-5b.

This discrepancy is handled by making p and q averages of the two prices and quantities. Thus arc elasticity e in figure 7-5a becomes

$$e = \frac{q_{11} - q_{12}}{p_1 - p_2} \frac{\dfrac{(p_1 + p_2)}{2}}{\dfrac{(q_{11} + q_{12})}{2}} = \frac{q_{11} - q_{12}}{p_1 - p_2} \frac{p_1 + p_2}{q_{11} + q_{12}} \qquad (7.24)$$

Using the numerical values given, for the arc elasticity over $T'_1 - T'_2$ in figure 7-5a we get:

$$e = \frac{-10}{1} \cdot \frac{199}{3,810} = -0.5223$$

The reader should now calculate the arc elasticity over $T''_1 - T''_2$ in figure 7-5b. The following should also be observed. Elasticities often are expressed in absolute values. Thus an elasticity of -0.5223 is simply called 0.5223, with the understanding being that in the case of a normal demand curve, demand changes are in the opposite direction from price changes. The key thing to remember here is that an elasticity which is smaller than 1 [or greater than -1 if we are observing signs (for instance, 0.75 or -0.75)] signifies that a given percentage change in price causes a smaller percentage change in quantity. The significance of all this for revenue is that when the demand is inelastic, a decrease in price means that even though demand might increase, it will not increase enough to prevent a fall in revenue. Analogously, when the demand is inelastic, an increase in price, means a percentage decrease in demand that is less than the price rise. Consider, for example, the situation with oil. Even when the price increased by almost 400 percent, demand fell only marginally, and producer revenues not only were maintained, but also increased drastically.

Before we leave this aspect of our subject, it should be observed that the most interesting problem involves starting out with the value of the elasticity and calculating changes in demand. For instance, if we assume that the (short-run) elasticity of oil is 0.1, then a 25 percent increase in price with all else constant means a percentage change in quantity of $\Delta q / q = n_p(\Delta p / p)$ $= -0.1\ (25) = -2.5$ percent, where n_p is the price elasticity demand.

Next we consider the income elasticity of demand. This concept is similar to the one just examined, and we are interested in the percentage change in demand, given a certain percentage change in income. Calling the income elasticity of demand n_i and income I, we have

$$\frac{\Delta q / q}{\Delta I / I} = n_i \tag{7.25}$$

Figure 7-6a shows a typical shift in the demand curve under the influence of an increase in income: With the price constant at $\bar{p}$, demand increases from q_0 to q_1 when income increases from I_0 to I_1. For instance, suppose we have an income elasticity of demand of $n_i = 1.1$ and $I_0 = 100$ and $I_1 = 150$. We thus get $\Delta q / q = n_i\ (\Delta I / I) = 1.1(50/100) = 0.55$. Here income increased by 50 percent, whereas demand increased by 55 percent. In this example, $I = I_1 - I_0 = 50$.

We can look at the effect of a parallel shift to the right of a linear demand curve on the price elasticity of demand. Using the definition of price elasticity given above, we note that, with an unchanged price, the absolute value of the price elasticity at B in figure 7-6b is smaller than that at A. This is so since, with $n_p = (\Delta q / q) / (\Delta p / p)$ at B the only change is in q, which is larger: $\Delta q / \Delta P$ is the same for D_0 and D_1 and, as mentioned, p is con-

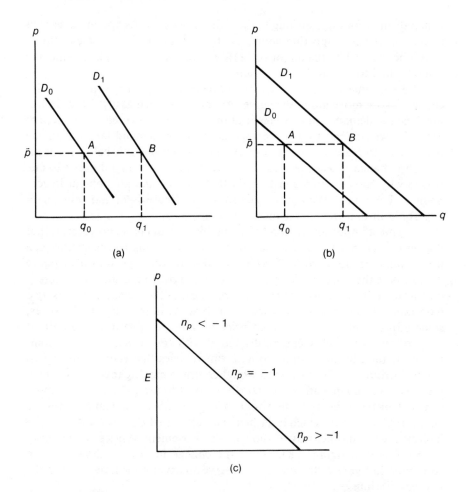

Source: Reprinted by permission of the publisher from Ferdinand E. Banks, *The Political Economy of Oil* (Lexington, Mass.: Lexington Books, D.C. Heath and Company, 1982), p. 94.

Figure 7-6. Elasticities on Linear Demand Curves

stant and equal to $\bar{p}$. [The expression *absolute value* is important. If n_p equals -0.3 at A and -0.2 at B, then the elasticity is actually larger at B if we concern ourselves with the sign; but the absolute value (the value considered without the sign) is smaller.] If the reader has any doubts about this matter, he or she should construct a numerical example or two, since this is a significant topic. The short-run price elasticity of oil is very low, while its medium-to long-run elasticity (as defined below) is somewhat larger. But because the demand curve for oil is shifting to the right under the effect of an increase in income in the oil-importing countries, the fall in demand that eventually would be induced by the major price rises which might take place some day is moderated to a considerable extent.

We continue by noting that on a linear demand curve, the elasticity changes as we move along the curve. As the reader can easily check, in figure 7-6c the demand curve is elastic above point E: A decrease in price, for example, would lead to a percentage decrease in demand larger than the percentage decrease in price. However, the curve is inelastic below point E.

The change in elasticity as we move along the demand curve is one reason why curves with a constant elasticity of demand are popular in economics. These have the appearance shown in figure 7-7a and resemble a hyperbola in that they never meet the p or q axis.

The most important reason for using this type of curve, however, is that it gives us a chance to put the concept of short- and long-run elasticities in a usable analytic framework. Take a situation of the type shown in figure 7-7b, where the price suddenly rises from p_0 to p_1. In the short run there is only a small decrease in demand, from q_0 to q_1, as the consumer moves up a (constant-elasticity) short-run curve from W to X. In the long run, however, some adjustment to the new high price may be possible in the sense that more of the commodity can be dispensed with. For example, if the commodity is fuel, in the medium to long run vehicles that economize on fuel can be introduced. Thus demand would continue moving toward q_2, which is on the long-run demand curve D_L. In terms of figure 7-7b, this movement can be depicted as going from X to Y; a large number of demand curves are being crossed, one of which is dashed and designated D_m (where m signifies intermediate). Another way of showing the movement of price and quantity can be found in figure 7-8a, where the change in price is shown in step form, and in figure 7-8b, where we observe an asymptotic movement to the new equilibrium q_2.

The situation with the elasticities is portrayed in figure 7-8c. Here, in order to avoid using negative numbers, absolute values are employed. For instance, as elasticity moves from n_s on curve D_S to n_L on curve D_L in figure 7-8b, where n_S and n_L are negative; in figure 7-8c these elasticities are shown in absolute terms (that is, they are positive) and are designated $|n_S|$ and $|n_L|$. As we move from X to Y in figure 7-7b we are crossing demand curves whose elasticities, in absolute terms, are between $|n_S|$ and $|n_L|$. Finally, let us reemphasize that under the pressure of increased income, a system such as the one shown in figure 7-7b is in motion to the right. Thus if we take the case of oil, where the absolute value of the short-run elasticity is low, when the price is raised by a large amount, demand falls by only a small amount (although in the medium to long run there are possibilities for a major fall in demand if the income remains constant). As we said, how-

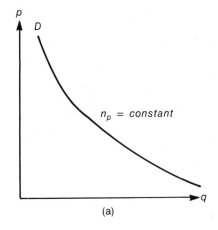

$n_p = constant$

(a)

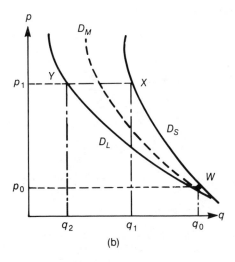

(b)

Source: Reprinted by permission of the publisher from Ferdinand E. Banks, *The Political Economy of Oil* (Lexington, Mass.: Lexington Books, D.C. Heath and Company, 1982), p. 96.

Figure 7–7. Constant-Elasticity Demand Curves and the Concept of Short- and Long-Run Elasticities

ever, the income does not remain constant: it increases and thereby substantially reduces the fall in demand. In figure 7–7b, for example, demand would not reach q_2 because the movement from q_1 to the vicinity of q_2 would take considerable time; meanwhile, the long-run demand curve would have shifted to the right.

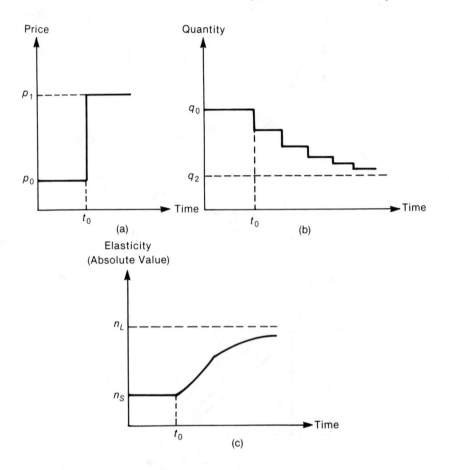

Source: Reprinted by permission of the publisher from Ferdinand E. Banks, *The Political Economy of Oil* (Lexington, Mass.: Lexington Books, D.C. Heath and Company, 1982), p. 97.

Figure 7-8. Movements to a Long-Run Equilibrium Following an Increase in Price

Next I propose to go through the foregoing exposition at a level commensurate to that employed earlier in this chapter. The discussion centers around figure 7-9, where the original price is p_0, and the long-run demand curve is d^*. If demand adjusted immediately to a fall in price to $\bar{p}$, demand would move to $\bar{d}$. Instead, in the very short run, with price $\bar{p}$, demand becomes only slightly larger than d_0. Later it arrives at d'_1, still later at d'_2, and as time goes by we get an asymptotic movement to $\bar{d}$. Another typical

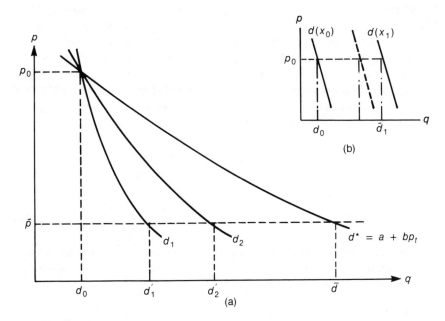

Note: These are constant elasticity curves—in logarithmic form in (b)—and the reader desiring a more rigorous introduction to them is referred to Ferdinand E. Banks, *The Political Economy of Oil,* 2nd ed. (Lexington, Mass.: Lexington Books, D.C. Heath and Company, 1982), pp. 131–132.

Figure 7-9. Demand Adjustment Following a Change in (a) Price and (b) Aggregate Variable Such as Income or Industrial Production

intermediate demand curve is d_2 (with d'_2 an intermediate consumption); and, as the reader can easily verify, it is possible to draw many such curves to either the right or the left of d_2.

The law of adjustment for this model is $d_t = d_{t-1} + \lambda(d^* - d_{t-1})$, where d^* is an equilibrium level of demand, and $0 < \lambda < 1$. When the *actual* value of demand at time t is equal to what is perceived as the equilibrium or desired value at time t, or $d_t = d^*$, our system is stationary. In the figure long-run demand is related to price by setting $d^* = a + bp_t$, and putting this into the adjustment equation. We can then get immediately by simple substitution $d_t = \lambda a + (1 - \lambda)d_{t-1} + \lambda bp_t$. When $d_t = d_{t-1} = d^*$, we have an equilibrium; this can be verified by putting this condition into the previous equation. This takes us back to the long-run demand curve $d_t^* = a + bp_t$, which functions conceptually as an equilibrium demand curve.

This is an example of the partial adjustment or stock adjustment hypothesis. In general we have a desired level of some variable, z_t^*, that depends on an explanatory variable x_t in a relationship such as $z_t^* = \phi_0 + \phi_1 x_t$. Frictions, delays, the inertia of habit, or cold calculation may make it impossible or undesirable to reach the new equilibrium in a single period, and so the actual change in z_t from the level of the previous period, or $z_t - z_{t-1}$, is made some fraction of the change needed to achieve the difference $z_t^* - z_{t-1}$. If we take λ to be this fraction, with $0 < \lambda < 1$, then the partial adjustment hypothesis can be written:

$$z_t - z_{t-1} = \lambda (z_t^* - z_{t-1})$$

or

$$z_t = (1 - \lambda) z_{t-1} + \lambda z_t^* \tag{7.26}$$

We can now substitute the expression $z_t^* = \phi_0 + \phi_1 x_t$ into the adjustment equation. The result is $z_t = \phi_0 \lambda + (1 - \lambda) z_{t-1} + \lambda \phi_1 x_t$, which is a perfect analogue of the earlier equation.

The Adaptive Expectations Hypothesis

We began the previous discussion with a change in price, without explaining how this change could have come about. Assuming the demand relationship shown in figure 7-9a, one explanation might be an increase in supply. Of course, for a typical raw material such as copper or aluminum, prices can be adjusted administratively, in response to market or nonmarket forces. We must also consider a possible change in demand in response to a change in industrial production. If this variable increases, then our entire demand curve shifts to the right. In such a situation we may be increasing demand because a larger production requires a larger input of a primary commodity; but because of this larger *flow* demand, stock demand could also increase, because it may be desirable to maintain a fixed relationship between the use of a primary commodity in the current production process and the amount held in inventory.

We will not try to distinguish between these two forces; instead we will simply postulate an increase in some variable x from x_0 to x_1, which shifts our demand curve to the right. This is shown in figure 7-9b. The assumption will also be some kind of elastic supply, so that price is constant. Once again we can observe a gradual displacement from d_0 to $\bar{d}$, where we have a partial adjustment situation such as we have already discussed. The only difference is that now consumption is related to industrial production (or a similar variable) and price is a parameter. If the relationship were linear we would have $d_t^* = h + wx_t$. Putting this in an adjustment equation of the previous type, with $d_t - d_{t-1} = \lambda(d^* - d_{t-1})$, we get $d_t = h\lambda + (1 - \lambda)d_{t-1} + w\lambda x_t$, as expected.

A similar approach to this type of problem begins with an expected change in a variable such as price, and relates the demand for a commodity to this expected value. We can begin by taking $d_t = h + wp_{t+1}^e$. Our problem now centers on how expectations are formed, and we will assume that we have adaptive expectation, or $p_{t+1}^e = p_t^e + \lambda(p_t - p_t^e)$. Using this in the previous equation gives $d_t = h + w[p_t^e + \lambda(p_t - p_t^e)]$. Next we lag $d_t = h + wp_{t+1}^e$ by one period, multiply the result by $(1 - \lambda)$, and subtract it from $d_t = h + w[p_t^e + \lambda(p_t - p_t^e)]$. As a result we get $d_t = h\lambda + (1 - \lambda)d_{t-1} + w\lambda p_t$. This is the same equation that we obtained using the adjustment hypothesis. The identification of this duality is largely due to Marc Nerlove. Although the econometric exploitation of these results takes place in the next section, if we fit a regression equation that provides a satisfactory explanation of d_t, we cannot say whether adjustment or expectation is at work. In some situations, however, economic theory may be able to provide an answer.

Some Econometrics of Consumption

If, in the analysis directly above, we had begun with a stochastic equation $d_t = \alpha + \theta p_{t+1}^e + u_t$, and then carried out the same manipulations, we would have ended up with:

$$d_t = \alpha\lambda + (1 - \lambda)d_{t-1} + \theta\lambda p_t + [u_t - (1 - \lambda)u_{t-1}] \quad (7.27)$$

u_t and u_{t-1} are stochastic variables, and the estimating equation is

$$d_t = \beta_0 + \beta_1 d_{t-1} + \beta_2 p_t + (\ )$$

Thus we get $\lambda = 1 - \beta_1$, $\alpha = \beta_0/(1 - \beta_1)$, and $\theta = \beta_2/(1 - \beta_1)$. The βs are obtained from the regression. A problem is raised by the stochastic term, however. One of the criteria placed on this term when ordinary least squares are used is no serial correlation, or $E(\epsilon_t, \epsilon_{t-1}) = 0$, where this expression is one of the off diagonal terms of the variance-covariance matrix. In the present analysis $E(\epsilon_t, \epsilon_{t-1}) = E[u_t - (1 - \lambda)u_{t-1}][u_{t-1} - (1 - \lambda)u_{t-2}]$, and if we assume no serial correlation among individual disturbances we get $E(u_t u_{t-1}) = E(u_t u_{t-2}) = E(u_{t-1}^* u_{t-2}) = 0$; but $E[-(1 - \lambda)u_{t-1}u_{t-1}] = (1 - \lambda)\,\mathrm{var}\,u_{t-1}$, and since $\mathrm{var}\,u_{t-1} \neq 0$, the composite stochastic term displays serial correlation. This means that least square estimates of α, θ, and λ are not consistent.

By way of completeness, two other demand equations should be referred to. Instead of adaptive expectations we could imagine a situation where extrapolative expectations prevailed, and thus we have $p_t^e = p_{t-1} + \phi\Delta p_{t-x}$. A simple lag structure might give $p_t^e = p_{t-1} + \phi\Delta p_{t-1} = p_{t-1} + \phi(p_{t-1} - p_{t-2})$. If consumption is a function of expected price we get right away $d_t = \alpha + \theta p_t^e = \alpha + \theta p_{t-1} + \theta\phi\Delta p_{t-1}$. The estimating equation for this case is $d_t = \beta_0 + \beta_1 p_{t-1} + \beta_2\Delta p_{t-1}$ and so $\alpha = \beta_0$, $\theta = \beta_1$, and $\phi = \beta_2/\beta_2$. As above, the βs are estimated using a suitable regression technique.

In an interesting and important contribution, Witherell (1967) seems to have attempted to combine adjustment and extrapolative expectations, ending up with a consumption equation in the following form: $d_t = \lambda\alpha + (1-\lambda)d_{t-1} + \lambda\theta p_{t-1} + \phi\lambda\theta\Delta p_{t-1}$. The estimating equation here is $d_t = \beta_0 + \beta_1 d_{t-1} + \beta_2 p_{t-1} + \beta_3\Delta p_{t-1}$. From these two equations we get $\lambda = 1 - \beta$, $\alpha = \beta_0/(1-\beta_1)$, $\theta = \beta_2/(1-\beta_1)$, $\phi = \beta_3/\beta_2$. Some questions could be raised about the theoretical soundness of this type of equation, but it is useful because it often gives good "fits."

Before looking at several econometric equations, we can say something about elasticities. A price elasticity, for example, is defined as

$$\epsilon_p = \frac{\partial q}{\partial p}\frac{p}{q} \qquad (7.28)$$

If we use regression results, we get $\partial q/\partial p$ from the regression equation, and for p and q we use the mean of the observations on price and quantity, or $\bar{p}$ and $\bar{q}$. This elasticity would normally be negative; but if demand is a function of the present price and expected price—for example, the price expected to prevail in the next period—we might have:

$$d_t = d_t(p_t, p_{t+1}^e) = a_0 + a_1 p_t + a_2 p_{t+1}^e \qquad a_1 < 0, a_2 > 0 \qquad (7.29)$$

Let us now take expectations as extrapolative, which means that we have $p_{t+1}^e = p_t + \phi(p_t - p_{t-1})$. Taking, $\phi \neq 0$, and putting this expression in the above equation gives:

$$d_t = a_0 + [a_1 + a_2(1 + \phi)]p_t - \phi a_2 p_{t-1} \qquad (7.30)$$

As is easy to show, we could have $[a_1 + a_2(1 + \phi)] \lessgtr 0$; everything depends on the magnitudes of a_1, a_2, and ϕ. Speculative effects could thus cause the relation between d_t and p_t to be positive instead of negative, which would mean a positive price elasticity. In any event, speculation based on conjectured future prices may play an important role in modifying elasticities. Cooper and Lawrence (1976) seem to have considered this possibility when dealing with certain primary commodity elasticities.

Then, too, we must distinguish between long and short term elasticities when we have dynamic equations. With an equation such as $d_t = \eta_0 + \eta_1 d_{t-1} + \eta_2 p_t$ we have a (long-run) equilibrium when $d_t = d_{t-1}$, and thus we get:

$$d_t = \frac{\eta_0}{1 - \eta_1} + \frac{\eta_2}{1 - \eta_1} p_t$$

Accordingly, the long-run elasticity is

$$\epsilon_L = \frac{\partial d_t}{\partial p_t} \frac{\bar{p}_t}{\bar{d}_t} = \frac{\eta_2}{1 - \eta_1} \frac{\bar{p}_t}{\bar{d}_t} \qquad (7.31)$$

On the other hand, the short-run elasticity is taken directly from $d_t = \eta_0 + \eta_1 d_{t-1} + \eta_2 p_t$, and is

$$\epsilon_S = \frac{\partial d_t}{\partial p_t} \frac{\bar{p}_t}{\bar{d}_t} = \eta_2 \frac{\bar{p}_t}{\bar{d}_t} \qquad (7.32)$$

In line with the theory we have discussed earlier, we expect the short-run elasticity to be smaller than the long-run: in the short run a price change changes demand slightly, but in the long run we may get a fairly large change in demand. This story is illustrated in figure 7–9. As pointed out above, η_2 should be negative if we do not have any unseemly speculative effects; and we should also have $0 < \eta_1 < 1$. If η_1 was in fact larger than unity, this would signify satiation on the part of buyers.

We can now present a few consumption equations that seem to follow from the above theory. Taking the case of tin in the United States we have:

$$d_t = 54.56 + 0.563\, d_{t-1} - 0.13246\, p_{t-1} + 0.050\, \Delta g + 9.97\, D$$
$$(0.186) \qquad (0.057) \qquad\qquad (0.014) \qquad (4.279)$$

$$\bar{R}^2 = 0.667 \qquad \epsilon_{PS} = -0.550 \qquad \epsilon_{PL} = -1.262$$

Here Δg is the change in inventories of durable goods in the United States, taken as an index, with $1963 = 100$. D is a dummy variable equal to unity for 1953 (and zero elsewhere), while standard errors are in parentheses. Next we present an equation for refined zinc in the United States:

$$d_t = 242.17 + 0.48998\, d_{t-1} + 2.171\, x_t + 12.736\, \Delta x_t + 44.66\, D$$
$$(2.49) \qquad\quad (1.31) \qquad (3.46) \qquad\quad (1.14)$$

$$\bar{R}^2 = 0.991 \qquad \epsilon_{XS} = 0.230 \qquad \epsilon_{XL} = 0.403$$

In this equation the dummy D is equal to unity for 1963–1965, and is zero elsewhere. The t ratios are in parentheses; x_t is an index of industrial production, and elasticities are taken with reference to this variable. (These elasticities, like income elasticities, are positive.) Finally, we take the case of the demand for primary copper in France. (Again, t ratios are in parentheses.)

$$d_t = 57.99 + \underset{(2.19)}{0.400}\, d_{t-1} - \underset{(2.53)}{0.310}\, p_t + \underset{(3.04)}{1.10}\, x_t$$

$$\bar{R}^2 = 0.925 \quad \epsilon_{PS} = -0.253 \quad \epsilon_{PL} = -0.421 \quad \epsilon_{XS} = 0.540 \quad \epsilon_{XL} = 0.900$$

Here we have both price elasticities and elasticities with respect to the variable x_t, which is the index of industrial production (1963 = 100). The logic of the deflation of the price by the wholesale price index has been examined at great length by Herfindahl (1959), among others. This deflation relates the cost of copper as an input to the price of the output in which it is used, thus saying something about its desirability relative to other inputs: if the price of the output, the input in question, and relative substitutes for the input doubled, we would expect no decline in the demand for the input per unit of output.

Just as the above elasticities were computed for individual countries, it would have been equally simple, in theory, to compute them for blocs of countries. The matter of projections will not be discussed here, but the above estimates are of interest primarily because they shed light on the general size of the elasticities for primary commodities; in line with our suspicions and intuition, these elasticities are quite low. The table 7–2 presents price and income elasticities for the most important commodity groups; the values in this table are weighted averages of the elasticities for individual products.

The significance of elasticities for such things as cartel forming has been referred to earlier in this book; for more on this matter, see the work of Herin and Wijkman (1976).

The Econometrics of Supply

Next we turn our attention to supply. The groundwork for this topic was carried out in the beginning of the previous section, and we obtain a dynamic equation for supply in the same way that we obtained one for consumption. First we relate equilibrium supply to price by $s^* = \beta + \pi p_t$; then we postulate an adjustment equation $s_t = s_{t-1} + \phi(s^* - s_{t-1})$, and by the usual manipulations we obtain $s_t = \phi\beta + \phi\pi p_t + (1 - \phi)s_{t-1}$. The

Table 7–2
Price and Income Elasticities for Major Commodity Groups

	Food	Metals and Minerals	Fuels
Price elasticity			
Low	-0.370	-0.140	-0.250
High	-0.540	-0.390	-0.250
Income elasticity			
Low	0.401	0.669	0.900
High	0.490	0.891	0.900

Source: Reprinted by permission of the publisher from Ferdinand E. Banks, *Scarcity, Energy, and Economic Progress* (Lexington, Mass.: Lexington Books, D.C. Heath and Company, 1977), p. 174.

adjustment hypothesis may make more sense for supply than for consumption, because changes in production invariably take time.

Another advantage of this type of equation, as we have seen, is that it permits us to regard supply as a function of a weighted average of present and former prices, which we might occasionally find useful. This can be shown simply by lagging the equation for s_t, and using it to replace s_{t-1} in the same equation. This gives us:

$$s_t = \phi\beta + (1 - \phi)[\phi\beta + (1 - \phi)s_{t-2} + \phi\pi p_{t-2}] + \phi\pi p_{t-1}$$

Repeated application of this procedure results in:

$$s_t = \beta + \phi\pi\Sigma(1 - \phi)^i p_{t-i} \qquad (7.33)$$

We have seen many expressions of this type earlier; in addition, the supply equation is, like its consumption analogue, a simple first-order difference equation. A device associated with multiplier theory will be used to clarify the dynamics of this equation. Our assumption is that we begin with an equilibrium (s_0, p_0), and that we have an increase in the price to p_1. Supply begins moving toward s_1, but in line with the adjustment hypothesis, we assume that this takes time. The supply curve in its usual form is shown in figure 7–10a, with the initial equilibrium at point A. The new equilibrium, which is on the long-run supply curve $s^* = \beta + \pi p_t$, is at (s_1, p_1).

This situation can also be studied with the aid of our supply difference equation. The beginning of our sequence is at point A' in figure 7–10b, where we have an equilibrium with $s_{t-1} = s_t = s_0$. The slope of FF is the coefficient of s_{t-1}, and is equal to $(1 - \phi)$. The intercept of FF is a function of the price, and so at the initial equilibrium it is a function of p_0. Next

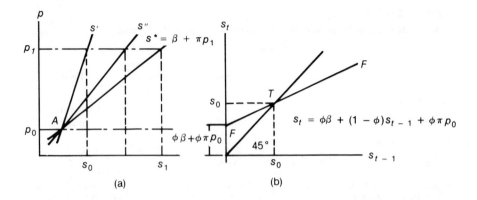

Source: Reprinted by permission of the publisher from Ferdinand E. Banks, *Scarcity, Energy, and Economic Progress* (Lexington, Mass.: Lexington Books, D.C. Heath and Company, 1977), p. 176.

Figure 7-10. (a) Supply Changes Given an Increase in Price and (b) the Initial Equilibrium in a Dynamic Supply Model

the price increases to p_1, and as a result FF shifts upward to $F'F'$. This arrangement is shown in figure 7-11a. The new equilibrium is s_1, but this is only reached asymptotically. In the first period the increase in production is TT', while in the second period it is $T'A$, etc. We thus have a multiplier sequence consisting of:

$$\phi\pi(p_1 - p_0)[1 + (1 - \phi) + (1-\phi)^2 + \ldots] = \frac{\phi\pi(p_1 - p_0)}{1 - (1 - \phi)}$$

$$= \pi(p_1 - p_0) \qquad (7.34)$$

This expression shows that $\phi\pi(p_1 - p_0)$ is equal to TT'; $\phi(1 - \phi)\pi$ $(p_1 - p_0)$ is equal to T^*W and so on. The asymptotic movement to the new equilibrium is also shown in figure 7-11b.

It could be argued from figure 7-11a that if producers were willing to supply the increase in demand $\Delta s = s_1 - s_0$ by reducing inventories, as well as from current production, inventory reduction would be $T'T''$ in the first period, $T'T'' - T^*W$ in the second period, and so on—assuming the full amount of Δs is supplied in each period, either from current production or from stocks.

We could also postulate that production could be increased by TT'' (to s_1) in the first period if consumers were willing to pay a price corresponding to the intercept F''. The reasoning here is that with $\phi\pi$ constant, every value

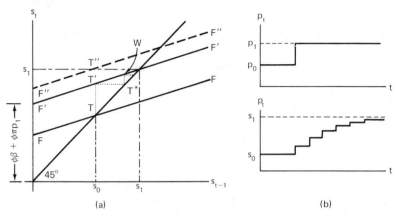

Source: Reprinted by permission of the publisher from Ferdinand E. Banks, *Scarcity, Energy, and Economic Progress* (Lexington, Mass.: Lexington Books, D.C. Heath and Company, 1977), p. 176.

Figure 7-11. Movement to the New Equilibrium

of the intercept corresponds to a price. The initial intercept corresponded to p_0; the increase in price to p_1 is associated with the intercept of $F'F'$ (and an increase in production of $FF' = TT'$ in the first period), and so there should be some price greater than p_1 which would cause producers to increase—or to attempt to increase—production by more than FF' in the first period.

Some Econometric Estimates

We shall now provide some supply equations for newly mined copper. The estimating equation corresponding to $s_t = \phi\beta + (1 - \phi)s_{t-1} + \phi\pi p_t$ is given by $s_t = \psi_0 + \psi_1 s_{t-1} + \psi_2 p_t + u_t$, where u_t is the error term. We thus see that $\phi = 1 - \psi_1$, $\beta = \psi_0/(1 - \psi_1)$, $\pi = \psi_2/(1 - \psi_1)$. If the lagging procedure used at the beginning of this section is applied to the estimating equation, we obtain a composite error term $v_t = u_t + \psi_1 u_{t-1} + \psi_1^2 u_{t-2} + \ldots$, but this expression can be lagged one period, multiplied by ψ_1, and subtracted from v_t *to yield* $u_t = v_t - \psi_1 v_{t-1}$. This indicates serially correlated error terms that the estimating procedure must take into account.

The logarithmic form of the estimating equation was chosen simply because it gave better fits than other types of estimating equations. This also simplifies the matter of obtaining supply elasticities, since if we have log $s_t = \psi_0 + \psi_1 \log s_{t-1} + \psi_2 \log p_t$, our short-run elasticity is $\epsilon_{SS} = \psi_2$, and the long-run elasticity (obtained employing the criterion that log $s_t =$

Table 7-3
Supply Equations for Primary Copper in Canada, Chile, and Zaire

Canada	$\log s_t = 0.21511 + 0.7949 \log s_{t-1} + 0.2365 \log p_{ut}$
	$\qquad\qquad\qquad\quad (6.876) \qquad\qquad (2.245)$
	$\bar{R}^2 = 0.8114 \qquad \epsilon_{SS} = 0.2365 \qquad \epsilon_{SL} = 1.16$
Chile	$\log s_t = 0.1236 + 0.7694 \log s_{t-1} + 0.2890 \log p_{t-1}$
	$\qquad\qquad\qquad\quad (8.412) \qquad\qquad (2.670)$
	$\bar{R}^2 = 0.920 \qquad \epsilon_{SS} = 0.2890 \qquad \epsilon_{SL} = 1.22$
Zaire	$\log s_t = 0.1444 + 0.7204 \log s_{t-1} + 0.1726 \log p_t + 0.031\,D$
	$\qquad\qquad\qquad\quad (8.854) \qquad\qquad (3.120) \qquad\qquad (2.869)$
	$\bar{R}^2 = 0.947 \qquad \epsilon_{SS} = 0.1726 \qquad \epsilon_{SL} = 0.625 \qquad D: 1959\text{-}62$

$\log s_{t-1}$) is $\epsilon_{SL} = \psi_2/(1 - \psi_1)$. The supply being examined here is the supply of newly mined copper in thousands of metric tons (copper content), expressed as an index with 1963 as the base year. The prices used here are the
price of refined copper on the London Metal Exchange, p_t, and the
American copper price, which is called p_{ut}. (This American copper price is
actually an average of several prices.) Both prices are in index form.

In the equations shown in table 7-3, we have t ratios in parentheses,
and D is a dummy variable that equals unity for the years given, and zero
otherwise. The basic data are annual observations over the period
1948-1967. The reader may also question the absence of time trend
variables in the above estimates, particularly since this book argued earlier
that for less-developed countries there is a secular expansion of raw material
production that has nothing to do with price. However, using time trend
variables introduced an unacceptable amount of multicollinearity into the
estimating equations.

Estimates of supply elasticities usually depend on who is doing the
estimating; table 7-4 shows estimates by United States Bureau of Mines of
the supply elasticities for several of the most important minerals. Other
organizations that are involved in determining supply elasticities for
primary commodities are UNCTAD and the World Bank, and the interested reader should examine their publications.

The Theory of Exhaustible Resources

At this point we question whether our econometric results might or should
be modified because we are, for the most part, dealing with nonrenewable

Table 7-4
Supply Elasticities for Metals

	Price Range	Elasticity
Aluminum	27–37	1.15
Copper	52–75	0.77
Nickel	128–200	2.03
Lead	14–20	1.84
Zinc	16–25	1.75

Source: U.S. Bureau of Mines estimates.

resources. (However, the techniques being discussed in this chapter also apply to such things as natural rubber, tea, wheat, and probably timber.) We need a rule for the optimal extraction of an exhaustible resource. I intend to derive (or perhaps I should say *rederive*) such a rule below, but I would like to make it clear that the connection between this rule and any econometric work is very tenuous indeed.

Stripped of unnecessary detail, the problem comes down to the following: if we have a unit of a resource in the ground that can be removed at a certain cost $\bar{M}_t$, which can be interpreted as a marginal cost, and this unit can be sold for a known price p_t, then the decision as to whether the unit should be removed turns on the rate of interest in the following way.

By removing the unit now and selling it at a price p_t, we obtain a *marginal profit* of $(p_t - \bar{M}_t) = MP_t$ for this one unit. If we put this marginal profit in a financial institution, our gain over one period is $MP_t(1 + r) - MP_t = rMP_t$, where r is the prevailing rate of interest.

If instead we leave the unit in the ground and extract it at the beginning of the next period, we realize a *capital gain* of $(p_{t+1} - \bar{M}_{t+1}) - (p_t - \bar{M}_t) = MP_{t+1} - MP_t$. If $rMP_t > (MP_{t+1} - MP_t)$, then the correct action is to remove the unit; but if by way of contrast we have $rMP_t < MP_{t+1} - MP_t$, then the unit of the resource is not extracted. Our equilibrium situation, which is the point at which we cease extracting the resource, comes when:

$$MP_{t+1} - MP_t = rMP_t.$$

that is:

$$\frac{MP_{t+1} - MP_t}{MP_t} = r,$$

or:

$$\frac{\Delta MP}{MP} = r. \qquad (7.35)$$

With zero extraction costs, the last expression reduces to $\Delta p/p = r$. We should also recognize that this last expression is a difference equation that, as time periods become very small, can be written as a differential equation with the solution

$$MP = MP(0)e^{rt}, \qquad (7.36)$$

or:

$$p - \bar{M} = MP(0)e^{rt}.$$

That is:

$$p = \bar{M} + MP(0)e^{rt}. \qquad (7.37)$$

Notice the difference between this expression and the usual profit-maximization rule, $p = \bar{M}$. In fact, $(p - \bar{M})e^{-rt}$ is the scarcity royalty on a unit of the resource, and can be interpreted as the opportunity cost of using the resource now instead of later (see chapter 6 of this book and *The Political Economy of Oil* for a more complete explanation of this concept).

We shall now look at the matter from another angle. Let us take a situation where we have some resource (such as conventional oil) that is being used now and that can be extracted at *zero* cost. Let us assume that we use A units per year, and that we have a stock of B units available. Next let us postulate an infinite potential supply of synthetic oil whose production requires nondepreciable capital worth K monetary units per barrel, but *no* other inputs. The rate of interest is r. (Notice that in this example the effect of price on future demand is sidestepped by specifying a yearly use rate of A.) With these data we thus see that the first units of synthetic oil will be required in $T = B/A$ years. Thus the efficiency price of conventional oil should be:

$$P(T) = Ke^{-r(t - T)}. \qquad (7.38)$$

This can also be interpreted as the *scarcity royalty* on conventional oil, since with a positive demand for some kind of oil after conventional oil has been exhausted, money must be accumulated to pay for the equipment that will produce synthetic oil. Thus we do not have $p = MC = 0$ in this case, but a positive price that yields revenue that can be put in a bank, or used to buy bonds. As a result, by the time the conventional oil has been exhausted, sufficient money has been accumulated to pay for the equipment to produce A units of oil per year, indefinitely. Note also that the per-period price of a barrel of synthetic oil is $P_s = rK$, which is the rental rate for the nondepreciable capital to produce this oil, since no other costs are assumed. This can be checked right away by noticing that:

$$\int_T^\infty rKe^{-r(t-T)}dt = K. \tag{7.39}$$

We also see that:

$$\frac{\partial P(t)}{\partial T} = -rKe^{-r(t-T)} \quad \text{and} \quad \frac{\partial P(t)}{\partial K} = e^{-r(t-T)} \tag{7.40}$$

If A decreases (and thus T increases), as is the case today, then $P(t)$ decreases; but as K increases (which is also happening today where oil is concerned), then $P(t)$ increases.

It is probably just as well to note that there are very many derivations of the foregoing results. Another, in fact, will be given next in the context of a slightly more difficult exercise. Still another can be found in *The Political Economy of Oil*.

Exhaustible Resources and Fixed Capital

Although the Hotelling analysis of the economics of exhaustible resources continues to fascinate economists, as Jacques Crémer (1979) recently noted, little or no effect has been put into understanding the significance of fixed capital for the production decisions of a mining firm, despite the well-known fact that mining-company directors cannot be as blasé about this issue as academic economists.

As Hotelling himself pointed out, "capital intensity in developing the mine . . . is a source of a need for steady production." This means simply that the necessity to make regular amortization and interest payments on the structures and machines used by the mining firm precludes the occasional closing of the mine because profits in a given period (such as a year) are substandard compared with expected profits in other periods—although according to the undiluted Hotelling analysis, this kind of behavior would be quite acceptable. It is also true that in the real world most mining output is sold on the basis of long-term contracts, with the mine committed to deliver a certain amount in every period regardless of the desire of management to restrict production in certain periods, and thus to save ore that could be sold for a larger profit in later periods.

Before turning these observations into simple algebra, it should be emphasized that despite the veritable deluge of papers on exhaustible resources, the microeconomic theory of the mining firm is extremely underdeveloped [although, admittedly, some recent work of Levhari and Liviatan (1977) clarify the basic issues]. For instance, if the reader refers to an important paper (1978) by John Helliwell in *Resources Policy,* and takes the time to generalize Helliwell's exposition, it becomes obvious that the analysis of Hotelling (and of his followers) has little, if anything, to do with actual mining installations. It is true, however, that anyone interested in the pricing and intertemporal allocation of energy resources such as oil and natural gas should devote a little attention to Hotelling's seminal article.

I will now construct a constrained maximum exercise that, with the help of Kuhn-Tucker theory, will give an insight into intertemporal production decisions in a situation where fixed capital is present, and the owner of this capital is committed to make periodic payments for the use of this capital—in this case interest payments since, for simplicity, I shall assume the equipment to be nondepreciable, and thus amortization need not be considered. I shall also, for pedagogic reasons, make the following assumptions: (1) the fixed capital is in place when the decision maker assumes his duties; thus his job is to say how much of the resource should be extracted in each period; (2) the amount of the resource is fixed, and we will deal with a fixed time-horizon situation that runs two periods, t and $t + 1$; thus the assumption must also be made that something is produced in each period, even had we decided to ignore the fixed capital (explicitly, $q_t, q_{t+1} > 0$).

I limit the analysis to two periods to make it easier for the reader to follow the discussion—in particular the algebraic portion; but the constraints used in this exposition to take into account the charges associated with fixed capital can be applied to more general discrete or continuous models with nonfixed time horizons, and the results will be similar to those presented here. Finally, I will assume that the ore prices p_t and p_{t+1} are known at the beginning of period t when the production decisions are made, and that output appears at the end of each period. This indicates that where output is concerned, the firm is operating in a situation characterized by pure competition; we do not have this situation in the capital market, however, since interest payments on the fixed capital cannot be accumulated, but must be paid every period. We can now go to the Lagrangian associated with our constrained maximum problem:

$$L = \frac{p_t q_t - c_t q_t}{(1 + r)} + \frac{p_{t+1} q_{t+1} - c_{t+1} q_{t+1}}{(1 + r)^2}$$

$$+ \lambda [R - (q_t + q_{t+1})] + \alpha_1 [p_t q_t - c_t q_t - r P_k K]$$

$$+ \alpha_2 [p_{t+1} q_{t+1} - c_{t+1} q_{t+1} - r P_k K]. \qquad (7.41)$$

The first two terms in equation 7.41 represent discounted quasi-rents for periods t and $t + 1$. Had we started with a nonlinear program, the sum of these expressions would have formed the objective function. As for the symbols, p is price, q is quantity, and c is unit variable cost. c is taken as a function of q, or $c_t = c_t(q_t)$. Note also that we have $c_t(q_t)q_t = C_t$, where C_t is the total variable cost of the output q_t. Thus the differentiation of C_t with respect to q_t, $C_t'(q_t) \equiv C_t'$, is the marginal cost of q_t.

The next expression follows from the constraint on resources: $q_t + q_{t+1} \leq R$. If for the present we ignore the last two terms of 7.41 and apply Kuhn-Tucker theory to the first three, we get:

$$\frac{\partial L}{\partial q_t} = \frac{p_t - C_t'}{(1 + r)} - \lambda \leq 0 \quad \text{and} \quad q_t \left[\frac{p_t - C_t'}{(1 + r)} - \lambda \right] = 0. \qquad (7.42)$$

$$\frac{\partial L}{\partial q_{t+1}} = \frac{p_{t+1} - C'_{t+1}}{(1 + r)^2} - \lambda \leq 0 \quad \text{and}$$

$$q_{t+1} \left[\frac{p_{t+1} - C'_{t+1}}{(1 + r)^2} - \lambda \right] = 0. \qquad (7.43)$$

$$\frac{\partial L}{\partial \lambda} = R - (q_t + q_{t+1}) \geq 0 \quad \text{and} \quad \lambda [R - (q_t + q_{t+1})] = 0. \qquad (7.44)$$

The expression $p - C'$ is a *marginal profit,* and since I have assumed $q_t, q_{t+1} > 0$, we get from the right-hand side of 7.42 and 7.43:

$$\frac{(p_{t+1} - C'_{t+1}) - (p_t - C_t')}{(p_t - C'_t)} = \frac{\Delta MP}{MP} = r. \qquad (7.45)$$

This is the well-known Hotelling rule. Observe also from 7.44 that if $q_t + q_{t+1} < R$, then Kuhn-Tucker theory says that $\lambda = 0$, and in each period we have $p = C'$. In other words, price is equal to marginal cost.

Now we turn to the last two constraints of 7.41. If we have nondepreciable capital that costs $P_k K$, the interest charge in each period will be $r P_k K$; and as mentioned earlier, this must be paid in each period. Thus in each period production must proceed to a point where, after variable costs are paid, enough revenue is left to pay interest charges—that is: $pq - c(q)q \geq rP_kK$. Next, if we apply Kuhn-Tucker theory to the entire Lagrangian, we get:

$$\frac{\partial L}{\partial q_t} = \frac{p_t - C'_t}{(1 + r)} - \lambda - \alpha_1\left[\frac{p_t - C'_t}{(1 + r)}\right] \leq 0 \quad \text{and}$$

$$q_t\left[\frac{p_t - C'_t}{(1 + r)} - \lambda - \alpha_1\frac{p_t - C'_t}{(1 + r)}\right] = 0. \tag{7.46}$$

$$\frac{\partial L}{\partial q_{t+1}} = \frac{p_{t+1} - C'_{t+1}}{(1 + r)^2} - \lambda - \alpha_2\left[\frac{p_{t+1} - C'_{t+1}}{(1 + r)^2}\right] \leq 0 \quad \text{and}$$

$$q_{t+1}\left[\frac{p_{t+1} - C'_{t+1}}{(1 + r)^2} - \lambda - \alpha_2\frac{p_{t+1} - C'_{t+1}}{(1 + r)^2}\right] = 0. \tag{7.47}$$

$$\frac{\partial L}{\partial \lambda} = R - (q_t + q_{t+1}) \geq 0 \quad \text{and} \quad \lambda[R - (q_t + q_{t+1})] = 0. \tag{7.48}$$

$$\frac{\partial L}{\partial \alpha_1} = p_tq_t - c_tq_t - rP_kk \geq 0 \quad \text{and} \quad \alpha_1[p_tq_t - c_tq_t - rP_kK] = 0. \tag{7.49}$$

$$\frac{\partial L}{\partial \alpha_2} = p_{t+1}q_{t+1} - c_{t+1}q_{t+1} - rP_kK \geq 0 \quad \text{and}$$

$$\alpha_2[p_{t+1}q_{t+1} - c_{t+1}q_{t+1} - rP_kK] = 0. \tag{7.50}$$

Some interpretations are now in order. If profit-maximizing behavior unconstrained by the financial commitments associated with the presence of fixed capital results in enough revenue in each period to pay the interest charge rP_kK, then we have $pq - cq \geq rP_kK$ for t and $t + 1$, and α_1 and α_2 are zero in accord with the Kuhn-Tucker theorem. From these equations we once again get the conventional Hotelling result.

If profit-maximizing behavior unconstrained by the financial commitments associated with the presence of fixed capital leads to an arrangement of the type shown in figure 7-1, however—where there is not enough revenue in period t to cover interest charges—then production must be increased in period t (and reduced in period $t + 1$) despite the fact that we reduce the total discounted profit (as given by the sum of the first two terms in expression 7.41). In this case we have $\alpha_2 = 0$, but $\alpha_1 > 0$. Thus, from the right-hand expressions in 7.46 and 7.47 we get:

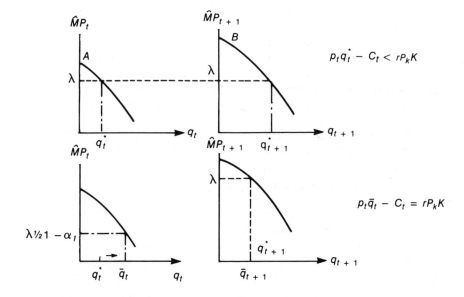

Figure 7–12. Profit Maximization in the Presence of a Constraint Provided by the Need to Pay Charges on Fixed Capital

$$\frac{p_t - C_t'}{(1 + r)} (1 - \alpha_1) = \lambda \qquad \text{or} \qquad \hat{M}P_t = \frac{p_t - C_t'}{(1 + r)} = \frac{\lambda}{1 - \alpha_1} \cdot \quad (7.51)$$

$$\frac{p_{t+1} - C_{t+1}'}{(1 + r)^2} = \hat{M}P_{t+1} = \lambda. \qquad (7.52)$$

In the notation used here, $\hat{M}P$ signifies *discounted* marginal profit. This arrangement is shown in the lower panel of figure 7–1. Notice also that in 7.51 we have an analogue of the expressions presented earlier in 7.35 and 7.37. This arrangement is shown in the lower panel of figure 7–12.

What we observe in figure 7–12 is a move away from the production pattern corresponding to maximum profit, to a scheme that involves less profit, but satisfies the last two constraints of equation 7.41. This displacement also gives the leveling-out effect in production referred to earlier. Note also that we have $\lambda > \lambda/(1 - \alpha_1)$, and since λ and λ and $\lambda/(1 - \alpha_1)$ are defined as the increase in total discounted profit given by an increase of one unit in the availability of the resource, if we were to obtain, without cost, an extra unit of the resource, it would be extracted in period $t + 1$.

Something should also be said about the sign of $1 - \alpha_1$. In the foregoing example $1 - \alpha_1$ is taken as greater than zero, but since we cannot put any a priori restrictions on α_1 other than specifying that it is nonnegative, it appears that $1 - \alpha_1$ could be less than zero. Given that interest charges must be paid in period t, then in this particular case it appears that production must be expanded past the point where $p_t = C_t'$—that is, to where $p_t < C_t'$. Unfortunately, however, this would not solve anything, since with $p_t < C_t'$, $MP_t < 0$, and thus $p_t q_t - c_t q_t$ is less than what it would have been with $p_t = C_t'$. Thus it is clear that at the optimum we must have $p_t \geq C_t'$, and as a result α_1 will have to be smaller than unity.

The Econometrics of Short-Run Prices

The background for this section can be found in the material on stocks and flows presented earlier in the chapter. There we related price changes to the difference between desired and actual inventories. Rather than take levels of inventories, we can relate price changes to changes in desired and actual inventories. In doing so, we might postulate the following equation.

$$\Delta p = p_t - p_{t-1} = \lambda (k \Delta d_{t-1} - I_{t-1}). \qquad (7.53)$$

In this expression d is demand and I is stocks. The estimating equation takes the form $\Delta p = \beta_0 + \beta_1 \Delta d_{t-1} + \beta_2 I_{t-1}$, from which we get $\lambda = -\beta_2$, $\beta_1 = \lambda k$, and so $k = -\beta_2/\beta_1$. We should also expect that $\beta_2 < 0$ and $\beta_1 > 0$. This construction will now be used to estimate a price equation for refined zinc in the United States. The flow variable here is actually disappearance, and thus represents zinc being used as a current input and for inventories. Theoretically, this situation is not ideal. We use annual data for the period 1953, and t ratios are in parentheses.

$$\Delta p = -0.885 + 0.0403 \, \Delta d_t - 0.195 \, \Delta I_{t-1} \qquad (7.54)$$
$$(1.476) \qquad\quad (3.13)$$

$$\bar{R}^2 = 0.5339 \qquad D.W. = 1.74$$

This equation is not particularly impressive, although it might be possible to improve it. Rather than speculate on its shortcomings, we can present a variant of this type of equation:

$$p = 74.46 + 4.55\frac{d_{t-1}}{I_{t-1}} + 0.348\frac{\Delta d_t}{\Delta I_t} + 5.24\,D \qquad (7.55)$$

$$(4.70) \qquad (1.95) \qquad (1.00)$$

$$\bar{R}^2 = 0.754 \qquad D.W. = 1.89$$

The dummy here is unity for 1967 and 1968 (during and after the severe strikes in the United States) and zero otherwise. One of the largest zinc companies in the world has used an equation like this for rough price forecasting and, at least until a few years ago, seemed quite satisfied. Figure 7–13 expresses some of the ideas found in this section.

Estimation and Some Other Problems

The estimation of commodity models has not presented as many problems as has the estimation of macroeconomic models, since for the most part—with commodity models—there are only a limited number of relevant

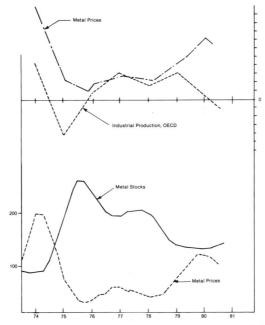

Figure 7–13. The Relationship of Metal Prices to Metal Stocks and Industrial Production

variables, and these either result in satisfactory equations or do not. In macroeconomics, on the contrary, there are so many variables available that, in principle, any type of equation can be estimated and any desired result can be obtained. Of course, irrelevant variables have been introduced in certain commodity models, but for the most part these frauds are so flagrant that even amatuer econometric sleuths are capable of detecting them.

Then, too, ordinary least squares or some version of the Cochrane-Orcutt technique will handle most commodity problems; and limited-information and full-information methods (as, for example, found in the TROLL system) should suffice for almost any conceivable estimating problem. However, a great deal of concern shown by econometricians for simultaneous equation bias and the like is wasted on commodity models; I am convinced that single equation estimates can do virtually everything that needs to be done in this area.

Where forecasting with commodity models is concerned, the situation is clear: there is no such thing as forecasting commodity prices, or any other prices, with an econometric model. This is what the record shows. It also shows that almost nothing can be forecast with macroeconometric models. The business journal *Euromoney* (and occasionally *Business Week*) makes a point of comparing econometric forecasts with reality, and these comparisons immediately suggest one thing: no one in his or her right mind would purchase the output of these models. The simple truth of the matter is that the attempt by established economists to convince students and noneconomists that forecasting with econometric models is a meaningful occupation is probably the greatest scandal in modern academic history— particularly since many of the present salesmen of econometric models are personally convinced of the lack of utility of these constructions. Moreover, the larger econometric models (economy wide models for policy analysis as well as forecasting) are and always have been hoaxes. The great danger, of course, is that results obtained from these inventions are taken seriously by honest men and women. (For example, the so-called evidence concerning the deleterious effect of social security on capital formation.) The simple fact of the matter is that the sooner econometrics is cut down to size, the better it will be for economics, economists, and perhaps even the world as a whole. As an example of what I mean here, I can cite the tactful withdrawal by its publishers of a particularly malodorous piece of Swedish "research" before it had a chance to offend the sensibilities of the profession. Unfortunately, however, similar concern was not shown for the Swedish taxpayers who financed this abomination or those economists who, against their will, had their work associated with it.

I have noticed, however, that many economists have started to react against some of the more flagrant excesses of econometrics. Professor Leontief, for example, has recently made very public his objection to the algebraic and econometric game playing that many of the leading journals insist upon forcing on their readers in place of issues that are vital to the survival of our civilization. This is one of the reasons why, in my note on the literature (beginning on page 315), the reader will find only a few of my key references associated with the so-called learned journals. This is because, as I tell my students, if you want to find out what is going on in the *real world,* then the journals are the wrong place to begin a voyage of discovery. As an example of what I mean, I can point out that only an insignificant amount of the new materials published in the journals every year finds a permanent home in the textbooks – because successful textbook writers understand that these materials are irrelevant for the education of economists.

8 Summary, Conclusions, and a Comment on the State of the World Economy

Since 1973–1974 the world economy has displayed many of the characteristics of the four worst economic downturns of the past two centuries: those taking place in the 1790s, 1850s, 1890s, and the 1930s. If we overlook incompetent political leadership and the unfortunate attempt by certain servants of the people to apply the crackpot rituals of monetarist economics to real life dilemmas, the cause of the present malaise can be traced directly to the energy price shocks of 1973 and 1979. In 1973, the growth of nonresidential fixed investment in four leading industrial countries (the United States, Japan, Germany, and the United Kingdom) left its long-run growth trend (of 4.9, 15.5, 6.5, and 5.8 percent respectively) and fell to a much lower trajectory, while simultaneously the inflation rate increased.[1] The productivity slide that is so pronounced in certain industrial countries can also be traced to the oil-price escalades of 1973 and 1979–1980, having its origin in two important phenomena. First, the reduction in spending on capital goods, which, in aggregate terms, has resulted in a decrease in the amount of equipment per employee and has thus directly affected productivity. On the psychological side, there has been a lowering of morale caused by people finding out that they are not, after all, going to be able to afford the goods and services that would enable them to live the creative and carefree lives of the Ewings or the Corleones. As a side effect, open unemployment in the OECD has increased by nearly ten million people in the past nine years, and now stands at 30.5 million. Not only is it probably accelerating but, as figure 8–1b indicates, for the first time since the Great Depression the growth of employment is stagnating.

In chapter 2 I attempted, among other things, to describe the way in which the world economy gradually returned to a kind of normality during 1975–1979, as producers and consumers adjusted to higher but stable oil prices; and to describe the devastating effects of the 1979–1980 oil-price rises on global economic activity in 1980–1981. (See figure 8–1a). These effects included a total income loss of about 5 percent for the OECD in 1980 and almost 8 percent in 1981. The situation in 1982 appears to be equally as dismal, although in reality it is much more serious than it looks because of the investment that did *not* take place over the past three years. Among other things, this means that even if income temporarily turns up for one

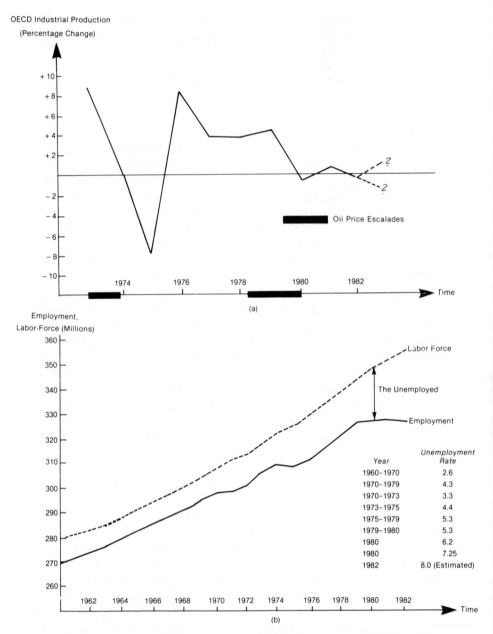

Figure 8-1. (a) The Change in OECD Industrial Production and (b) Unemployment in the OECD

reason or another in the near future, a sizable fraction of the unemployed will remain without jobs unless there is a decline in *real* remuneration that parallels the fall in productivity. As unemployment begins to approach crisis proportions, there is some indication this is already taking place— through both a stagnation in the rate of increase of wages and salaries and the depressant effect of price increases on purchasing power.

It has been made clear to me that not all economists agree with either the tone or emphasis of the above analysis. Only a few years ago a number of so-called experts took great pains to point out that abnormalities in world price movements were apparent before 1973, which is a useless and irrelevant observation in the light of subsequent aggregate price developments; while other economists have insisted that the world economy is capable of progressing as it did during the golden age of the 1950s and 1960s with a substantially lower input of energy per unit of output. The arguments supporting this provocative contention invariably reveal a deficient insight into the nature of energy *conservation* and *substitution,* so as a complement to the material presented in chapter 2 I will give a straightforward interpretation of these important phenomena.

Conservation, as it is practiced today, means, to a considerable extent, depriving the productive sector of the economy of irreplaceable energy inputs, which in turn leads to a fall in macroeconomic activity and thereby to an involuntary lowering of standards and expectations. Conservation comes about not just through turning down thermostats or driving fewer miles, but as a reaction to the decline or collapse of current and expected incomes caused by falling production. Among other things it means people conserving on food, medical and dental treatment, and the education of their children. In Denmark, for example, many decision makers (and would-be decision makers) were intent upon showing the rest of the world that modern economies can function on a much smaller diet of traditional energy materials; and until recently Denmark had the highest unemployment rate in northern Europe, a freight of foreign borrowing that has almost spiraled out of control, and in 1980 and 1981 a more rapid decline in domestic demand than any other OECD country. These are a few of the economic costs of the Danish conservation miracle in quantitative terms; and they have also had a deleterious effect on the social harmony existing in Denmark since, as might be expected, many of the residents of that country are not particularly overjoyed by the possibility that it might someday regress to the status of a banana republic.

A similar, but somewhat more complicated, story can be told about substitution. Substitution does not just mean substituting coal for oil, wood for coal, or rubbish for wood. It also means a change in the pattern of consumer and investment goods produced by many industrial countries and, along with this, an alteration in the structure of production. More specifically, in the OECD countries it means the disappearance or scaling down of a number of vital economic activities that helped finance the high economic and cultural levels that many of us have been privileged to enjoy over the past two decades or so. On the other hand, substitution and conservation are manifestations of technological progress, and in certain respects technological progress is moving faster than ever. But up to now, in most industrial countries, technological progress has had the embarrassing proclivity to be labor saving and energy using, which is one of the reasons that the growth of employment is now displaying an ominous tendency to flatten out. It should be made clear, though, that the *net* effect of technological progress may turn out to be energy saving. This is so because the increased amount of energy used by robots, word processors, and automatic-banking systems could be more than compensated for by the decreased demand for such things as electricity and fuel by people whose incomes have been reduced as a result of being displaced by these remarkable innovations.

For instance, industrial production has fallen over most of the past year or two in the United States, while Sweden lost 40,000 industrial jobs last year as industrial production fell by approximately 4 percent. This creeping deindustrialization is what conservation and substitution are really all about. Moreover, not only are many of the lost industrial jobs gone forever, but many of the service jobs, because they will pay less, that will replace them are going to be viewed as unsatisfactory by the people taking them. Note here that, in most countries, wages and salaries in the service sector are directly or indirectly tied to the buying power generated in the traditional industrial sector, although unfortunately in Sweden this simple fact of economic life has escaped the notice of many amateur economists who, typically, think that factories can be replaced by child-care centers. It should also be understood that many of the so-called high-technology industries that are supposed to eventually replace activities such as metal processing, steel making, and ship building will *not* materialize in places like Scandinavia. This is because these high-technology industries will also be established in low-wage, nonunionized parts of the world that now, as was not the case in the fairly recent past, are also in possession of a large and growing reservoir of technical and managerial skills.

I would like to conclude this portion of the discussion with the following observation. Too much attention is being paid to the benefits from saving energy and not enough to the costs. Given the production that is being lost as a result of conservation and substitution, it could be argued that the implicit value of a barrel of oil is actually many hundreds of dollars. What has been almost completely forgotten, it seems, is that for most people the oil-supply and oil-price dilemmas are only important in terms of overall economic conditions; and not even the Hudson Institute physicists claim to be in possession of workable prescriptions for alleviating the macroeconomic discomforts (for example, unemployment) that have accompanied the fall in oil consumption.

In chapters 3, 4, and 5 I examined some of the alternatives to oil, and a few remarks on these matters are in order at the present time. The world may be on its way to an energy technology largely based on coal, but I believe that this is the wrong choice for the immediate future. It is also true that in the short-run increasing amounts of gas and, possibly, nuclear power are going to be used. My research indicates that large quantities of gas should be consumed—if possible, so much that a large increase in coal usage can be delayed until the environmental problems associated with coal are at least partially overcome. As it happens, however, if Western Europe and Japan are to raise their consumption of natural gas by substantial amounts in the near future, then a fairly high percentage of this gas will have to be purchased from the USSR. This immediately raises a political dilemma, since certain people in both the United States and Europe are of the opinion that the energy dependence of Western Europe on the USSR should not be increased. They are also of the opinion, although they do not say so, that the USSR should be isolated economically; and that without Western technology the Soviet economy could not function.

The simple truth here is that Western Europe is *not* dependent on the USSR for its natural gas or for any other commodity or service. On the contrary, it is the USSR that is dependent upon Western Europe for the hard currency that is required to pay for the large amounts of technology the USSR desires to import for at least the remainder of the 1980s. As is well known, copious amounts of gas can be found in other parts of the world, to include the North Sea; and when the European Gas Network is complete it will be capable of introducing and distributing a great deal of gas from non-Soviet sources to many Western European countries. But from a strictly business point of view, the USSR is the optimal supplier of this superb energy resource, since it is willing to sell large quantities at a price appreciably below that being requested by its competitors. I conclude that the more of this gas purchased the better; and once the tap is fully opened on Soviet gas, and it shows signs of staying open, it may be possible to make use of pipelines through the southern part of the USSR to market some of the huge gas reserves of Iran. If so, the Western European (and the world) energy picture would certainly assume a saner composition, particularly from an environmental point of view.

As indicated above, the main disadvantages associated with coal are environmental. Unless I am mistaken, after the first oil-price shock I was one of the first economists in the world to propose that coal should be returned to its former position as the world's most important energy resource. This option still makes sense in the long run, but if attempts were made to raise coal consumption by a factor of two or three in the next twenty years, the results might well be catastrophic. At present, the atmosphere over many industrial countries simply cannot tolerate a large increase in sulphates and other noxious elements that originate with the burning of coal. Moreover, the efficiency of pollution-suppression devices has been grossly oversold both by commercial interests associated with the coal industry, and by certain bureaucrats whose knowledge and expertise is unfortunately vastly inferior to the grandeur of the positions they occupy. In addition, because of the present climate of recession, wavering profits, and indifference or ignorance, efforts are being made everywhere to avoid making investments of the type needed to make coal environmentally acceptable, and unless I am mistaken, many of these efforts are going to be successful.

What about nuclear energy? With the exception of hydro-systems, nuclear generators are, in my opinion, the most inexpensive way to obtain large amounts of electrical power. The problem is that successful nuclear operations tend to increase the respectability of the breeder reactor, which is conceivably the most dangerous piece of machinery ever devised. Moreover, given the deficient intelligence and imagination of many of our political masters, the possibility that either the plutonium community *or* a more light water intensive community can be made safe within the framework of conventional democratic practices may be very small. For instance, a majority of Swedish legislators agree that, along with unemployment, the widespread distribution and use of narcotics represents the greatest danger facing Swedish society today; but even so, these ladies and gentlemen have not been able to do anything about the large-scale marketing of hard drugs that takes place less than 100 meters from the front doors of the Swedish Parliament, openly, to a considerable extent in broad daylight, and often within sight of the police. Accordingly it is not unthinkable that this type of well-meaning but perplexed statesman and stateswoman are completely incapable of dealing with the security problems of an energy source, such as plutonium, where there is absolutely no margin for error. Furthermore, a finite danger also exists that in order to avoid providing the same gratuitous liberties to terrorists in possession of plutonium that are now accorded drug wholesalers and incompetent Gothenburg economists, parliamentarians in countries using the breeder might lean toward a general restriction of civil liberties.

Chapter 6 consisted of a long exposition of the economics of nonfuel primary commodities in which it was made clear that the same thing is now true of the market for oil that has long been true of the market for metals and minerals: a lack of demand due to a failing world economy. In fact, in relative terms, the recent oil surplus was less than the surplus of iron ore, and the *real price* of nonfuel minerals—taken as an aggregate—is lower today than it has been for the past twenty years. Not only that, bad vibrations from the growing world economic crisis periodically causes convulsions on these markets. On February 8, 1982, the day after a major U.S. television network presented a recapitulation of the stock market crash of 1929, both the stock markets and the Chicago and New York commodity exchanges were hit by a wave of selling. Unfortunately, though, some economists refuse to place the correct interpretation on events of this nature. A recent Nobel Prize winner has ventured that the fall in the price of primary commodities lengthens and aggravates the recession, when, in fact, the chain of causality moves in the opposite direction: the falling price of primary commodities is simply confirming trends that exist in the manufacturing, commercial, and financial sectors of the major industrial countries.

It also appears that primary commodities have lost their attraction for speculators. One of the reasons for this is the high interest rates prevailing over the past year. These high rates have made it inopportune to finance and hold large inventories of various metals and minerals; bonds and certain other financial assets are much more attractive. The major losers in this drama are the countries of the Third World that depend for their livelihood on the production of primary commodities; and some of these countries are attempting to form producers' organizations of the OPEC type. The major tin exporters, apparently led by Malaysia, have invested more than a half-billion dollars in stockpiling tin in an attempt to support the price of tin. As far as I can tell, this exercise will only be successful in the long run if the cooperation of some of the larger tin-consuming countries can be obtained, and in particular the United States. But with no tin being produced in the United States, I doubt whether the present U.S. government is inclined to be sympathetic to a "TINPEC." Here I should mention that the International Cocoa Organization *has* managed to keep the price of cocoa from crashing by maintaining large inventories of cocoa. The question must be asked, though, just how long it will be possible to continue this practice, particularly when the main cocoa-growing countries (Ghana, the Ivory Coast, and Brazil) seem unable or unwilling to reduce their production.

Something should also be said here about gold, since in *The Political Economy of Oil,* I took the liberty of predicting rising prices for this commodity. As far as I am concerned, that prediction still holds. The recent decline in the gold price is directly traceable to the unnaturally high level of interest rates prevailing in the world economy in general (and the United States in particular), which makes the holding of gold an uneconomical proposition. (Of lesser importance is the departure from the market of OPEC purchasers, and the dumping of gold by the USSR.) In turn, the structure of U.S. (and world) interest rates is due to the not-quite-realistic economic policies of President Reagan, who has introduced a succession of large budget deficits for the purpose of financing unproductive investments in armaments, and whose tax reductions have—at least up to now—promoted private consumption rather than capital formation. Short-term interest rates in the United States are now about 16 percent, while long-run rates are 14-15 percent, and since the present rate of inflation is 6-7 percent, the real rate of interest is in the vicinity of 8 percent. But the average real rate of return on such things as machinery is traditionally about 4.5-5 percent in the United States, and so investment has stagnated. It is clear that this situation is not only intolerable, but, from a macroeconomic point of view, very unstable; and unless an adjustment takes place it will mean even less investment, a continuation—and perhaps acceleration—of the decline in industrial production, and employment rates approaching those experienced during the great depression.

Since it is now becoming common knowledge that unadulterated Reaganomics has no more future than pure *or* diluted Thatcheromics, I see the adjustment consisting of a softening up or gradual repudiation of at least two main planks in President Reagan's game plan. The rearmament program will be reduced to sensible proportions—hopefully along the lines suggested by people like General Maxwell Taylor; and the more grotesque antics of monetarism will be abandoned—although, unfortunately, before the bankruptcy of monetarist economics is recognized in every nook and cranny of the academic world. If these matters are handled correctly, the U.S. economy could again display some of the strong growth tendencies witnessed during 1976-1978. What will not happen, however, is a resumption of the growth rates experienced before the first oil-price shock, because the kind of world in which a performance of that nature was possible no longer exists.

Two more important topics need to be mentioned: exchange rates and inflation. Where the first is concerned, I think that it is now the belief of a large majority of bankers and central bankers that the introduction of fully flexible exchange rates was a major blunder. Ironically enough, one of the reasons that nothing is being done to offset this fiasco is that certain influential economists and politicians are of the (false) opinion that a free market in international currencies is a logical extension of free domestic markets in goods and services. One of the more interesting effects of flexible (or fluctuating) exchange rates is the way they have disguised movements in the *real price* of oil. Since the dollar has appreciated relative to almost all other currencies over the past year by *more* than inflation rates, the real price of oil has increased for Sweden, Germany, France, Japan, and many other countries even though the *money price* of oil has been constant, or nearly constant. Put another way, the purchasing power of oil has steadily increased for its exporters: just as the dollar of an American tourist buys more in France this year than last year, the same is true for an OPEC country buying French machinery with dollars. This is an important factor in explaining why, despite appearances, it is comparatively easy for some OPEC countries to freeze or decrease their exports of oil.

As for inflation, I think I can say that in *The International Economy: A Modern Approach,* I said just about all there is to say on that topic that I feel is important. In particular, I am pleased to have pointed out that highly successful economies like Germany and Japan have shown that a rapidly growing money supply does *not* mean excessive inflation; and that the key element in any anti-inflation program is reestablishing or, in some cases, establishing, the link between productivity and remuneration. What I did not point out, however, is that as things now stand a modicum of inflation may be preferable to the deflation that certain governments seem intent upon bringing about (where deflation here does not mean a declining price level, but an annual rise in prices slightly above that existing in the 1960s).

To understand this, consider the effect of deflation on homeowners. These people have gone into considerable debt to buy their homes, believing that due to inflationary conditions their salaries will also increase in such a manner as to enable them to make mortgage and amortization payments. More important, they believe that their homes are a liquid asset that, in an emergency, can be easily sold for at least its purchase price or used as a security against which to incur additional debt. Deflation makes these anticipations less realistic, and by decreasing the value of this asset often causes them to reduce their spending.

The business sector in many countries is in an analogous situation. Many businesses now carry enormous amounts of debt—debt that was acquired in order to purchase machines and structures. (Consider, for example, the major airlines). The positive *net worth* of these firms is maintained only by the larger anticipated revenues that will result from the sale of the goods and/or services produced by these physical assets. A fall in the price (or, in some cases, the *expected* price) of these goods and services leads to a writing down of the value of the physical assets held by corporations, either on their balance sheets or in the thinking of the financial institutions that have lent these corporations money. (This is because as the profit-making capacity of assets decline, the market value of these assets must reflect this change. Once again the reader can think of the airlines with their enormously expensive inventories of planes and servicing facilities whose resale value tumbles when people decide to ride busses or trains instead of airplanes.) At this point, two things are likely, both of them unpleasant. Creditors can start foreclosing, and the financial panic that was called into the wings when debt started to be measured in units of one trillion, is sung on stage by a chorus of bankruptcies. Or, less drastically, creditors threaten to foreclose, or even start asking the directors of firms embarrassing questions about their financing in unfriendly or sarcastic tones. For individual companies this is the signal for investment to decrease, and the discharging of employees to increase; and for the economy as a whole the onset or deepening of a recession. Here I can remind the reader that, as implied above, one of the main contributing elements to this type of predicament is a fall in consumer demand on the part of those individuals whose consumption behavior is sensitive to their wealth. An important group here is, of course, people whose wealth is in the form of real estate (that is, homeowners who eventually realize that in a deflationary situation the value of their homes and lots have substantially decreased and as a result their net worth is significantly reduced). Considering again the airlines example employed above, a slowing down in the rate of inflation has meant a fall in the price of air travel; but this fall has been insufficient to maintain airline revenues because the consumption of travel services, which to a great extent is a luxury, has fallen even more rapidly than consumption in general.

Let me close by emphasizing that I am *not* saying that a high rate of inflation is preferable to the kind of price increases experienced during the 1950s or 1960s, which I have chosen to call deflation, because they are so much lower than the price rises we have seen during the past few years. I *am* saying that attempting to check inflation by monetary means in the presence of the kind of consumer and corporation debt that exists today, almost everywhere in the world, is begging for a catastrophe. Monetarism, like champagne, is good for pleasure—especially the pleasure of academic economists trying to add a few millimeters to their bibliography; but it is very bad for business.

Note

1. Ferdinand Banks, *The Political Economy of Oil* (Lexington, Mass.: Lexington Books, D.C. Heath and Company), figure 2–2.

A Note on the Literature
and a Bibliography

Perhaps the best reading on energy subjects is to be found in the journals *Energy Policy* (Butterworth Scientific Limited), and the *Energy Journal*. The most important journal for nonfuel minerals is *Resources Policy*. Other important journals are the *OPEC Bulletin* and the *Chemical Economy and Engineering Review (Tokyo)*. These reviews contain material on oil products and petrochemicals as well as petroleum. Important periodicals for keeping up to date on energy matters are *Business Week* and *L'Expansion*. In my work on this book I have also found the *Engineering and Mining Journal* and the *Oil and Gas Journal* useful.

The best introductory article on oil is Flower (1978), but I have attempted to expand on his exposition in this book and especially in the second printing of *The Political Economy of Oil*. An important survey of oil, coal, gas, and uranium is found in Folie and McColl (1978). An interesting analysis of the first oil crisis is that of Charles Issawi (1978). Pierre Desprairies (1981) has a vision of the development of the oil price that I share, as does Edward Erickson (1981). Also very interesting is Murcier and Boissonnat's interview (1982) with Robert Lattés called 'Pétrole: la grande illusion,' in *L'Expansion*. Provocative scenarios dealing with the demand for energy materials have been constructed by Steve Salant, and Martin Hoffmeyer and Axel Neu. An important book on coal, although limited by its concentration on the U.S. market, has been written by Richard F. Gordon (1978); but Dr. Martin Zimmerman of MIT is now doing the most comprehensive work on the economics of coal. An important book on enhanced oil recovery has been written by Ray Dafter (1981).

Very good examinations of the markets for nonfuel minerals can be found in the books by Carmine Nappi (1979) and Satyadev Gupta (1982); see also the work of P.C.F. Crowson (1982). Where microeconomic and market theory applicable to nonfuel minerals is concerned, perhaps the most important contributions are those of Robert Pindyck. Outstanding work on copper has been done by Raymond Mikesell (1979). Straightforward discussions of commodity problems can be found in Rex Bosson and Bension Varon (1977), Paul Hallwood (1979), Richard Cooper and Robert Lawrence (1975), Varon and Kenji Takeuchi (1974). For models of cartel behavior see Robert Pindyck (1977, 1978), and also Michael Folie and Alistair Ulph (1980); but the OPEC point of view is well presented by Salah El Serafy (1981) in a nontechnical exposition. A very interesting introduction to the important oil-transfer problem, as well as some other macroeconomic problems associated with industrial raw materials, can be found in the work of Michael Schmid. I have also attempted to survey various aspects of these issues in *The Political Economy of Oil* (second printing) and *The International Economy: A Modern Approach* (second printing). For matters concerning the substitutability between energy and other production factors see Dale Jorgenson (1978) and Berndt and Wood (1977).

For the examination of resource problems from history of doctrine point of view the reader should consult G.F. Ray, T.J.C. Robinson, and Harrie Paul Hendrikx. The resources portion of Anthony Fishers new book is also quite useful; and the same is true of Dasgupta and Heal (1979), which is also an excellent advanced microeconomics text.

The economics of the Soviet energy sector has been well surveyed by Krylov (1979) and Jonathan Stern (1980), but also on a somewhat smaller scale by Joseph Martellaro (1981). Excellent analyses of the mineral and energy potential of Australia are now available. See, in particular, the work of Stephen Derrick, Daina McDonald, and Phyllis Rosendale (1981), Michael Folie and Greg McColl (1980), David Gallagher and McColl (1981), Greg McColl (1981), and Edward Shann (1981). Shann treats in some detail the Australian version of the Dutch Disease. The background to this ailment and, in my opinion, the best understanding of its origin is to be found in the work of Paul Rayment—for the most part integrated into various publications of the Economic Commission for Europe. But see also the seminal paper by Robert Gregory (1976). The best elementary treatment of uranium economics is by Marian Radetzki (1981).

For the theory of resource extraction I recommend the work of John Hartwick and Murray Kemp. Their production is so extensive, and all inclusive, that there would be little point in listing it here. Among the many interesting extensions of this particular topic see Florian Sauter-Servaes (1979). The best work on primary commodity econometrics is now being done by Anthony Owen and David Hojman and, in comparison with much of the other work being done on this topic, it is also honest. Very important and provocative work on ultimate resource availability has been carried out by Ken Walker (1979) and F.E. Trainer (1982); but for a broader view of this topic, the work of Nicholas Georgescu-Roegen is a must.

An interesting aspect of commodity politics can be found in the work of Harry Johnson (1976) and Helge Berge (1977). For many of the quantitative as well as philosophical and political issues involving energy, the forthcoming book edited by J. Ferricelli and J.B. Lesourd is to be recommended; and a very useful elementary, but very informative, book on energy and the environment has been written by Earl Cook (1976). Perhaps the best up-to-date summary of energy-policy issues can be found in the transactions of the 1982 meeting of the International Association of Energy Economists, edited by Paul Tempest, which also contains some of the conclusions of the summary report of the Stanford Energy Modeling Forum titled *World Oil*.

There are many excellent intermediate microeconomic texts available now (such as Call and Holahan, Lancaster, Mansefield, and R.L. Miller), but at the elementary-intermediate level I have found Laidler (1981) and Quirk (1976) very useful. Probably the best intermediate macro book ever written is Dornbusch and Fisher (1981), which among other things contains an interesting reference to the macroeconomic effects of the 1973 energy price shock; but certainly the best introduction to the significance of debt is to be found in the very nontechnical book, *Paper Money,* by Adam Smith. A very clear and thorough book on international economics is that of Kreinin (1979).

Adelman, M.A. "Energy-Income Coefficients and Ratios: Their Use and Abuse," *Energy Economics* (January 1980).

Alchian, A., and Allen, W. *Exchange and Production Theory in Use.* Belmont, Calif.: Wadsworth, 1964.

Al-Janabi, A. "Opec Reserves, Production and Exports." *OPEC Weekly Bulletin* (January 1977).

———. "Estimating Energy Demand in the OPEC Countries." *Energy Economics* (April 1979).

Altenpohl, D. *Materials in World Perspective.* Berlin: Springer-Verlag, 1980.

Amacher, R.C., and Sweeney, R.J. "International Commodity Cartels and the Threat of New Entry: Implications of Ocean Mineral Resources." *Kyklos* 29 (1976).

Artus, J. "Potential and Actual Output in Industrial Countries." *Finance and Development* (1979).

Bailly, P.A. "The Problems of Converting Resources to Reserves." *Mining Engineering* (1976).

Bambrick, S. *Australian Minerals and Energy Policy.* Canberra: Australian National University Press, 1979.

Banks, F.E. "An Econometric Model of the World Tin Economy: A Comment," *Econometrica,* 1972.

———. *The World Copper Market: An Economic Analysis.* Cambridge: Ballinger, 1974.

———. "Laying Hamlets Ghost in Commodities." *New Scientist* (August 1976).

———. *Scarcity, Energy, and Economic Progress.* Lexington, Mass.: Lexington Books, D.C. Heath and Company, 1977.

———. *The International Economy: A Modern Approach.* Lexington, Mass.: Lexington Books, D.C. Heath and Company, 1979a.

———. *Bauxite and Aluminum: An Introduction to the Economics of Non-Fuel Minerals.* Lexington, Mass.: Lexington Books, D.C. Heath and Company, 1979b.

———. *The Political Economy of Oil.* Lexington, Mass.: Lexington Books, D.C. Heath and Company, 1980.

———. "The Price of Oil in 1985" CEER Review (April–May, 1982).

Basevi, G. and Steinherr, A. "The 1974 Increase in Oil Prices: Optimum Tariff or Transfer Problem." *Weltwirtschaftliches Archiv* (August 1976).

Baumol, W.J. *Economic Dynamics: An Introduction,* 3rd ed. New York: Macmillan, 1970.

Bell, G. "The OPEC Recycling Problem in Perspective." *Columbia Journal of World Business* (Fall 1976).

Berge, H. "En Ny Ekonomisk Världsordning?" *Ekonomisk Revy* (March 1977).

Bergsten, C.F. "A New OPEC in Bauxite." *Challenge* (July–August 1976).

Berndt, E.R., and Wood, D.O. "Engineering and Econometric Approaches to Industrial Energy Conservation and Capital Formation." Resources Paper no. 16, University of British Columbia (December 1977).

Björk, O. "Utvecklingen på den Internationella Oljemarknaden." Department of Economics, University of Stockholm, Skrift Nr 1978:5 (1978).

Bosson, R., and Varon, B. *The Mining Industry and the Developing Countries.* New York: Oxford University Press, 1977.

Brennan, M.J. "The Supply of Storage," *American Economic Review* 48 (1958).

Bronfenbrenner, M. *Macroeconomic Alternatives.* Arlington Heights, Ill.: AHM Publishing Corporation, 1979.

Brooks, D., and Andrews, P.W. "Mineral Resources, Economic Growth, and World Population." *Science,* 5 July 1974.

Brown, M., and Butler, J. *The Production, Marketing, and Consumption of Copper and Aluminum.* New York: Praeger, 1968.

Brown, W.M. "Can OPEC Survive the Glut." Hudson Institute (October 1981).

Carman, J. "Comments on the Report Entitled 'The Limits to Growth.' " Address at the University of Brunswick, 1972.

Carmoy, Guy de. "Nuclear Energy in France: An Economic Policy Overview." *Energy Economics* (July 1979).

Charles River Associates. Policy Implications on Producer Country Supply Restrictions: The World Aluminum/Bauxite Market. Cambridge, Mass., 1977.

Chevalier, J.M. *Det Nye Spill om Oljen.* Trondheim (1974).

Choucri, N. "Analytical Specification of the World Oil Market." *Journal of Conflict Resolution* (June 1979).

Clarke, R., and Hocking, D.M. "Natural Gas: Pricing, Export, and Taxation Policy." Centre for Policy Studies, Monash University, 1980. Stencil.

Clower, R.W. "An Investigation into the Dynamics of Investment." *American Economic Review* 64 (1954).

Cook, E. *Man, Energy, Society.* San Francisco: W.H. Freeman, 1976.

Cooper, R., and Lawrence, R.Z. "The 1972-73 Commodity Boom." *Brookings Papers on Economic Activity* no. 3 (1975).

Cox, C.C. "Futures Trading and Market Information. *Journal of Political Economy* (December 1976).

Cranston, D.A., and Martin, H.C. "Are Ore Discovery Costs Increasing?" *Canadian Mining Journal* (1973).

Cremer, J. "On Hotellings Formula and the Use of Permanent Equipment in the Extraction of Natural Resources," *International Economic Review* (June 1979).

Crowson, P.C.F. "Investment and Future Mineral Production," *Resources Policy* (March 1982).

Dafter, R. "Winning More Oil." London: Financial Times Business Information Ltd., 1981.

Dasgupta, B. "Oil Prices, OPEC, and the Poor Oil Consuming Countries." in Paul Rogers, ed. *Futute Resources and World Development.* New York: Plenum, 1976.

Dasgupta, P. and Heal, G. *Economic Theory and Exhaustible Resources.* Cambridge: Cambridge University Press, 1979.

Davidson, P. "Fiscal Problems of the Domestic Crude Oil Industry." *American Economic Review* (March 1963).

Derrick, S., McDonald, D., and Rosendale, P. "The Development of Energy Resources in Australia: 1981 to 1990." *Australian Economic Review,* no. 3 (1981).

DeSouza, G.R. *Energy Policy and Forecasting.* Lexington, Mass.: Lexington Books, D.C. Heath and Company, 1981.

Desprairies, P. "A Note on International Energy Prospects." French Institute of Petroleum, Paris, 1981. Stencil.

Donges, J.B. "UNCTAD's Integration Program for Commodities." *Resources Policy* (March 1979).

Doran, C.F., and Hopkins, J. "Three Models of OPEC Leadership and Policy in the Aftermath of Iran." *Journal of Policy Modeling* (1969).

Dornbusch, R. and Fisher, S. *Macroeconomics* (second edition) New York: McGraw Hill, 1981.

Dorr, A. "International Trade in the Primary Aluminum Industry." Ph.D. diss., Pennsylvania State University, 1975.

Dowell, R. "Resources Rent Taxation." Australian Graduate School of Management, 1978. Stencil.

El Serafy, S. "Absorptive Capacity, the Demand for Revenue, and the Supply of Petroleum." *Journal of Energy and Development* (Autumn 1981).

———. "Oil and the World Economy." *Finance and Development* (March 1982).

Erickson, E.W. "United States Energy Policy." *Cato Journal* (Fall 1981).

———. "The World Price of Oil." North Carolina State University, 1982. Stencil.

Erickson, E.W., and Grennes, T.J. "Arms, Oil, and the American Dollar." *Current History* (May–June 1979).

Ferricelli, J., and Lesourd, J.-B. *Econometrics and the Energy Crisis.* Paris: Economica, 1982.

Fisher, A. *Resource and Environmental Economics.* Cambridge: Cambridge University Press, 1981.

Fisher, F.M. *Supply and Costs in the U.S. Petroleum Industry.* Washington, D.C.: Resources for the Future, 1964.

Fisher, F.M.; Cootner, P.H.; and Baily, M.N. "An Econometric Model of the World Copper Industry." *Bell Journal of Economics* (Autumn 1972).

Fisher, L.A., and Owen, A.D. "An Economic Model of the U.S. Aluminum Industry." *Resources Policy* (September 1981).

Flower, A.R. "World Oil Production." *Scientific American* (March 1978).

Folie, M. "An Analysis of the Policy for Developing Australia's Oil Reserves." *Australian Quarterly,* no. 3 (1978).

———. "An Economic Appraisal of the Australian Oil and Gas Industry." *Resources Policy* (June 1980).

Folie, M., and McColl, G. *The International Energy Situation Five Years After the OPEC Price Rises.* Sydney: Centre for Economic Research, 1978.

———. "The Australian Coal Industry—An Economic Appraisal." *Resources Policy* (March 1980).

Folie, M., and Ulph, A.M. "Outline of an Energy Model for Australia." CRES Working Paper, 1976.

———. "The Use of Simulation to Analyze Exhaustible Resource Cartels." Canberra: Australian National University, 1978.

———. "Energy Policy for Australia." In R. Webb and R. Allen, eds., *Australian Industrial Policy.* London: Allen and Unwin, 1980.

Fox, W.A. *The Working of a Tin Agreement.* London: Mining Journal Books, 1974.

Fritsch, B. "The Zencap-Project: Future Capital Requirements of Alternative Energy Strategies Global Perspectives." Fifth World Congress of the International Economic Association, Tokyo (September 1977).

———. Über die Partielle Substitution von Energie, Ressourcen und Wissen." Paper presented to the Vereins für Socialpolitik Gesellschaft für Wirtschafts-und Sozialwissenschaften, Mannheim (1979).

Gallagher, D.R., and McColl, G.D. "Resource Scarcity? Australian Coal in a World Perspective." Brisbane, Australia, May 1981. Stencil.

Gälli, A. "The Foreign Trade of the OPEC States." *Intereconomics* (November–December 1979).

Gately, D. "The Possibility of Major Abrupt Increases in World Oil Prices by 1990." Discussion Paper Series, New York University, Faculty of Arts and Sciences, Department of Economics (May 1978).

———. "OPEC Pricing and Output Decisions." Paper prepared for the Conference on Applied Game Theory, Institute for Advanced Studies, Vienna, 13–16 June 1978.

———. "The Prospects for OPEC Five Years after 1973/74." *European Economic Review,* no. 12 (1979).

Georgescu-Roegen, N. *The Entropy Law and the Economic Process.* Cambridge, Mass., 1971.

———. *Energy and Economic Myths.* New York: Pergamon, 1976.

Gilbert, C.L. "The Post War Tin Agreements." *Resources Policy,* June 1977.

Gordon, R.L. *Coal in the U.S. Energy Market.* Lexington, Mass.: Lexington Books, D.C. Heath and Company, 1978.

Gregory, R. "Some Implications of Growth in the Minerals Sector." *Australian Journal of Agricultural Economics* (1976).

Griffith, E.D., and Clarke, A.W. "World Coal Production." *Scientific American* (January 1979).

Grillo, H. "The Importance of Scrap." *The Metal Bulletin,* Special Issue on Copper (1965).

Gupta, S. *The World Zinc Industry.* Lexington, Mass.: Lexington Books, D.C. Heath and Company, 1982.

Habenicht, H. "Processing Mineral Raw Materials." *Intereconomics* 9/10 (1977).

Haefele, W. "Global Perspectives and Options for Long-Range Energy Strategies. Keynote Address at the Conference on Energy Alternatives, East-West Center, Honolulu, Hawaii (1979).

Hallwood, P. *Stabilization of International Commodity Markets.* Greenwich, Connecticut: JAI Press, 1979.

Hammoudeh, S. "The Future Price Behavior of OPEC and Saudi Arabia: A Survey of Optimization Models." *Energy Economics* (July 1979).

Harlinger, H. "Neue Modelle Für die Zukunft der Menschheit." IFO-Institut fur Wirtschaftsforschung, Munich (February 1975).

Hashimoto, H. *Market Prospects for Aluminum and Bauxite.* World Bank, 1978.

Hawkins, R.G. *The Demand for Energy in Australia.* Ph.D. diss., Australian National University, Canberra, 1976.

Helliwell, J.F. "Effects of Taxes and Royalties on Copper Mining Investment in British Columbia." *Resources Policy* (March 1978).

Hendrikx, H.P. "Resources Rent and Changing Cost Conditions in the Mineral Industries." Department of Economics, University of Queensland, 1981.

Herfindahl, O.C. *Copper Costs and Prices: 1879*-1957. Baltimore: Johns Hopkins University Press, 1959.

Herin, J., and Wijkman, P.M. *Den Internationella Bakgrunden.* Stockholm: Institut for Internationella Ekonomi, 1976.

Hippel, F. von, and Williams, R. "Solar Technologies." *Bulletin of the Atomic Scientists* (November 1975).

Hoffmeyer, M., and Neu A. "Zu den Entwicklungsaussichten der Energiemärkte." *Die Wirtschaft,* Heft 1 (1979).

Hojman, D. "The IBA and Cartel Problems." *Resources Policy* (December 1980).

———. "An Econometric Model of the International Bauxite-Aluminum Economy." *Resources Policy* (June 1981).

Hotelling, H. "The Economics of Exhaustible Resources." *Journal of Political Economy* (April 1931).

Hu, S.-Y.D. "The Copper Commodity Model and Energy Issues." Unpublished diss. University of Pennsylvania, 1978.

Hughes, H., and Singh, S. "Economic Rent: Incidence in Selected Metals and Minerals Resources." *Resources Policy* (June 1978).

International Economic Studies Institute. *Raw Materials and Foreign Policy.* Washington, D.C., 1976.

Issawi, C. "The 1973 Oil Crisis and After." *Journal of Post Keynesian Economics* (Winter 1978-1979).

Janssen, E.R. "Le Prix du Petrole Brut et des Produits Pétroliers et leur Evolution en Europe." *Recherches Economiques de Louvain* (March 1978).

Johnson, C.J. "Cartels in Minerals and Metals Supply." *Mining Congress Journal* 62 (1976).

Johnson, H. "Commodities: Less Developed Countries' Demand and Developed Countries' Response." Unpublished manuscript, 1976.

Jorgenson, D.W. "The Role of Energy in the United States Economy." *National Tax Journal* (September 1978).

Jorgenson, D.W., and Hudson, E. "Economic Analysis of Alternative Energy Growth Patterns, 1975-2000." In D. Freeman et al., eds., *A Time to Choose.* Cambridge, Mass.: Ballinger, 1974.

Josefsson, M. *Den Internationella Arbetsfördelningen.* Särtryck ur SOU 1977:16. Stockholm, 1977.

Kapitza, P. "Physics and the Energy Problem." *New Scientist* (October 1976).

Kellog, H. "Sizing up the Energy Requirements for Producing Primary Metals." *Engineering and Mining Journal* (April 1977).

Kemp, A.G., and Crichton, D. "Effects of Changes in UK North Sea Oil Taxation." *Energy Economics* (October 1976).

————. "North Sea Oil Taxation in Norway." *Energy Economics* (January 1979).

Köhler, K. "Rohstoffpreisindizes: Methodik und Aussagefähigkeit." *Bremer Ausschuss für Wirtschaftforschung* 20 (1976).

Kolbe, H., and Timm, H.J. "Die Bestimmungsfaktorn der Preisentwicklung auf dem Weltmarkt fur Naturkautschuk-Eine Okonometrische Modellanalyse," no. 10. Hamburg: IIWWA Inst. für Wirtschaftforschung, 1972.

Kreinin, M.E. *International Economics: A Policy Approach* New York: Harcourt Brace Jovanovich, 1979.

Krylov, C.A. *The Soviet Economy.* Lexington, Mass.: Lexington Books, D.C. Heath and Company, 1979.

Kymn, K.O. "A Contribution Toward a Correct Understanding of the Welfare Analysis of Foreign Supply Interruptions." *Energy Communications* 4 (1978).

Laidler, D. *Introduction to Microeconomics.* Oxford: Phillip Allen, 1981.

Laulajainen, R. "The U.S. Frash Industry, 1955-1975." *Sulphur* (March-April 1977).

Lecomber, R. *The Economics of Natural Resources.* London: Macmillan, 1979.

Levhari, D., and Liviatan, N. "Notes on Hotellings Economics of Exhaustible Resources." *Canadian Journal of Economics.* (May 1977).

Leontief, W. *The Future of the World Economy.* New York: United Nations, 1977.

Levy, H., and Sarnat, M. "The World Oil Crisis: A Portfolio Interpretation." *Economic Inquiry* (September 1975).

Levy, W.J. "The Years That the Locust Hath Eaten: Oil Policy and OPEC Development Prospects." *Foreign Affairs* (Winter 1978).

Lloyd, B., and Wheeler, E. "Brazil's Mineral Development: Potential and Problems." *Resources Policy* (March 1977).

Lovering, T.S. "Mineral Resources from the Land." In Preston Cloud, ed., *Resources and Man.* San Francisco: W.H. Freeman, 1969.

Lütkenhorst, W., and Minte, H. "Probleme des Monetären und Realen Transfers." In M. Tietzel, ed., *Die Energiekrise: Funf Jahre Danach.* Bonn: Neue Gesellschaft GmbH, 1978.

————. "The Petrodollars and the World Economy." *Intereconomics* (March-April 1979).

Lybeck, J.A. *The Step One Model of the Swedish Economy.* Stockholm: Stockholm School of Economics, 1979. (Recalled).

MacAvoy, P.W. "Economic Perspective on the Politics of Economic Commodity Indexing." The Institute of Government Research, University of Arizona, 1977.

MacKay, G.A. "Uranium Mining in Australia." In J.T. Woodcock, ed., *International Resource Management*. Canberra: Australian Institute of Mining and Metallurgy, 1978.

Madigan, A.L. "Oil Is Still Cheap." *Foreign Policy* (Summer 1979).

Malenbaum, W. *Material Requirements in the U.S. and Abroad in the Year 2000*. National Technical Information Service, Report PB 219-6/5/PB (1977).

Manners, G. *The Changing World Market for Iron Ore, 1950–1980*. Baltimore: Johns Hopkins University Press, 1971.

Martellaro, J.A. "Soviet Energy Resources: Present and Perspective." *Economia Internazionale* (Spring 1981).

McColl, G.D. "Prospects for the Australian Coal Exports." Working Paper no. 23. Sydney: Centre for Economic Research, University of New South Wales, June 1981.

McCulloch, R. "Global Commodity Politics." *Wharton Magazine* (Spring 1977).

McKie, J.W. "Market Structure and Uncertainty in Oil and Gas Exploration." *Quarterly Journal of Economics* (November 1960).

McNichol, D.L. *Commodity Agreements and Price Stabilization*. Lexington, Mass.: Lexington Books, D.C. Heath and Company, 1978.

Mikdashi, Z. *The International Politics of Natural Resources*. Ithaca, N.Y. Cornell University Press, 1976.

Mikesell, R.F. *The World Copper Industry*. Baltimore: Johns Hopkins University Press, 1979.

Mitchell, J. "Oil and the Outlook for the Eighties." London, 1981. Stencil.
———. "Fragmentation and Competition in the Energy Market." Presented to the Oxford Energy Seminar, 1981. Stencil.

Murcier, A., and Boissonnat, J. "Petrole: la grande Illusion." *L'Expansion 2/15 (April 1982)*.

Nappi, C. *Commodity Market Controls*. Lexington, Mass.: Lexington Books, D.C. Heath and Company, 1979.

Neu, A.D. "Entkoppeling von Wirtschaftswachtum und Energieverbrauch—Eine Strategie der Energiepolitik?" *Kieler Diskussions Beitrage*, no. 52 (1981).

Owen, A.D. "Elasticity of Supply for the U.S. Uranium Industry." *Resources Policy* (December 1980).

Page, W.P., and Kymn, K.O. "Cartel Policy and the World Price of Oil: An Explanation?" *Energy Communications* (1978).

Pearce, D., and Rose, J. *The Economics of Natural Resource Depletion*. London: Longmans, 1976.

Percebois, J. "Energie, Croissance et Calcul Economique." *Revue Economique* (May 1978).

———. "A Propos de quelques Concepts Utilises en Economie de l'Energie." In *Economies et Societes,* no. 3, (July–September 1978).

———. "Is the Concept of Energy Intensity Meaningful?" *Energy Economics* (July 1979).

Petersen, F.M. "A Model of Mining and Exploring for Natural Resources." *Journal of Environmental Economics and Management,* no. 5 (1978).

Petersen, U., and Maxwell, S.R. "Historical Mineral Production and Price Trends." *Mining Engineering* (January 1979).

Pindyck, R.S. "Cartel Pricing and the Structure of the World Bauxite Market." *Bell Journal of Economics* (Autumn 1977).

———. "Gains to Producers from the Cartelization of Exhaustible Resources." *Review of Economics and Statistics* (May 1978).

———. "OPEC's Threat to the West." *Foreign Policy* (Spring 1978).

———. *The Structure of World Energy Demand.* Cambridge, Mass.: MIT Press, 1979.

Prain, R. *Copper: The Anatomy of an Industry.* London: Mining Journal Books, 1875.

Quirk, J.P. *Intermediate Microeconomics.* Chicago: Science Research Associates, 1976.

Radetzki, M. "Falling Oil Prices in the "80s." Stockholm, 1981. Stencil.

———. *Uranium: A Strategic Source of Energy.* London: Croom Helm, 1981.

Ray, G.F. "Energy Economics: A Random Walk in History." *Energy Economics* (July 1979).

Rayment, P.B.W. "On the Analysis of the Export Performances of Developing Countries." *Economic Record* (June 1971).

———. "The Homogenity of Manufacturing Industries with Respect to Factor Intensity: The Case of the United Kingdom." *Oxford Bulletin of Economics and Statistics* (August 1976).

———. "Petrol Prices, Conservation, and Macro-Economic Policy." Economic Commission for Europe, February 1980.

Rees, R. *Public Enterprise Economics.* London: Weidenfeld and Nicolson, 1976.

Renton, A. "A Bigger Bonanza." *New Scientist,* 23 September 1976.

Reza, A.M. "Analysis of the Supply of Oil." *Energy Journal* (June 1981).

Robinson, T.J.C. "Classical Foundations of the Contemporary Theory of Non-Renewable Resources." *Resources Policy* (December 1976).

Rogers, P. "The Role of Less Developed Countries in World Resource Use." In P. Rogers and A. Vann, eds., *Future Resources and World Development.* New York: Plenum, 1976.

Rose, S. "Third World Commodity Power is a Costly Illusion." *Fortune* (November 1976).

Ross, M.H., and Wiliams, R.H. "Energy Efficiency: Our Most Underrated Energy Resource." *Bulletin of the Atomic Scientists* (November 1976).

Rostow, W.W. *Getting from Here to There: A Policy for the Post-Keynesian Age.* New York: Macmillan, 1978.

Russell, R.W. "Governing the World's Money: Don't Just Do Something, Stand There." *International Organization* (Winter 1977).

Rustow, D.A., and Mugno, J. *OPEC: Success and Prospects.* New York: New York University Press, 1976.

Salant, S.W. "Exhaustible Resources and Industrial Structure: A Nash-Courant Approach to the World Oil Market." *Journal of Political Economy,* 1976.

Sauter-Servaes, F. "Der Übergang von einer Erschöpfbaren Ressource zu einem Synthetischen Substitut." Paper presented to the Vereins für Socialpolitik Gesellschaft für Wirtschafts-und Sozialwissenschaften, Mannheim, 1979.

Schmid, M. "A Model of Trade in Money, Goods, and Factors." *Journal of International Economics,* no. 6 (1976).

Shann, E.W. "The Analysis of Mining Sector Growth." Paper presented at the Tenth Congress of Economists, Canberra, 24–28 August 1981.

Siebert, H. "Erschopfbare Ressourcen." *Wirtschaftsdienst,* no. 10 (1979).

Slesser, M. *Energy in the Economy.* London: Macmillan, 1978.

Smith, B. "Bilateral Monopoly and Export Price Bargaining in the Resource Goods Trade." *The Economic Record* (March 1977).

Smith, G., and Shink, F. "International Tin Agreement: A Reassessment." U.S. Treasury Department, OASIA, Research Discussion Paper no. 75/18 (1975).

Solow, R. "Richard T. Ely Lecture: The Economics of Resources or the Resources of Economics." *American Economic Review* (May 1974).

Spengler, J. "Population and World Hunger." *Rivista Internazionale di Scienze Economiche E. Commericali* (December 1976).

Stern, J.D. *Soviet Natural Gas Development to 1990.* Lexington, Mass.: Lexington Books, D.C. Heath and Company, 1980.

Stobaugh, R., and Yergin, D. "After the Second Shock: Pragmatic Energy Strategies." *Foreign Affairs* (Spring 1979).

Takeuchi, K. "CIPEC and the Copper Earnings of Member Countries." *The Developing Countries* (February 1972).

Timm, H.J. "Kurzfristige Internationale Rohstoffpreisentwicklung und Konjunkturschwankungen," HWWA Institute für Wirtschaftsforschung, Hamburg, March 1976.

Tobin, J. "Liquidity Preference as Behavior Toward Risk," *Review of Economic Studies* (February 1958).

Trainer, F.E. "Potentially Recoverable Resources." *Resources Policy* (March 1982).

Tuve, G.L. *Energy, Environment, Populations and Food: Our Four Interdependent Crises.* New York: Wiley Interscience, 1976.

Ulph, A.M. "World Energy Models: A Survey and Critique." *Energy Economics* (January 1980).

Ulph, A.M., and Folie, M. "Gains and Losses to Producers from the Cartelization of Exhaustible Resources." Canberra: CRES Working Paper, 1978.

Vann, A., and Rogers, P. *Human Ecology and World Development.* New York: Plenum, 1974.

Varon, B., and Takeuchi, K. "Developing Countries and Non-Fuel Minerals." *Foreign Affairs* (April 1974).

Vedavalli, R. *Market Structure of Bauxite/Alumina/Aluminum and Prospects for Developing Countries.* Washington, D.C.: World Bank, 1977.

Verleger, Philip. "The U.S. Petroleum Crisis of 1979." *Brookings Papers,* no. 2 (1979).

Walker, K.J. "Minerals Consumption Implication of a Fully Industrialized World." *Resources Policy* (December 1979).

Zimmerman, M.B. *Long-Run Mineral Supply: The Case of Coal in the United States.* Ph.D. diss., Massachusetts Institute of Technology, 1975.

———. "Modeling Depletion in a Mineral Industry: The Case of Coal." *Bell Journal of Economics* (Spring 1977).

———. "Estimating a Model of the U.S. Coal Supply." *Materials and Society,* no. 2, 1978.

Index

About the Author

Ferdinand E. Banks is Social Sciences Faculty Research Fellow at the University of Uppsala. In 1981, he held a Tore Browaldh Foundation Fellowship for travel and research in Australia and Germany; and during 1980 he was visiting professor at the Centre for Policy Studies, Monash University, Melbourne. In 1978–1979, he was professional Fellow in economic policy at the Reserve Bank of Australia and visiting professor in the Department of Econometrics, the University of New South Wales. Dr. Banks attended Illinois Institute of Technology and Roosevelt University, receiving the B.A. in economics. After serving with the U.S. army in the Orient and Europe, he worked as a systems and procedures analyst and engineer. He received the M.Sc. and Fil. Lic. from the University of Stockholm, and the Fil. Dr. from the University of Uppsala, where he has taught since 1971.

Dr. Banks was a lecturer at the University of Stockholm for five years; was senior lecturer in mathematics, economics, and statistics at the United Nations African Institute for Economic and Development Planning, Dakar, Senegal; and has been consultant lecturer in macroeconomics and input-output analysis for the OECD at the Technical University of Lisbon, Portugal. From 1968 to 1971 he was an econometrician and primary commodity economist for the United Nations Commission on Trade and Development (UNCTAD) in Geneva, and he has also been a consultant on planning models and the steel industry for the United Nations Industrial Development Organization (UNIDO) in Vienna. His previous books are *The World Copper Market* (1974); *The Economics of Natural Resources* (1976); *Scarcity, Energy, and Economic Progress* (1977); *The International Economy: A Modern Approach* (1979); *Bauxite and Aluminum: An Introduction to the Economics of Nonfuel Minerals* (1979); and *The Political Economy of Oil* (1980). He has also published fifty-two articles and notes in various journals and collections of essays. He is currently working on the Western Europe–Soviet Union natural-gas transaction, the futures market for oil and oil products, and a book on coal, *The Political Economy of Coal* (Lexington Books, forthcoming).